Harnessing Indigenous Knowledge and Place-Based Learning for Sustainability

Ahmad Qablan
https://orcid.org/0000-0002-2780-9796
United Arab Emirates University, UAE

Asma Abdallah
https://orcid.org/0000-0003-1028-7618
Sharjah Education Academy, UAE

Ahmed Alkaabi
United Arab Emirates University, UAE

IGI Global
Scientific Publishing
Publishing Tomorrow's Research Today

Vice President of Editorial	Melissa Wagner
Director of Acquisitions	Mikaela Felty
Director of Book Development	Jocelynn Hessler
Production Manager	Mike Brehm
Cover Design	Jose Rosado

Published in the United States of America by
IGI Global Scientific Publishing
701 East Chocolate Avenue
Hershey, PA, 17033, USA
Tel: 717-533-8845 | Fax: 717-533-7115
Website: https://www.igi-global.com E-mail: cust@igi-global.com

Library of Congress Cataloging-in-Publication Data

Names: Qablan, Ahmad, 1974- editor | Abdallah, Asma, 1966- editor | Alkaabi, Ahmed, 1984- editor
Title: Harnessing indigenous knowledge and place-based learning for sustainability / edited by: Ahmad Qablan, Asma Abdallah, Ahmed Alkaabi.
Description: Hershey, PA : IGI Global Scientific Publishing, [2026] | Collection of essays by Usep Suhud and others. | Includes bibliographical references and index. | Summary: "This book explores how Indigenous knowledge systems and place-based education can advance sustainability, equity, and environmental stewardship. Through global perspectives, it bridges traditional ecological wisdom with contemporary research to reimagine education and community development for more resilient and sustainable futures"-- Provided by publisher.
Identifiers: LCCN 2025054470 (print) | LCCN 2025054471 (ebook) | ISBN 9798337376509 hardcover | ISBN 9798337376516 paperback | ISBN 9798337376523 ebook
Subjects: LCSH: Traditional ecological knowledge | Sustainability--Study and teaching | Sustainable tourism | Sustainable development--Mexico
Classification: LCC GN476.7 .H37 2027 (print) | LCC GN476.7 (ebook)
LC record available at https://lccn.loc.gov/2025054470
LC ebook record available at https://lccn.loc.gov/2025054471

British Cataloguing in Publication Data
A Cataloguing in Publication record for this book is available from the British Library.

Table of Contents

Detailed Table of Contents

This chapter investigates Indigenous Knowledge Systems (IKS) and transformative sustainability education, focusing on the fusion of cultural legacy with contemporary educational practice. It analyses the relational, interdependent, and place-based wisdom of Indigenous epistemologies and its contributions to educational reforms aimed at nurturing ecological awareness and the adaptive capacity of the community. With global illustrations from the worldwide, the chapter expresses the need to combine Indigenous and place-based approaches in formal and non-formal education, within ethical bounds, in alignment with UNESCO's ESD 2030 and SDG 4. It acknowledges the prevalence of cultural appropriation; the cultural and epistemological silences, and the active avoidance of cultural appropriation; and it offers varied avenues to achieve effective knowledge co-production and a change in educational institutions. The chapter reiterates the necessity of innovatively integrating in education for the attainment of sustainability learning that is culturally relevant and internationally responsive.

The idea of indigenous knowledge systems is based on the ancient connection with the land and culture and serves as the main pillars of the reconsideration of the sustainability education in the context of new environmental and social issues of our current world. The chapter explores the opportunities of teaching sustainability through the incorporation of the Indigenous epistemologies and place-based pedagogies to alter abstract and theory-based teaching towards the experiential, moral, and community-based learning. Using the literature on the world and current case studies, the chapter proves that land-based learning, Indigenous storytelling, and community-school partnerships contribute to the increase in ecological awareness, cultural identity, and agency of learners. It also emphasizes ethical activity, knowledge sovereignty, and policy congruence as the main preconditions of meaningful integration stating that Indigenous ways of knowing must be used as central reference points when forming more resilient, fair, and sustainable futures.

The United Arab Emirates' shift toward a diversified, post-oil economy is reflected in its national curriculum. This chapter argues that the UAE curriculum represents a hybrid sustainability epistemology integrating ethical–religious formation, place-based ecological knowledge, scientific literacy, and technological advancement. Sustainability is embedded across subjects rather than taught independently: Islamic Studies and Moral Education promote ethical responsibility, social and heritage subjects connect learning to local ecological contexts, and science and innovation address climate change, energy transitions, and emerging technologies. The chapter further shows how the curriculum operates as an institutional system that organizes and reproduces sustainability knowledge in alignment with national priorities. While this model demonstrates coherence across school types, it also highlights tensions between global frameworks, local contexts, and development goals. Overall, the chapter frames curriculum as a hybrid knowledge infrastructure in rapidly modernizing societies.

Place-Based Education (PBE) relates education to the local settings, cultures and communities to enhance sustainability and environmental consciousness. Based on the Indigenous knowledge systems, PBE promotes experiential and land-based learning that can develop environmental stewardship, critical thinking, and social responsibility. The chapter examines the theoretical perspectives, pedagogical approaches and policy models supporting PBE in the world and its contribution to decolonizing education and the promotion of Education for Sustainable Development (ESD). It brings forth the successful school-community partnerships and new teaching strategies through case studies, both global and local. The chapter wraps up by discerning the important measures in incorporating PBE into the curriculum that can help develop climate resilience and sustainability literacy.

This chapter examines how sustainability is institutionalized within the governance architecture of the UAE education system, defined as the interaction of national strategic visions, centralized policy design, regulatory oversight, and leadership practices shaping education planning and implementation. National frameworks UAE Centennial 2071 and Net Zero 2050 place education within long-term economic and environmental planning supporting post-oil diversification and national development. In this context, sustainability extends beyond environmental protection to include economic transformation, technological modernization, and institutional resilience. While global agendas such as SDG 4 and 16 influence the policy environment, the UAE model reflects selective alignment rather than direct transfer. Centralized authorities translate national priorities into regulatory frameworks and planning requirements aligning schools with national goals. The chapter highlights how strategic vision and regulatory coordination support sustainability governance in a rapidly evolving education system.

This chapter, "Social Justice, Self-Determination, and Sovereignty in Educational Reform," examines how Indigenous knowledge systems and place-based pedagogies contribute to equitable and sustainable education. Through the lens of decolonial and critical pedagogical inquiry, this chapter explores the interconnections between social justice, self-determination, and educational sovereignty, contending that Indigenous communities' rights to design and govern their own education systems hold the key to sustainable and resilient learning. The authors provide examples from Māori schools in Aotearoa/New Zealand, Tribal Colleges in North America, and community-led models in India and Latin America to show that sovereignty enables cultural revitalization, ecological stewardship, and social empowerment. The chapter identifies links to SDGs 4, 13, and 15 but also recognizes constraints including policy, funding inequities, and epistemic tokenism.

Twice exceptional (2e) learners' students who combine high intellectual or creative potential with learning or developmental challenges represent an underserved group and a key example of sustainable, equitable inclusion. This chapter explores how schools can shift from theoretical equity to evidence-based practice. In alignment with UN SDG 4, it argues that sustainable inclusion must recognize full cognitive diversity, especially learners whose profiles defy traditional categories. Due to the masking effect, many 2e students remain unidentified, leading to academic, social, and emotional challenges. Drawing on research in gifted and special education, the chapter highlights differentiated instruction as an effective approach. It integrates major developmental and talent models to explain the cognitive, emotional, and ecological dimensions of 2e learners and concludes with recommendations showing that embracing cognitive diversity is essential for sustainable human development.

Outdoor learning and indigenous education are highly interconnected concepts. Since indigenous education involves the transmission of indigenous knowledge, it is naturally place-based and cannot be effectively carried out solely within a classroom setting. By observing and engaging with the environments in which indigenous culture has been formed, students can better concretize indigenous knowledge and experience a more meaningful learning process. In this regard, the present study aims to offer educators ideas for implementing outdoor learning activities and to provide a resource to support them in designing educational practices. To achieve this goal, an AI-supported tool named IndigenTerra has been developed to localize the proposed activity designs. This tool modifies the designed activities according to the location and the age group of the students. It is anticipated that IndigenTerra will represent an example of the integration of technology within outdoor learning and indigenous education.

In this chapter, we offer a panoramic view of the English language education stance adopted in the United Arab Emirates across the years. The chapter specifically highlights how English language education is gradually evolving towards a sustainable ideology, by focusing on equitable approaches that are now prevalent in the field, such as that of translanguaging. The chapter ends with specific instructional practices that can be implemented by instructors and policy makers that will help to fulfill educational sustainable goals

The study aimed to examine the challenges social studies teachers face in implementing global citizenship education in two primary schools in Bangladesh. This study employed a qualitative case study approach, focusing on two elementary schools: one in an urban area and the other in a rural area. The primary data for this study were acquired through in-depth interviews with four social studies teachers and two head teachers from two schools. Additionally, four classroom observations and two focus group sessions with students were conducted to gain more clarity. This study analyzed data thematically. This study identifies seven core challenges regarding knowledge of GC and GCED, integrating GCED into facilities and infrastructure, classroom management and teaching skills, training and professional development, assessment of GCED, and a heavy workload. So, policymakers and school authorities need to focus on solving these challenges to make students appropriate global citizens.

This chapter explores how chemistry education can become more meaningful by connecting learning to students' local environments, cultures, and real-life experiences. It highlights place-based education (PBE) as an approach that encourages students to engage with community issues and sustainability challenges. By integrating socio-scientific issues and ethno-chemistry, the chapter shows how traditional knowledge and cultural practices can be linked to scientific concepts. It also presents a community-engaged, project-based learning model where students collaborate with local communities to address real problems. This approach enhances understanding, strengthens relevance, and fosters responsible, reflective, and environmentally aware learners

Preface

INTRODUCTION

Education today stands at a critical crossroads. As societies face environmental degradation, cultural homogenization, technological disruption, and widening inequities, the challenge is no longer whether education should respond, but how it can do so in meaningful and transformative ways. *Roots of Resilience: Place-Based Leadership for Sustainable Education* addresses this urgency by positioning education as a contextual, relational practice grounded in place. The volume brings together interdisciplinary perspectives that examine how place-based leadership—rooted in local cultures, communities, and ecologies—can drive sustainable education. It extends sustainability beyond environmental concerns to include social, cultural, ethical, and technological dimensions, emphasizing the integration of indigenous knowledge, community engagement, inclusive practices, and emerging technologies. At its core, the book argues that educational transformation must begin with reconnecting to place through understanding local histories, honoring cultural identities, and fostering responsive and ethical leadership.

In today's global landscape, education is increasingly influenced by frameworks such as the United Nations' Sustainable Development Goals. However, many systems remain disconnected from learners' lived realities. This gap is especially visible in rapidly developing contexts like the United Arab Emirates, where modernization intersects with cultural preservation. Education must prepare students for a global, digital future while sustaining local identity and environmental responsibility. This book offers a localized yet globally relevant perspective, demonstrating how place-based leadership can bridge global sustainability agendas and local implementation. It also addresses key challenges, including climate change, cultural preservation, artificial intelligence integration, inclusive education, and policy-practice gaps, contributing to broader discussions on adaptive and sustainable education systems.

The volume targets a diverse audience, including educational leaders, policymakers, practitioners, researchers, and graduate students. It is particularly relevant for those seeking to implement context-responsive reforms, embed inclusive practices, and advance sustainability-focused research. Teacher educators and professional development facilitators will also benefit from its insights into preparing future-ready educators, while interdisciplinary readers will find value in its connections between education, environment, technology, and culture.

Collectively, the chapters construct a coherent narrative of place-based leadership for sustainability. Early chapters establish foundational concepts, including the integration of indigenous knowledge and hybrid epistemologies in curriculum design. Subsequent contributions explore technological applications, leadership practices, and policy implementation, highlighting the importance of translating sustainability into practice. The volume further examines inclusive education, teacher leadership, and innovative intersections between artificial intelligence and indigenous practices. Additional chapters address STEM education, smart technologies, and climate-responsive schooling, emphasizing the role of learning environments in promoting equity and sustainability.

The later chapters expand the discussion to broader societal dimensions, including sustainable mobility, supply chains, and social justice. They highlight the importance of equity, self-determination, and global citizenship, enriched by international case studies that provide comparative insights. Together, these contributions offer a comprehensive and multidimensional understanding of sustainability in education.

CHAPTER OVERVIEW

The opening chapter examines the relationship between Indigenous knowledge systems and transformative sustainability education, emphasizing culturally grounded learning, ecological awareness, and ethical knowledge integration within global educational reform efforts.

The second chapter explores how land-based learning, storytelling, and community partnerships support sustainability education by strengthening ecological consciousness, cultural identity, and learner agency.

The third chapter analyzes sustainability within the curriculum framework of the United Arab Emirates, highlighting the integration of ethical, religious, scientific, and place-based perspectives in national educational development.

The fourth chapter investigates place-based education as a framework for experiential and community-centered learning that promotes environmental stewardship, critical thinking, and sustainability literacy.

The fifth chapter examines how sustainability is institutionalized within the governance structures of the UAE education system through strategic planning, policy coordination, and long-term national development frameworks.

The sixth chapter addresses issues of social justice, self-determination, and educational sovereignty, demonstrating how Indigenous-led educational models contribute to equity, cultural revitalization, and ecological resilience.

The seventh chapter focuses on neurodiversity and twice-exceptional learners, arguing that inclusive and differentiated educational practices are essential for sustainable and equitable human development.

The eighth chapter presents the relationship between Indigenous education and outdoor learning while introducing an AI-supported educational tool designed to localize and adapt sustainability-oriented learning activities.

The ninth chapter traces the evolution of English language education in the United Arab Emirates toward more sustainable and equitable approaches, including translanguaging and culturally responsive instructional practices.

The tenth chapter investigates the challenges faced by primary school social studies teachers in implementing global citizenship education in Bangladesh, identifying institutional, pedagogical, and resource-related barriers.

The eleventh chapter reimagines chemistry education through place-based and sustainability-oriented approaches that connect scientific learning with local cultures, socio-scientific issues, and community engagement.

The concluding chapter explores students' perceptions of science education in the UAE and examines how relevance, trust, and interactive pedagogy shape place-based sustainability learning and ethical engagement with Indigenous knowledge.

CONCLUSION

In conclusion, this book advances the field by reframing leadership as context-sensitive and place-responsive, integrating indigenous knowledge into mainstream discourse, and bridging theory with practice. It promotes inclusive and equitable education while offering interdisciplinary insights that connect education to global challenges. Ultimately, it calls for a shift from standardized systems to resilient, place-based learning ecosystems, encouraging educators and leaders to act as co-creators of sustainable futures grounded in local contexts.

Ahmad Qablan
United Arab Emirates University, UAE

Asma Abdallah

Sharjah Education Academy, UAE

Ahmed Alkaabi
United Arab Emirates University, UAE

Acknowledgment

As the editors of *Roots of Resilience: Place-Based Leadership for Sustainable Education*, we extend our sincere gratitude to the institutions, scholars, and practitioners whose support, expertise, and collaboration made this book possible. This volume represents a collective effort grounded in a shared commitment to advancing sustainable, context-responsive educational leadership.

We would first like to acknowledge the invaluable institutional support that enabled the conceptualization and development of this work. The commitment of educational institutions to fostering research, innovation, and scholarly dialogue has provided a strong foundation for this publication. Such environments not only nurture academic inquiry but also encourage the exploration of new ideas that address the evolving needs of education systems in a rapidly changing world.

This book is a reflection of meaningful international collaboration. We are deeply appreciative of the diverse group of contributors who have brought their knowledge, research, and professional experiences to this volume. Scholars and education professionals from various regions have enriched this work by offering multiple perspectives on place-based leadership, sustainability, equity, and community engagement. Each chapter represents a unique contribution, collectively forming a comprehensive and globally relevant body of knowledge.

We would also like to express our sincere appreciation to the peer reviewers whose critical insights and constructive feedback played a vital role in strengthening the quality of this book. Their dedication to maintaining high academic standards has enhanced the coherence, depth, and rigor of the chapters. While their contributions may not always be visible, their impact is deeply embedded in the final outcome of this publication.

Our gratitude also extends to the broader academic and professional community whose ongoing commitment to advancing educational leadership and sustainability continues to inspire and inform this work. The collective efforts of scholars, practitioners, and policymakers in exploring innovative, context-responsive approaches have significantly contributed to shaping the perspectives presented in this volume and reinforcing its relevance to current educational challenges.

Finally, we acknowledge that this book stands as a collective achievement. It is the result of shared vision, intellectual collaboration, and sustained effort among contributors, reviewers, editors, and institutional partners. We are profoundly grateful for the spirit of cooperation and dedication that has shaped this journey.

We hope that *Roots of Resilience: Place-Based Leadership for Sustainable Education* will contribute meaningfully to the field, inspiring educators, leaders, and researchers to engage with place-based approaches and to advance sustainable practices within their own educational contexts.

Chapter 1
Bridging Tradition and Innovation:
Indigenous Knowledge in Transformative Sustainability Education

Sovanna Huot
http://orcid.org/0009-0006-8335-442X
Department of Education, National Institute of Education (NIE), Phnom Penh, Cambodia

Phearun Tep
http://orcid.org/0009-0000-2490-677X
Department of English, University of Delhi, New Delhi, India

ABSTRACT

This chapter investigates Indigenous Knowledge Systems (IKS) and transformative sustainability education, focusing on the fusion of cultural legacy with contemporary educational practice. It analyses the relational, interdependent, and place-based wisdom of Indigenous epistemologies and its contributions to educational reforms aimed at nurturing ecological awareness and the adaptive capacity of the community. With global illustrations from the worldwide, the chapter expresses the need to combine Indigenous and place-based approaches in formal and non-formal education, within ethical bounds, in alignment with UNESCO's ESD 2030 and SDG 4. It acknowledges the prevalence of cultural appropriation; the cultural and epistemological silences, and the active avoidance of cultural appropriation; and it offers varied avenues to achieve effective knowledge co-production and a change

DOI: 10.4018/979-8-3373-7650-9.ch001

in educational institutions. The chapter reiterates the necessity of innovatively integrating in education for the attainment of sustainability learning that is culturally relevant and internationally responsive.

I. INTRODUCTION

There is a growing recognition worldwide that a value and worldview shift is necessary, and, as Kimmerer (2013) so beautifully states, Indigenous ways of knowing provide a moral and a spiritual framework that advocates living as kin with the world beyond the human, fostering respect, reciprocity, and responsibility, which is what the world needs now. In the same vein, Barnhardt and Kawagley (2005) argue the merging of Indigenous and Western ways of knowing is not simply additive, but rather synergistic creates new pedagogical spaces that value the integration of both the contemporary and the traditional.

Likewise, UNESCO (2020), ESD 2030 Roadmap, advocates for education for sustainable development frameworks that embrace multiple ways of knowing and, as innovations for education, incorporate Indigenous ways of knowing as integral to the fostering of sustainable ecological and resilient systems. This chapter seeks to position itself at the merging of these discourses which aims to connect the innovative pedagogies of contemporary sustainability education with the Indigenous traditions of historical wisdom. By defining this merging as an epistemological reconciliation and a pedagogical shift, the chapter aims to engage with the global conversation on how education can serve both cultural restoration and ecological preservation (Huot, 2026). The case for the merging of Indigenous knowledge with contemporary educational innovation rests on the fact that such frameworks have proven to be more effective manifestations of contextual and ethically informed sustainable outcomes. As Berkes (2017) explains, Indigenous knowledge systems embody thousands of years of integrated, cumulative learning from the observation and interaction with local ecosystems, knowledge that is especially relevant today in the context of climate adaptation and the conservation of biodiversity. These systems have exemplified, and practiced, long-term sustainability that aligns contemporary ecological principles such as resilience, reciprocity, and adaptive management (Fernández-Llamazares & Álvaro Cabeza, 2017).

Western scientific paradigms have, in contrast, historically repeated the same patterns of abstraction and universalism as the rest of the sciences, and have for the most part, inappropriately distanced, the human-environment interactions that are the focus of their research. This is the gap through which modern education systems have operated, to avoid fostering a sense of ecological belonging and responsibility that is so often needed (Gwekwerere, 2011). Bridging this gap by incorporating Indigenous worldviews, provides a counterbalance to these movements by re-establishing the importance of relationality, holism, and responsibility to place, which are the most valuable in reconstructing sustainability as a practice, rather than an abstraction (Huot & Em, 2025; McGregor et al., 2020).

Moreover, Sterling (2001) notes that learning about sustainability must now develop into transformative education, which, in addition to knowing, fosters the ability to initiate systemic change and cultural renewal. The combination of holistic ecological understanding and creative pedagogy represents a double movement: the retention of foundational knowledge and the use of contemporary pedagogical technologies and transdisciplinary inquiry approach to engage learners in real-world challenges of sustainability (Lotz Sisitka et al., 2015). The rationale, from a global policy perspective, conforms to UNESCO's Reimagining Our Futures Together: A New Social Contract for Education (2022) which articulates pluralism of knowledge, transcultural approaches, and collaborative knowledge creation as the cornerstone of sustainable societies. Thus, the integration of IKS in educational innovation becomes a necessity, not merely a want, in the creation of learning systems that are culturally pluralistic and ecologically restorative (Huot & Hok, 2025a). Indigenous knowledge and local knowledge systems, despite global commitments to inclusive and sustainable education, continue to be marginalized, tokenized, and inserted into formal education systems in a piecemeal and superficial way. Current research (2020–2025) shows that although sustainability frameworks such as ESD 2030 advocate for the integration of diverse knowledge systems, the advocacy in practice is poorly and unevenly applied, continuing the dominant Western approach (Lin et al., 2021; UNESCO, 2023).

Chandra (2014) and Redvers et al. (2020) suggest that the education for sustainability still adopting Eurocentric paradigms that still favour empirical and instrumental reasoning compared to relational and spiritual connections with the nature. Yunkaporta (2019) explains that the most marginal indigeneity epistemologies are not due to irrelevant, but due to the neglect of systemic gaps that undermines the community and oral ways of knowledge as non-scientific. Bartlett et al. (2012) and Zeyer (2022) illustrate that this systemic marginalization continues to perpetuate epistemic injustice by excluding the critical role of Indigenous peoples and narrowing the scope of diversity within sustainability education. More recently, several studies indicate that educational institutions continue to use Indigenous emblems and stories without any real interaction or co-authorship (Bartlett et al., 2012; Shahjahan & Seinn, 2025). This so-called performative inclusion, which is often seen as cultural add-ons, is more likely to maintain than to change colonial structures. As Zurba et al. (2021) put it, the custodians of ecological knowledge, Indigenous peoples, often face an epistemological gap within the ambitious goals of global sustainability.

Moreover, the prominence of both a rigid technocratic and empirical approach and some forms of sustainability education may further downplay the importance of ethics and the relational glue that is necessary to hold the socio-ecological system together. As Manns et al. (2025) and Zurba et al. (2021) explain, the absence of, or inadequate, consultation with Indigenous and local knowledge, sustains the fragmentation of collective learning by marginalizing the community and the outcome's sustainability. Consequently, education systems have the potential to reinforce the extractive and anthropocentric rationales driving the current ecological crisis (Chinn, 2007; Lotz Sisitka et al., 2015). Hence, the central concern of this chapter is the neglect of the epistemic dimensions of the sustainability education movement. Educational reforms, in the absence of authentic Indigenous and local knowledge, will continue to lack a significant degree of potential change. Therefore, in this chapter, and considering the described circumstances, this study addresses two objectives. One is to examine the role of Indigenous knowledge in transformative sustainability education, with a focus on the relational, ethical, and philosophical dimensions of the human-nature connection. The second is modeling Indigenous and other innovative pedagogies, in the form of a guide, as a model for educators and policymakers, and a model for re-visioning sustainability education that is holistic, contextual, cross intercultural, and anticipatory. These objectives represent the promise of transdisciplinary pluralism and the decolonization of education for sustainable development (Leal Filho et al., 2019; Lotz Sisitka et al., 2015; UNESCO, 2023).

This chapter makes the case for choosing, as a primary, not peripheral, framework of transformative learning, Indigenous Knowledge Systems. Following Gruenewald (2003) and, more recently, Chinn (2007), this study argues that sustainable education must also embrace decolonization, not only of the structures of power and of the epistemic silencing, but also of the rein habitation of the relationship among people, and place, and culture. The chapter articulates a conceptual pathway that connects Indigenous relational epistemologies and one type of emergent pedagogy, thus contributing to the theorization and the practice of transformative sustainability learning. It builds on Sterling (2001) advocacy for re-integrative education by integrating systems thinking, moral imagination, and ecological conscience.

This research aims to construct an integrated and workable model for educators, policymakers, and scholars, drawing on worldwide case studies and emerging research in both Indigenous studies and sustainability education. This chapter articulates a re-envisioned model of sustainability education as a living and adaptive system, integrated with both ancient and modern systems of knowledge and fostering socially, ecologically, and culturally diverse adaptive futures.

II. THEORETICAL AND CONCEPTUAL FRAMEWORK

2.1 Indigenous Knowledge Systems (IKS)

Indigenous Knowledge Systems (IKS) are acknowledged as holistic and dynamic systems, and as such, IKS embodies a living system. In her own words, Battiste (2019) explains that indigenous knowledge is not a system of mere cultural artifacts. It is a system of cognition that exists within the web of interrelationships of land, people, and the spiritual plane. While many Western epistemologies are abstract and universal, IKS speaks to the systems of relationality, interdependence, and collectivism (Dei, 2011; McGregor et al., 2020; Zurba et al., 2021). Relationality is often the central pillar of IKS, which is the concept that knowledge is a product of, and because of, relationships, be it with people, other-than-human beings, ancestors, or the land (Wilson, 2020). Knowledge is relational because it is created, validated, and lived through interconnectedness rather than isolation (Kovach, 2009). Relational views of the world create a paradigm of learning as a process of ethical involvement and reciprocity, rather than a quest for control or mastery (Kimmerer, 2013).

The principle of interdependence enables all beings to collaborate as co-signers of an ecological web of life. Indigenous ecological knowledge, as Berkes (2017) argues, mirrors adaptive systems thinking long before systems theory entered the Western lexicon. Profound knowledge of the dynamics, resilience, and sustainability of ecosystems was constructed by Indigenous peoples through centuries of systems observation and experimentation, coupled with the oral tradition. Seasonal rituals, harvestings, and rotations of land, and the transcendental harvest, all of which reinforce ecological balance, reflect interdependence (Fernández-Llamazares & Álvaro Cabeza, 2017). Intergenerational community learning, collective and experiential, constitutes another of the bases of Indigenous knowledge systems (IKS). Participation in the life of the community as Barnhardt and Kawagley (2005) observe, constitutes the central aspect of learning in Indigenous societies. Elders and knowledge keepers function as social and cultural surrogate living libraries in the teaching and learning process. The social, collective, and cultural attributes of this which pedagogy elicits an inspiration. Freire (1970) is often cited as the architect of this pedagogy, characterized by dialogue as the process of knowledge co-creation via reflection and action (praxis).

To conclude, IKS is a radical critique of abstract, vertical relations of power with knowledge and the vertical relations of power with knowledge and the vertical relations of power as a fractal theory of knowledge and knowledge systems. The vertical relations of power with knowledge fractal, and power with knowledge. Knowing, and the vertical relation of knowledge encourages a re-thinking of the relations of power, knowledge, and the vertical relations of knowledge with a systems, fractal, and power lens. The vertical relations of power with knowledge, and the vertical relations of knowledge, advocate an understanding of knowledge as an ecological system, a system of the relations of power, knowledge, and an ecology of systems, a fractal, inter-relation of systems differentiated by a unique system of the relations of the vertical, knowledge, and systemic, ecological inter-ordered paradigm. This shift in thinking encourages a radical transformation and regenerative, and a re-conceptualization of purpose, of education.

2.2 Place-Based and Transformative Sustainability Education Theories

As a pedagogical approach, the theory of place-based education (PBE) has strong connections with Indigenous epistemologies. Defining place-based education, Sobel (2004, 2005) states it as knowledge construction that begins with the local knowledge and entails the understanding of the various landscapes, cultures, and communities that the learners are a part of. PBE is understood as a critical pedagogy of place, and in this context, education must fully engage in the two processes of decolonization, challenging the existing structures of power, and rein habitation, restoring the relationships and order of people to a given place (Huot & Chheang, 2025). The pillars of PBE are context-based, participatory, and experiential learning, which aligns with experiential learning cycle, in which all components, concrete experience, reflection, conceptualization, and experimentation, reinforce one another (Kolb, 2015). As Smith and Sobel (2010) state, when learners undertake activities within their local setting, whether it is land restoration, community mapping, or storytelling, they develop a relationship to their environment that is emotional and ethical, a phenomenon that cultivates stewardship and agency.

Like PBE, Transformative Sustainability Education (TSE) focuses on similar goals, but with the addition of fostering radical changes in world views, values and behaviours (Sterling, 2024; Wals, 2010). From Fleming (2018) transformative learning theory, Sterling (2001) and Sterling (2024) argue that for sustainability education to be effective, it must engage the whole person "cognitive, affective, and spiritual", so that they can achieve integrated reflection and depth of awareness. TSE conceptualizes learning as progressive, comprising an element of unlearning that aids the breaking down of barriers to understanding and the development of a more sophisticated empathy and critical awareness of the interdependencies of the global and the local (Lotz Sisitka et al., 2015).

More recent studies Leal Filho et al. (2019), Lotz Sisitka et al. (2015), and Wals (2011) stress the importance of context, collaborative inquiry, and ethical reflection, three transformative learning constructs, and these are also prominent in Indigenous pedagogies. Thus, PBE and TSE together offer an integrated pedagogical framework that respects global sustainability and local context. From a theoretical perspective, both frameworks challenge and move away from reductionist and technocratic models to enhance ecological and emotional literacy and civic participation. They also bridge the gap between UNESCO (2020) call for pluralism in the understanding of knowledge systems and the need for sustainably oriented education that is rooted in a particular place.

2.3 Integrative Conceptual Model: Bridging IKS and Innovative Pedagogies

With the threads weaved together, the Integrative Conceptual Model for Transformative Sustainability Learning (ICTSL) is a blending of frameworks for Indigenous Knowledge Systems (IKS), place-based education, and transformative sustainability education. The model represents a framework of systems and structure at its most dynamic within the three intersecting domains of the model:

1. Epistemological Integration (Ways of Knowing): This is the recognition and legitimization of layered and diverse epistemologies. It builds on the works of Chinn (2007) and Dei (2011), positing that the inclusion of diverse and relational, empirical, and spiritual epistemologies strengthens sustainability learning. Educators as cultural mediators foster the dialogue between Indigenous and scientific rationality (Bartlett et al., 2012; Zeyer, 2022).
2. Pedagogical Innovation (Ways of Teaching): This dimension, based on PBE and TSE theories, speaks to the experiential, active, and reflexive pedagogy. Potential learning activities include community co-teaching, place-based learning (land-based), digital storytelling, and cross-generational teaching-learning. Wals (2010, 2011) refers to these practices as transgressive learning, the type of education that seeks to change dominant paradigms and, in the process, enhance collective imagination and creativity.
3. Transformative Praxis (Ways of Being and Acting): A definitive domain concerned with the learning of the engaged, action-oriented ethical nurturing of relational and socio-ecological care. This domain is in step with Freire (1970) and the liberatory education of the oppressed, and Cajete (1999) learning as ceremony, as an act of honouring life's interrelatedness. Such a praxis-oriented perspective supports learners transiting knowledge, and head learning, to embodied, lived, or, the sustainability of Indigenous values of reciprocity, respect, and responsibility, in practice.

The envisioned model of ICTSL presents the relational dynamic of IKS and contemporary pedagogies as synergistic and cyclical. Knowledge is seen to be, and is, flowing in and out of the community's wisdom, the inquisitive classroom, and the engagement with the discourse of a sustainable future; thus, the flow of knowledge is innovative, culturally and ecologically relevant (Huot et al., 2025). This prospect, grounded in the Indigenous relational epistemologies, applied in a transformative and place-responsive manner, the integrative framework, shifts the educational focus from the mere transmission of information to the cultivation of wisdom.

III. BRIDGING TRADITION AND INNOVATION IN EDUCATION

3.1 Convergence Between Traditional Ecological Wisdom and Modern Sustainability Tools

The integration of traditional ecological wisdom and contemporary innovations in sustainability may be considered one of the most significant and urgently needed changes in today's education. Indigenous knowledge should not be considered primitive or antiquated; instead, it should be seen as a vital, active, and living source of knowledge that can offer solutions to contemporary problems (Dei, 2011). Therefore, this convergence is not an act of revival, but an act of renewal. It is the process of utilizing the knowledge of our ancestors to respond to the most pressing global sustainability challenges of the 21st century. Moreover, for centuries Indigenous and local communities practiced various sustainable forms of land and biodiversity stewardship and community governance long before the term sustainability science became part of Western academia (Berkes, 2017). Rotational farming systems, sacred groves, and seasonal harvesting rituals are a few examples that illustrate an understanding of ecology based on reciprocity, and not exploitation. Kimmerer (2013) notes, to know the world through Indigenous eyes is to know the world as a gift. A gift that requires gratitude and responsibility, not ownership.

Modern-day sustainability initiatives using digital tools, monitoring, AI, renewable tools, and GIS, though useful in planning and building governance frameworks for climate change adjustment (Sterling, 2001), for the most part, remain unframed for ethics and local context, elements critical for transformative change (Bartlett et al., 2012; Zeyer, 2022). The more recent literature tends to highlight the complementarity of the two paradigms rather than the cross-paradigmatic stance. An example of such a dual epistemic approach is the theory of Two-Eyed Seeing, or Etuaptmumk, articulated by Bartlett with his colleagues (2012). It suggests that one side as the value of Indigenous knowledge and the other as the value of Western knowledge, and both together give us a more complete view of the world.

As discussed by Zurba et al. (2021), the value of Indigenous and Western systems of knowledge, in totality, fosters a more rounded perception of the world.

This pluralistic view offers a ground for the knowledge of Indigenous peoples and Western systems of knowledge, a vital aspect for planning for climate change. The combination of traditional ecological approaches and modern digital tools fosters ecological literacy that is beyond cultures and disciplines. The combination of Indigenous seasonal calendars and AI in climate modelling, for instance, will enhance the accuracy of the model and cultural relevance of climate adaptation planning (Shahjahan & Seinn, 2025). Likewise, Indigenous land-use histories guide the drone mapping of ecosystems to reveal culturally significant areas and biodiversity (McGregor et al., 2020).

Nevertheless, integration efforts should avoid extractive and tokenistic approaches. As Tuck and Yang (2012) explain, the use of Indigenous knowledge without the community's approval leads to the appropriation of the knowledge and reproduces what they call epistemic violence. Thus, convergence of this kind requires what Cajete (1999) refers to as ethical relationality, which, in this instance, means positive interdependence, and, in this case, also means respect for Indigenous sovereignty and the positive interdependence relational framework.

The synthesis of ancient knowledge and contemporary innovations in sustainability encourages a revolution in the field of education. It encourages the rethinking of the term sustainability, not as a technical challenge, but as a matter of, and critique, of a culture, a civilization, and the societal structures that exist. Wals (2011) aptly puts it, the future of learning for sustainability depends not only on cognitive innovation but on recovering the moral and ecological intelligence that Indigenous traditions have long sustained.

3.2 The Integration of Local Knowledge and Technology in Curriculum Development

Current sustainability curricula strive to differ from the abstract global perspectives evident in the United Nations' Sustainable Development Goals (SDGs), and seek to adopt, and implement, community-centred and place-based methodologies (Leal Filho et al., 2019). It is no longer acceptable to include Indigenous Knowledge Systems (IKS) in curriculum development as a form of cultural enhancement. Rather, it is a necessity for the pedagogy of social change that is required in order to facilitate the transformation of students' perceptions and actions concerning the sustainability problems they face. Barnhardt and Kawagley (2005) contend that education should link students to the ecological and cultural foundations of their communities. This is fulfilled when local knowledge functions as the content and the methodology. That is, learning that derives from the ethics, practices, and rhythms of a given community.

In sustainability education, there is an ancient and resilient sustainable resource management system in the falaj irrigation systems on the Arabian Peninsula and the qanat systems in Persia, which are paired with Indigenous hydrological knowledge and satellite data on water scarcity (Abo-Khalil, 2024). In this sense, technology can function more as a cultural interface than as a replacement for the older systems of technologies. McGregor et al. (2020) posit that digital storytelling combined with oral history is one way to document and revitalize Indigenous ecological memory. Similarly, virtual and augmented reality has the potential to foster empathy and ecological mindfulness through immersive learning experiences that help students engage with the environments of their ancestors (Lotz Sisitka et al., 2015).

Integrating Indigenous and non-Indigenous techno knowledge entails knowing the power dynamics and epistemic hierarchies involved (Shahjahan & Seinn, 2025). The authors note that the peripheries of 'dominant' Western academia systematically ignore and construe non-Western epistemologies in a contextual manner, thus perpetuating the asymmetry which continues to 'colonialize' the recognition of Indigenous knowledge (Huot & Chheang, 2025). The authors' curriculum redesign indicates the need to address the epistemic injustice stemming from the Western dominant academic structures, where Indigenous peoples are wholly regarded as equal co-designers, authors, and knowledge collaborators in the co-creation and legitimization of the structures. Some models of Indigenous epistemologies and pedagogies demonstrate this such as the following:

1. Community-Based Co-Design: Educational programs that are collaboratively designed with Indigenous communities, where curricula include and centre the local histories, languages, and priorities of the community (Chinn, 2007; Dei, 2011).
2. Participatory Use of Technology: The deployment of digital mapping or citizen science in a way that honours and respects the community's ownership of the data and knowledge (Sterling, 2024).
3. Dialogic Pedagogy: The fusion of Freirean (1970) dialogical methods with Indigenous oral pedagogies, where learners are positioned as co-inquirers and pedagogical agents rather than as passive receivers.

When the aim of technology is not to supplant but to bolster the Indigenous voice, it becomes a tool for reconciliation and regeneration. In this sense, Dei (2011) argues, knowledge integration succeeds only when Indigenous epistemology is not an add-on to Western science, but a rebalancing of what counts as knowledge itself.

In the UAE, sustainability education is an intentional example of the integration of the two extremes of the fusion of heritage-based knowledge and advanced technologies. The Zayed Sustainability Prize and The Sustainable Schools Initiative (SSI) incorporate certain Bedouin ecological wisdom, such as water and sand dune management, and traditional wayfinding, along with the teaching of renewable energy and AI climate modelling (Abo-Khalil, 2024). This curricular integration illustrates the implementation of the UAE Vision 2071 principle of sustainability through innovation grounded in heritage. observes that the Emirati education reforms now advocate for heritage-informed innovation, whereby the cultural dimension of the identity of the UAE is reflected in the preservation and conservation of the environment. With the Smart Desert Farming project, where students use satellite systems to monitor irrigation and Indigenous practices of soil management, the UAE exemplifies the integration of tradition with innovation (Chheang & Huot, 2026).

New Zealand's education system includes extensive examples of integration of Indigenous knowledge systems. The guardianship (Kaitiakitanga) of the environment from Māori cosmology is taught at school level (Te Marautanga o Aotearoa) and as part of environmental education, for example in the initiative Enviroschools (Macfarlane et al., 2007). The Enviroschools program also reframes environmental education as cultural restoration. Students, for example, use Māori lunar planting calendars (*maramataka*) and meteorological data to plant and monitor the growth of trees on sacred sites with drones, and use GIS to map and document their ancestors' lands (Reihana et al., 2021). This approach to education emphasizes the relational and spiritual aspects of sustainability and education, and the principles of peace, reciprocity, and responsibility.

Canada's incorporation of Mi'kmaw Elders Albert and Murdena Marshall's Two-Eyed Seeing (Etuaptmumk) approach into Indigenous STEM Bartlett et al. (2012) has been particularly notable. This approach has been utilized by the Universities of Cape Breton and Saskatchewan for teaching environmental science and climate change. Students learn geospatial technologies and ecological field techniques through land-based learning with Elders (Cirkony et al., 2023; Zeyer, 2022). These programs develop epistemic humility by recognizing the limits of a single perspective. Two-Eyed Seeing, as Bartlett et al. (2012) and Zeyer (2022) state, facilitates qualification experiences in contrast to the quantitative science. Therefore, the Canadian model is an example of an educational ecosystem where innovation and tradition as interdependent sources of value.

The examples above illustrate that to bridge both tradition and innovation, adjustments to the curriculum alone will not suffice, and that a more profound shift in the educational philosophy is necessary. Indigenous Knowledge Systems continue to serve as an ethical guide and centre that promotes learning through the lenses of respect, reciprocity, and relational accountability. At the same time, innovation in technology further expands the methodological boundaries of sustainable education, allowing local wisdom to globalize. This merging of both local and global practices converts education to what Sterling (2024) refers to as a regenerative learning ecosystem.

Here, multiple knowledge systems integrate to create a future that is socially and ecologically balanced. It also responds to UNESCO (2023) advocacy for epistemic pluralism, as it integrates cultural diversity, which is necessary for the survival of the planet, into the education for sustainable development (ESD). Successful integration, as illustrated in the UAE, New Zealand and Canada shows that: (1) indigenous communities spearhead the educational design and validation; (2) technology is used as a mediator for dialogue rather than a controller and (3) the desired learning outcomes focus on transformational change rather than knowledge acquisition.

The integration of modern innovation with Indigenous knowledge systems reframes education as cultural renewal and stewardship of the planet. As Wals (2011) states so poignantly, the way out of unsustainability cannot be innovated without first remembering how to belong. It is obvious, then, that the combination of tradition and innovation is much more than an educational practice; as represented in Table 1.

Table 1.

Country	Concept	Innovation	Application	Outcomes
United Arab Emirates (UAE)	Bedouin Ecological Wisdom (Falaj system, Desert Adaptation)	Blending heritage-based sustainability with renewable energy, AI-driven farming, and smart water management	Sustainable Schools Initiative (SSI); Smart Desert Farming projects	Enhanced cultural identity, climate adaptability, and applied eco-literacy among students
New Zealand (Aotearoa)	Māori Concept of Kaitiakitanga (Guardianship of Nature)	Integration of Māori cosmology with STEM learning and environmental science through digital tools (GIS, drones)	Enviroschools Programme; National Curriculum (Te Marautanga o Aotearoa)	Deepened place-based awareness, ecological ethics, and biocultural conservation practices
Canada	Two-Eyed Seeing (Etuaptmumk) – Mi'kmaw Epistemology	Co-learning model uniting Indigenous land-based knowledge with scientific and technological methodologies	University-based land-learning programs; STEM-integrated sustainability education	Fostered epistemic humility, intergenerational learning, and reconciliation through education

IV. IMPLEMENTATION AND PRACTICE

4.1 Strategies for Incorporating Indigenous and Place-Based Learning in Education

Regarding formal and non-formal education embracing Indigenous Knowledge Systems (IKS) and place-based learning, the practice needs to be far removed from superficial, one-off inclusion. There needs to be a shift from content insertion to epistemic transformation (Sterling, 2024). This means that Indigenous knowledge systems must interactively become part of the foundational philosophy of education, not merely as an addition. In formal education, such as in schools and universities, initiatives often begin with curriculum localization, where the learning outcomes are geared towards specific local ecosystems and cultural systems. As Gruenewald (2003) and Smith (1999) point out, this is a process that, in terms of place-based pedagogy, facilitates the articulation of academic learning to the social and cultural practices of the targeted community. For instance, instead of isolating the concept of sustainability as an abstract idea, schools can teach it in relation to local ecosystems and Indigenous land management practices and community stories within the context of a science and social studies integrated curriculum.

Recent developments demonstrate that participatory curriculum models have positive results on developing learners' environmental awareness. For example, in Canada, Two-Eyed Seeing has been institutionalized in the science co-teaching model, where Elders and scientists co-design and co-deliver lessons (Cirkony et al., 2023; Zeyer, 2022). Equally relevant is the incorporation of ecological activities alongside Māori principles such as kaitiakitanga (guardianship) and whakapapa (genealogy) in the Aotearoa New Zealand Enviro-schools Program (Macfarlane et al., 2007).

Involvement of Indigenous peoples' voices in the implementation of such initiatives in informal education, including community education, heritage education, and non-governmental organizations (NGOs) is encouraged. Lotz Sisitka et al. (2015) describe the combination of knowledge, action, and community well-being as the integration using research and community-based service-learning, which model best the participatory approach. In the UAE for instance, local councils conduct heritage-based environmentally focused programs that promote the Global Education for Sustainable Development (ESD) objectives by incorporating traditional crafts, camel herding, and oasis restoration (Abo-Khalil, 2024).

Wals (2011) identifies three interrelated guiding principles to learning for the future: (1) the creation of equitable learning environments where different knowledge systems can engage and be appreciated equally, which he terms as dialogic learning; (2) learning by doing which promotes engagement of learners with the environment, the community, and local practices, which is also termed experiential and land-based education; and (3) constructive and collaborative cycles of reflection and co-evaluation, which involve educators, community members, and learners actively. These guidelines nurture a relational sense of education, drawing from Wilson (2020). With this perspective, all learning is accountable to the relationships that make it possible: people, place, and spirit.

4.2 Community Partnerships and Teacher Education

Collaboration with educational institutions and Indigenous community and local organization partnerships is central to implementation. Dei (2011) argues that the meaningful integration of IKS is inclusive of relational pedagogy, that is, pedagogy, elapsing over time, where trust, reciprocity, and respect are hegemonic in appreciating the Indigenous governance structures. The significance of community school partnerships is even more pronounced in the context of place-based learning. Within the UAE, partnerships involving the Ministry of Education and local universities and local heritage councils have pioneered projects in which students create digital stories about the traditional desert practices, integrating oral histories and climate science (Abo-Khalil, 2024). These partnerships enable communities to be viewed as active participants in the cocreation of knowledge rather than as passive study subjects.

Equally important is teacher education. Bartlett et al. (2012) and Zeyer (2022) posit that in addition to being culturally knowledgeable, educators must possess critical reflexivity and, therefore, the ability to suspend their assumptions around knowledge and authority. Indigenous Teacher Education Programmes (ITEPs) in Canada and *Te Kotahitanga* in New Zealand are exemplary programs in decolonial teacher education that foreground the integration of relational ethics and local knowledge (Macfarlane et al., 2007).

Integrating Indigenous Knowledge Systems (IKS) into teacher training requires innovative approaches based on IKS integration. Educators must find ways to integrate Indigenous teachings (like storytelling, seasonal observation, and intergenerational mentorship) with technology and experiential learning. For instance, augmented reality (AR) tools can be used to immerse learners in places like ancestral lands, or landscapes and systems of traditional resource management so learners can connect Indigenous knowledge systems to contemporary learning (Lotz Sisitka et al., 2015). Another fundamental principle of implementation is participatory curriculum development. For example, Tuck and Yang (2012) state that the curriculum, in this case, should be designed with Indigenous peoples, and not for them. This also aligns with the UNESCO (2020) Education for Sustainable Development Roadmap, that emphasizes the need for equitable partnerships in the co-creation and dissemination of knowledge for culturally sustaining education, aimed at the equitable and systemic outcome of Sustainability.

4.3 Representation, Ownership, and Respect: Ethical Considerations

The process of incorporating Indigenous knowledge is highly ethical. As Smith (1999) and Zurba et al. (2021) remind, ethical knowledge and Indigenous knowledge systems, dovetail such that Indigenous peoples have the right to control the knowledge, and how it is used, represented, and shared. While framing the idea of *Authentic Representation*, the paradox of the monetization and romanticization of the portrayal of Indigenous cultures must always be opposed. It must be emphasized that each of the various systems of knowledge emerged from a particular ecosystem and history. For example, the merging of the Māori models of the universe and the Māori frameworks, with other Indigenous systems, does a disservice to the local meanings and the knowledge systems. This ethically consequential implementation of knowledge requires, at the very least, a contextually relevant adaptation, co-authorship, Kahawai cultural depiction, and validation of Chinn (2007).

A critical aspect of Indigenous knowledge is the communal and intergenerational nature of knowledge ownership. Battiste (2019) and Wilson (2020) highlight that educational settings that incorporate Indigenous knowledge must obtain permissions consistent with the cultural protocols of the respective Indigenous community. Therefore, the Indigenous community should have ownership and control over educational resources, data, and digital content that derive from community knowledge, not educational institutions or publishing companies.

Moreover, culturally ethical practice involves the spiritual and ceremonial aspects of knowledge. Kimmerer (2013) mentioned that the information contained in the Indigenous cultures is relational and there is a relationship that aligns with the information. Because of this, there is a need to establish and maintain pedagogical practices that acknowledge and respect the relationships and connections to specific places and cultures so that practices such as land-based recognition at the opening of the class, the use of local languages, and the instruction of Elders, become standard practice (Huot & Hok, 2025b). Recent literature has suggested multiple methods to implement these principles:

- Zurba et al. (2021) suggest the development of Ethical Partnership Agreements (EPAs) which secure terms of partnerships, attribution, and benefit sharing between educational institutions and Indigenous communities.
- Shahjahan and Seinn (2025) suggest the development of Community Review Boards, which are intended to assess and vet the curricula for cultural safety and other aspects of Indigenous content.
- UNESCO (2020) proposes Digital Sovereignty Guidelines that secure online collections of Indigenous knowledge and ensure they are governed by Indigenous protocols.

This transformative approach, based on the co-creation of new pathways, alongside the ethical and historical injustices of the past, offers the potential for a sustainable and just future. As Wals (2011) succinctly puts it, education for sustainability that is not education for equity and for respect is education for no future.

The Indigenous and place-based education is a constantly evolving, transforming process, and not a checklist of activities. Effective practice comes from relational engagement, the ethics of accountability, and the ethics of reciprocal learning (Huot & Em, 2025). When schools, communities, and policymakers work together under these guiding principles, education becomes, and should be, more than a vehicle for the transmission of knowledge. It becomes a living, breathing, culturally revitalized system. The integrated heritage-innovation approach of the UAE; New Zealand's bicultural model of education; and Canada's Two-Eyed Seeing exemplify that sustainable learning is possible when knowledge is integrated rather than imposed; communities are equal partners and not subjects; and the land is both a teacher and a resource. This holistic approach reaffirms that education for sustainability is, in essence, a cultural and ethical journey characterized by the integration of humility, the reciprocity of all living things, and the respect for the multiple ways of knowing that sustain life on the planet, as illustrated in Table 2.

Table 2.

Context	Strategies	Stakeholders	Programs
Formal Education (Schools & Universities)	Curriculum localization; Co-teaching with Elders; Integration of Indigenous epistemologies into STEM; Experiential and land-based learning modules.	Ministry of Education, Teachers, Indigenous Elders, Curriculum Designers, Students.	Two-Eyed Seeing (Canada); Enviroschools (New Zealand); Sustainable Schools Initiative (UAE).
Non-Formal Education (Community and NGO Programs)	Community-based workshops; Heritage education; Eco-cultural restoration; Youth environmental stewardship projects.	Local councils, NGOs, Elders, Cultural organizations, Environmental educators.	Oasis Restoration & Desert Heritage Projects (UAE); Community Eco-Camps (Canada).
Teacher Education and Professional Development	Training in Indigenous pedagogies; Reflexivity and decolonial methods; Integrating technology with traditional knowledge transmission.	Teacher colleges, Universities, Indigenous Knowledge Keepers, Education Ministries.	Indigenous Teacher Education Programs (Canada); Te Kotahitanga (New Zealand).
Policy and Institutional Implementation	Creating inclusive policies; Ethical partnership frameworks; Establishing cultural advisory boards and community review panels.	Government agencies, Policy-makers, Universities, Indigenous councils, UNESCO affiliates.	ESD 2030 Framework; UAE Green Education Strategy; UNESCO Local Knowledge Initiatives.

V. CHALLENGES AND OPPORTUNITIES

There are challenges in integrating Indigenous Knowledge Systems (IKS) with contemporary sustainability education that are epistemological, institutional, and ethical. One of the biggest challenges is the difference in worldviews regarding Indigenous knowledge systems, which are described as holistic, relational, and place-bound, and the Western systems of knowledge, which are based on analytical, compartmentalized, and universalist structures or paradigms. As Dei (2011) indicates, the difference in worldviews is not only academic, but is also deeply political and indicates the marginalization of Indigenous ways of knowing and the power streamlined dominance to the western systems of knowledge.

According to Bartlett et al. (2012) and Zeyer (2022), mainstream curricula based on western objectivity and standardization fail to consider relational ways of knowing that views knowledge in the context of relationships, community, and ecosystem, and spirituality. Such a difference in worldviews can lead to what is called "epistemic friction" where Indigenous knowledge and ways of knowing are simplified and relegated to the background as mere cultural embellishments. The tokenistic inclusion of Indigenous peoples describes the mere inclusion of Indigenous ways of knowing and is aimed to achieve symbolic recognition of diversity without altering the existing paradigm (Shahjahan & Seinn, 2025).

Policy restrictions aggravate these tensions. Despite a number of countries signing the United Nations Declaration on the Rights of Indigenous Peoples (UNDRIP) and the Education for Sustainable Development (ESD) 2030 Roadmap, they continue to lack coherent and actionable relevant frameworks. National curricula still focus on economic growth and STEM innovation as in the case of the Arab Gulf region, while neglecting Indigenous ecological knowledge, and thus ignoring a critical source of sustainability wisdom (Qablan et al., 2023). The systemic adaptation of place-based learning models is often slowed down by the bureaucracy that lacks teacher training and lacks Indigenous consultation mechanisms (Huot & Hok, 2025b).

The other remaining issue is that of cultural misappropriation and misrepresentation. The rising academic and policy interest in Indigenous knowledge, Smith (1999) points out, runs the risk of perpetuating the colonial patterns of extraction, using knowledge without the consent of the community, and without the cultural context of the community. The digitization of Indigenous resources, often on publicly accessible global platforms, remains a risk to unprotected Indigenous knowledge and the privacy of Indigenous peoples (UNESCO, 2020). Additionally, the knowledge that is acquired and modified in a Western way tends to lose its essence, value and sacredness. In order to lessen these risks, Wals (2011) and Zurba et al. (2021) suggest what can be termed as epistemic humility, which involves acknowledging the absence of a dominant, ultimate knowledge system. This kind of humility involves the need for a certain type of co-governance, wherein Indigenous communities are not merely consulted, but are policy and practice co-authors.

Innovative policies integrating Indigenous knowledge and learning from local places into global educational frameworks have great potential. The UNESCO Roadmap for Education for Sustainable Development (ESD) and the 2030 Agenda for Sustainable Development, which include Sustainable Development Goal (SDG) 4) state inclusive and equitable quality education and demand an education system that honours and integrates the local cultures, local knowledge, and the sustainability of the ecosystems (UNESCO, 2020). Among the priority areas of the ESD 2030 framework, policy advancement, learning environment transformation, educator empowerment, youth empowerment, and local action acceleration, indigenous educational paradigms seamlessly integrate (Sterling, 2024). For example, Indigenous pedagogy is an example of transformative learning and experiential learning, which is what SDG 4.7 aims to achieve, learning that promotes sustainable living, cultural diversity, and global citizenship (Wals, 2011). Thus, to align sustainability education policy to UNESCO's intercultural framework, national and regional institutions should: (a) implement co-management boards composed of education ministry officials and Indigenous councils; (b) integrate local sustainability practices grounded in ecologically-informed traditional practices; (c) adopt intergenerational mentorship frameworks that honour oral traditions and informal science and champion intergenerational mentoring; and (d) advocate education policy reform that incorporates Indigenous pedagogy.

Examples of this type of alignment can already be found in international practice. In New Zealands, Te Aho Matua has integrated the Māori educational frameworks of whanaungatanga (relationships) and manaakitanga (care) as foundational pillars of its curriculum (Macfarlane et al., 2007). Within Canada, the Truth and Reconciliation Commission's Calls to Action (2015) has prompted the legal adoption of land-based learning and supportive government pedagogy (Cirkony et al., 2023; Zeyer, 2022). In the UAE, the Green Education Partnership Framework (2023) supports the model of integrating sustainability with the preservation of cultural heritage and, by formalizing the inclusion of Indigenous ecological knowledge, could further advance (Abo-Khalil, 2024). The cases above demonstrate the necessity to incorporate Indigenous Knowledge Systems (IKS) with Education for Sustainable Development (ESD) and Sustainable Development Goals (SDG) frameworks in the interest of sustainability.

Indeed, as Zurba et al. (2021) express, Indigenous-led initiatives to sustainability governance are crucial to the preservation of biodiversity and the promotion of epistemic justice; two of the most critical goals of the global educational reform movement. The most significant for the impact of Indigenous and place-based education will be the sustained capacity-building activities to be undertaken throughout the educational, institutional, and community spheres. Scholars identify three interrelated components: capacity development, institutional transformation, and the co-production of knowledge (Lotz Sisitka et al., 2015; Wals, 2011).

Capacity Development: Building capacity demands starting with educators. Modules inclusive of cultural epistemology, Indigenous methodologies, and participatory curriculum design must be incorporated into teacher education programs (Bartlett et al., 2012; Zeyer, 2022). Along with teacher professional development efforts, Indigenous-led workshops, and immersion programs, investments in professional development create and sustain the pedagogical confidence necessary for the integration of multiple cultures:

(a) **Institutional Transformation:** Institutions must look past inclusion and toward true transformation. As Sterling (2024) notes, it is paradoxical that sustainability is achieved with unchanged institutional arrangements. A decolonial governance framework, with Indigenous people in leadership and curriculum design and development roles, is essential for universities and ministries. Furthermore, the reconciliation of Indigenous epistemic values, specifically, reciprocity, collective responsibility, and stewardship, with institutional mission statements, is likely to help shift education toward relational sustainability.

(b) **Knowledge Co-Production and Networking:** Sustainable integration requires both global and regional networks that enable the flow of knowledge from and between Indigenous communities, scholars, and policymakers. Examples of the UNESCO LINKS Programme (Local and Indigenous Knowledge Systems) and the GLOBE Program (Global Learning and Observations to Benefit the Environment) illustrate how knowledge co-production can strengthen community systems and scientific systems (UNESCO, 2020).

Ethical governance of collaborative digital platforms can foster multilingual and intercultural content sharing and the preservation of oral histories, ecological knowledge, and teaching materials. These platforms must be governed by Indigenous data sovereignty frameworks to guarantee the control and ownership of digital content (Battiste, 2019; Smith & Sobel, 2010). Lastly, regional centres of excellence focused on Indigenous sustainability education can be transformative. These centres can be based in the Gulf, Oceania, or North America and can be interdisciplinary research and teaching training and policy innovation hubs.

Although the epistemology divides, policy stagnation, and the cultural appropriation risk can be perceived as the barriers, they are also the indicators of the potential areas for transformative change. Each barrier calls for greater reflexivity, dialog, and collaboration between Indigenous pedagogies, communities, and education systems. Indigenous knowledge, when aligned with the principles of

UNESCO's ESD 2030 and SDG 4, should no longer be perceived as a complement to sustainability education. Rather, it is the moral and ecological underpinning of it. Through the distribution of governance, teacher advocacy, and the respect for epistemic pluralism, the combination of Indigenous and place-based education will lead societies to genuine transformative sustainability defined by justice, equity, and the wisdom of the land.

VI. CONCLUSION

Integrating Indigenous Knowledge Systems (IKS) with transformative sustainability education is both an innovative merging of disciplines and a radical shift in societal conceptions of what knowledge is, learning, and what the future may hold for humanity. It has been documented throughout this chapter the vital contributions of Indigenous Wisdom strengthened by the relational, interdependent, and reverent approaches to all of Creation, in envisioning the reformation of sustainability education as a living process in real-time, within contexts, and in particular communities. Fusing this vision with contemporary advancements in education, such as digital learning, systems thinking, and ecological design, has the potential to develop learning environments that are culturally relevant and responsive in a global context.

What stands out from this synthesis is that the achievement of sustainability is not found in technical knowledge alone. It is found in the moral imagination, cultural persistence, and the strong sense of place. Indigenous epistemologies indicate that knowledge is not a thing to be used up, rather it is a relationship that needs to be cultivated and cared for. When this relational understanding is synthesized with existing modern educational systems, a transformative learning paradigm is created. One that equips learners not just to understand the sustainability challenges, but also to live sustainably and fully in their own communities.

A value-driven education system that is flexible and inclusive can help balance cultural heritage and innovation. Educational institutions and governing bodies need to understand that culturally informed education and scientific advancement are collaborative pathways toward resilience. This balance is possible when curricula are designed with Indigenous perspectives, teachers are trained to promote ethical intercultural learning, and communities are collaboratively integrated into the educational governance process. Renewed cultures and transformed global sustainable educations are mutually possible.

Future research should develop models that assess the learning outcomes of sustainability education longitudinally. Indigenous pedagogy is often unmeasured and here is an absence of longitudinal assessments that measure the impact of Indigenous pedagogy on community health, ecological care, and knowledge sharing across generations. There is a need for policymakers to reform the structures that allow for the

REFERENCES

Abo-Khalil, A. G. (2024). Integrating sustainability into higher education challenges and opportunities for universities worldwide. *Heliyon, A Cell Press Journal, 10*(9), 1–13. DOI: 10.1016/j.heliyon.2024.e29946

Barnhardt, R. A. Y., & Kawagley, O. A. (2005). Indigenous knowledge systems and Alaska Native ways of knowing. *Anthropology & Education Quarterly, 36*(1), 8–23. DOI: 10.1525/aeq.2005.36.1.008

Bartlett, C., Marshall, M., & Marshall, A. (2012). Two-eyed seeing and other lessons learned within a co-learning journey of bringing together indigenous and mainstream knowledges and ways of knowing. *Journal of Environmental Studies and Sciences, 2*(4), 331–340. DOI: 10.1007/s13412-012-0086-8

Battiste, M. (2019). *Decolonizing education.* Association of Canadian University Presses. DOI: 10.59962/9781895830972

Berkes, F. (2017). *Sacred ecology.* Routledge., DOI: 10.4324/9781315114644

Cajete, G. (1999). *Native science: Natural laws of interdependence.* Clear Light Pub.

Chandra, D. V. (2014). Re-examining the importance of indigenous perspectives in the western environmental education for sustainability: "From tribal to mainstream education". *Journal of Teacher Education for Sustainability, 16*(1), 117–127. DOI: 10.2478/jtes-2014-0007

Chheang, S., & Huot, S. (2026). Safeguarding digital classrooms: Cybersecurity implications of AI-driven tools in Cambodian education. [JEILT]. *Journal of Education Innovation & Language Teaching, 2*(1), 23–45. DOI: 10.63381/jeilt.v2i1.68

Chinn, P. W. U. (2007). Decolonizing methodologies and indigenous knowledge: The role of culture, place and personal experience in professional development. *Journal of Research in Science Teaching, 44*(9), 1247–1268. DOI: 10.1002/tea.20192

Cirkony, C., Kenny, J., & Zandvliet, D. (2023). A two-eyed seeing teaching and learning framework for science education. *Canadian Journal of Science, Mathematics and Technology Education = Revue Canadienne de l'Enseignement des Sciences, des Mathématiques et de la Technologie, 23*(2), 340–364. DOI: 10.1007/s42330-023-00276-z

Dei, G. J. S. (2011). *Indigenous philosophies and critical education.* Peter Lang Group AG. DOI: 10.3726/978-1-4539-0131-1

Fernández-Llamazares, Á., & Cabeza, M. (2017). Rediscovering the potential of indigenous storytelling for conservation practice. *Conservation Letters*, *11*(3), 1–12. DOI: 10.1111/conl.12398

Fleming, T. (2018). Mezirow and the theory of transformative learning. In Wang, V. (Ed.), *Critical theory and transformative learning* (pp. 120–136). IGI Global Publishing., DOI: 10.4018/978-1-5225-6086-9.ch009

Freire, P. (1970). *Pedagogy of the oppressed* (M. B. Ramos, Trans.). *Continuum.*

Gruenewald, D. A. (2003). The best of both worlds: A critical pedagogy of place. *Educational Researcher*, *32*(4), 3–12. DOI: 10.3102/0013189X032004003

Gwekwerere, Y. (2011). Bridging culture, indigenous and scientific ways of knowing. *Cultural and Pedagogical Inquiry*, *6*(2), 1–4. DOI: 10.18733/C3GW24

Huot, S. (2026). Emerging voices from Cambodia: International student journeys and educational aspirations in a globalized academic landscape. In Irham, I., Miftahul, H., & Lalu Nurul, Y. (Eds.), *International student narratives in globalized higher education* (pp. 363–398). IGI Global Scientific Publishing., DOI: 10.4018/979-8-3373-4212-2.ch014

Huot, S., & Chheang, S. (2025). Unwriting empire: Decolonizing Khmer language, literature, and cultural memory. *The Context Journal, 12*(6), 280–293. https://doi.org/https://doi.org/10.5281/zenodo.17061523

Huot, S., & Em, S. (2025). The integration of English language teaching with environmental sustainability: A comprehensive review of pedagogical strategies and global impacts. [CJESS]. *Cambodian Journal of Educational and Social Sciences*, *2*(2), 83–95. DOI: 10.69496/cjess.v2i2.68

Huot, S., & Hok, C. (2025a). Enhancing English language teaching for multilingual learners: Leveraging heritage languages in the classroom. *Journal of Social and Educational Research*, *4*(1), 87–95. DOI: 10.5281/zenodo.15698963

Huot, S., & Hok, C. (2025b). Unpacking the struggle: Linguistic, pedagogical, and cognitive barriers in Cambodian students' English essay writing. *Context (Ibadan)*, *12*(4), 149–168. DOI: 10.5281/zenodo.15807401

Huot, S., Loch, S., & Rin, R. (2025). Land, life, and the sacred: Reimagining development through tribal worldviews. In Sayak, P., Susmita, H., & Nitesh, T. (Eds.), *Psychology of media influence: Shaping thoughts, behavior and society* (pp. 448–462). Mitra Press.

Kimmerer, R. W. (2013). *Braiding sweetgrass: Indigenous wisdom, scientific knowledge and the teachings of plants*. Milkweed.

Kolb, D. A. (2015). *Experiential learning: Experience as the source of learning and development* (2nd ed.). Pearson Education.

Kovach, M. (2009). *Indigenous methodologies: Characteristics, conversations, and contexts*. University of Toronto Press.

Leal Filho, W., Vargas, V. R., Salvia, A. L., Brandli, L. L., Pallant, E., Klavins, M., Ray, S., Moggi, S., Maruna, M., Conticelli, E., Ayanore, M. A., Radovic, V., Gupta, B., Sen, S., Paço, A., Michalopoulou, E., Saikim, F. H., Koh, H. L., Frankenberger, F., & Vaccari, M. (2019). The role of higher education institutions in sustainability initiatives at the local level. *Journal of Cleaner Production*, *233*, 1004–1015. DOI: 10.1016/j.jclepro.2019.06.059

Lin, J., Stoltz, A., Aruch, M., & Rappeport, A. (2021). Decolonization and transformation of higher education for sustainability. *Journal of Comparative & International Higher Education*, *13*(Summer), 134–156. DOI: 10.32674/jcihe.v13iSummer.3255

Lotz Sisitka, H., Wals, A. E. J., Kronlid, D., & McGarry, D. (2015). Transformative, transgressive social learning: Rethinking higher education pedagogy in times of systemic global dysfunction. *Current Opinion in Environmental Sustainability*, *16*, 73–80. DOI: 10.1016/j.cosust.2015.07.018

Macfarlane, A., Glynn, T., Cavanagh, T., & Bateman, S. (2007). Creating culturally safe schools for māori students. *Australian Journal of Indigenous Education*, *36*(1), 65–76. https://doi.org/https://doi.org/10.1017/s1326011100004439. DOI: 10.1017/S1326011100004439

Manns, M., Heiss, J., Morohashi, J., & Loupis, L. (2025). Harnessing indigenous knowledge for education for sustainable development. *Journal of Education for Sustainable Development*, *19*(2), 157–167. DOI: 10.1177/09734082251360875

McGregor, D., Whitaker, S., & Sritharan, M. (2020). Indigenous environmental justice and sustainability. *Current Opinion in Environmental Sustainability*, *43*, 35–40. DOI: 10.1016/j.cosust.2020.01.007

Qablan, A., Alblooshi, K. M., & Alkaabi, F. A. (2023). Education for sustainable development (ESD) and school leadership. In Abdallah, A. K., & Alkaabi, A. (Eds.), *Restructuring Leadership for School Improvement and Reform* (pp. 378–398). IGI Global., DOI: 10.4018/978-1-6684-7818-9.ch019

Redvers, N., Schultz, C., Vera Prince, M., Cunningham, M., Jones, R., & Blondin, B. (2020). Indigenous perspectives on education for sustainable healthcare. *Medical Teacher*, *42*(10), 1085–1090. DOI: 10.1080/0142159X.2020.1791320 PMID: 32657230

Reihana, K. R., Wehi, P. M., Harcourt, N., Booth, P., Murray, J. M., & Pomare-Pieta, M. (2021). Indigenisation of conservation education in New Zealand. *Pacific Conservation Biology*, *27*(4), 493–504. DOI: 10.1071/PC20060

Shahjahan, R. A., & Seinn, A. P. (2025). Towards 'decolonizing' engineering education: An integrative review. *Cogent Education*, *12*(1), 1–15. DOI: 10.1080/2331186X.2025.2541075

Smith, G. A., & Sobel, D. (2010). *Place-and community- based education in schools*. Routledge.

Smith, L. T. (1999). *Decolonizing methodologies research and indigenous peoples*. University of Otago Press.

Sobel, D. (2004). 2005). In *Place-based education: Connecting classrooms & communities* (2nd ed.). The Orion Society.

Sterling, S. (2001). *Sustainable education: Re-visioning learning and change*. Green Books for the Schumacher Society.

Sterling, S. (2024). *Learning and sustainability in dangerous times: The stephen sterling reader*. Agenda Publishing., DOI: 10.2307/jj.13473649

Tuck, E., & Yang, K. W. (2012). Decolonization is not a metaphor. *Decolonization*, *1*(1), 1–40. DOI: 10.25058/20112742.n38.04

UNESCO. (2020). *Education for sustainable development: A roadmap*. United Nations Educational, Scientific and Cultural Organization (UNESCO). DOI: 10.54675/YFRE1448

UNESCO. (2023). *New report and guidelines for indigenous data sovereignty in artificial intelligence developments*. UNESCO. https://www.unesco.org/en/articles/new-report-and-guidelines-indigenous-data-sovereignty-artificial-intelligence-developments

Wals, A. E. J. (2010). *Message in a bottle: Learning our way out of unsustainability*. Wageningen University & Research (WUR).

Wals, A. E. J. (2011). Learning our way to sustainability. *Journal of Education for Sustainable Development*, *5*(2), 177–186. DOI: 10.1177/097340821100500208

Wilson, S. (2020). *Research is ceremony: Indigenous research methods*. Fernwood publishing.

Yunkaporta, T. (2019). *Sand talk: How indigenous thinking can save the world.* Text Publishing.

Zeyer, A. (2022). Teaching two-eyed seeing in education for sustainable development: Inspirations from the science environment health pedagogy in pandemic times. *Sustainability (Basel), 14*(10), 1–12. DOI: 10.3390/su14106343

Zurba, M., Petriello, M. A., Madge, C., McCarney, P., Bishop, B., McBeth, S., Denniston, M., Bodwitch, H., & Bailey, M. (2021). Learning from knowledge co-production research and practice in the twenty-first century: Global lessons and what they mean for collaborative research in nunatsiavut. *Sustainability Science, 17*(2), 449–467. DOI: 10.1007/s11625-021-00996-x

KEY TERMS AND DEFINITIONS

Cultural Heritage: Cultural Heritage includes the community and the individual expressions of knowledge of a community, including traditions, languages, arts, rituals, and environment. It is a collection of community memory and is a bridge between the past and the present. It provides the moral and the cultural foundations for sustainable for sustainable development and education.

Indigenous Knowledge Systems (IKS): These systems exemplify the intricate, cross-generationally segmented knowledge, value, and practice systems Indigenous groups develop through their distinctive, prolonged, and interactive engagements with their natural and cultural environments. These systems also reflect the distinct relational worldview and collective experience of the specific Indigenous community. Aspects of IKS include ecological knowledge and wisdom, spirituality, social (community) structuring, and community ethics of sustainability of resilience.

Innovation in Education: Educational innovation encompasses the creation and the application of new strategies of teaching and new technologies and new institutional practices to improve the effectiveness and the inclusiveness of learning. In the context of sustainability, this innovation means the combination of contemporary, including digital, media, cross- and inter-disciplinary, and experiential forms of learning, with the teaching and learning practices of both contemporary and traditional and Indigenous knowledge systems.

Place-Based Learning: This type of learning integrates the local environment and community with the learning process. It ties learning to the local context and lived experiences of learners, encouraging them to explore and interact with their environment to develop ecological awareness, cultural identity, and civic engagement.

Policy Integration: Policy integration deals with the incorporation of the cross-cutting principles of sustainability and the Indigenous paradigms of education into the design of education systems, structures of governance, and the frameworks of institutional leadership. It harmonizes the cross-sectional integration of the environmental, cultural, and education domains to provide the learning systems that promote equitable and sustainable transformation of society.

Sustainability Education: This type of education embodies a transformative learning experience and focuses on empowering individuals and communities with the requisite knowledge, skills, and values that defend the sustainability of the environment, equity of the society, and viability of the economy. It motivates learners to think rationally, and act and participate responsibly in the realization of sustainable communities.

Transformative Pedagogy: Transformative pedagogy is concerned with teaching techniques that promote, in addition to learning the content of a subject, the importance of critical thinking, self-consciousness and action, and social awareness. It helps learners to think of the alternative to the dominant systems and structures, re-evaluate their attitudes towards the ecosystem, and promote attitudes that favour sustainability and equity.

Chapter 2
Land, Culture, and Learning:
Integrating Indigenous Wisdom Into Transformative Sustainability Education

Yuldoshev Rustam Elmurodovich
Karshi State University, Uzbekistan

Huseynova Abira Amanovna
Oliy Majlis of the Republic of Uzbekistan, Uzbekistan

Ravikumar R. N.
http://orcid.org/0009-0009-3705-1681
Marwadi University, Rajkot, India

Azizjonova Asilabonu
Termez University of Economics and Service, Uzbekistan

S. Aarthi
http://orcid.org/0009-0006-9064-2091
Marwadi University, Rajkot, India

Rustamov Shavkat Shuxratovich
Bukhara State University, Uzbekistan

ABSTRACT

The idea of indigenous knowledge systems is based on the ancient connection with the land and culture and serves as the main pillars of the reconsideration of the sustainability education in the context of new environmental and social issues of our current world. The chapter explores the opportunities of teaching sustainability through the incorporation of the Indigenous epistemologies and place-based pedagogies to alter abstract and theory-based teaching towards the experiential, moral, and community-based learning. Using the literature on the world and current case studies, the chapter proves that land-based learning, Indigenous storytelling, and community-school partnerships contribute to the increase in ecological awareness, cultural identity, and agency of learners. It also emphasizes ethical activity, knowl-

DOI: 10.4018/979-8-3373-7650-9.ch002

edge sovereignty, and policy congruence as the main preconditions of meaningful integration stating that Indigenous ways of knowing must be used as central reference points when forming more resilient, fair, and sustainable futures.

1. INTRODUCTION

The growing environmental degradation, biodiversity, climate instability, and growing socio-cultural inequities have increased the need for sustainability education as a critical response. Standard systems of education, which are mostly based on theory-focused and technocratic paradigms have been failing to consider the highly relational and place-based character of sustainability issues. This chapter establishes Indigenous knowledge systems as the basis and not subordinate to sustainability education in that they are understood to be holistic and that land, culture and learning are inter-connected (Kioupi & Voulvoulis, 2019). The indigenous worldviews, formed on the basis of centuries of life experience and intergenerational transmission, offer ethical and ecological information that is required to promote resilience, stewardship, and a sense of responsibility to the group. By presupposing place-based learning and Indigenous epistemologies, the chapter proposes a radical change in the sustainability education that is no longer based on abstract global narratives but rests on the local pedagogies based on cultural responsiveness (Cruz, 2020). The chapter incorporates both theoretical understanding, empirical literature, and current case studies in order to show that education based on land and culture can create ecological awareness, cultural identity, and ethical involvement. In general, the introduction provides the justification of rethinking sustainability education based on Indigenous wisdom in order to enable just, resilient, and sustainable futures.

1.1 Sustainability Education in Global Crises

The global environmental crises like climate change, deforestation, water scarcity, and ecological degradation are getting more and more intertwined with cultural erosion, social marginalization, and loss of traditional knowledge system. Sustainability education has been advocated as an intervention tool to combat these problems but prevalent models tend to focus more on scientific rationalism and adherence to policy rather than on experience and sense-making (McCarty & Lee, 2014). These practices often ignore the social and moral aspects of sustainability leading to disjointed knowledge and little behavioral change. Simultaneously, Indigenous cultures on the global level are being endangered by the forces of globalization, land displacement, and epistemic marginalization which adds to the undermining of the environmental stewardship by communities. Such convergent crises are evidence of the insufficiency of sustainability education models that are not yet contextualized with the cultural backgrounds and local truths (Levitan & Johnson, 2020). There is a need to do a reframing sustainability as both ecological and cultural is an ecological and cultural project. By implementing the Indigenous worldviews, education systems will be able to solve ecological problems and, at the same time, restore the cultural knowledge base, enhance the resilience of communities, and promote the moral relationship between people and nature.

1.2 Indigenous Knowledge in Sustainability Education

Place-based epistemologies are indigenous knowledge systems that are exercised on the basis of relationality, reciprocity and respect to living systems. They combine ecological knowledge with cultural, spiritual and social values in contrast to the reductionist frameworks, which provide a holistic basis of sustainability education. These systems include extensive expertise of biodiversity, cyclic processes, resource use and adaptive resilience, which have been acquired over years of engagement with particular landscapes (Nesterova, 2020). They are important with respect to environmental wisdom, but their ethical character of stewardship and common good. Through the incorporation of Indigenous knowing in sustainability education, students get to engage with other ways of knowing that challenge anthropocentric and extractive ways of knowing. It further promotes culturally reduced learning activities which are relevant to the local settings and realities on the ground. It is worth noting that Indigenous knowledge is significant to educational equity, since it affirms marginalized epistemologies and supports decolonizing pedagogies. Thus, integrated into a meaningful way, such knowledge systems can help in improving critical thinking, environmental responsibility and moral interaction among learners.

1.3 Chapter Scope and Structure

The chapter critically discusses the opportunities for incorporating Indigenous knowledge and place-based learning into sustainability education to achieve transformative learning outcomes. Its vision entails theoretical perspectives, international and Indian scenarios, pedagogical forms, and the contemporary community-based instances. This chapter is written in a progressive way of building up the insight, the first section is the conceptual background of the literature, the synthesis of the literature, then the methodological implications, the empirical examples and the analysis. It also adds to the body of existing knowledge by not placing Indigenous knowledge as an extra resource but as one of the key epistemological tools, which must be utilized in explaining how humans and nature interact. Moreover, the chapter provides lessons that can be applied to practice by teachers, policy makers and institutional leaders who would wish to implement culture-based sustainability curriculum. The chapter brings to a sensible and conscientious approach to knowledge integration addressing the dangers of ethical participation, institutional issues and usurpation. In general, this would contribute to the statement of a clear, culturally sensitive model of sustainability education based on the global concerns and local conditions.

2. BACKGROUND STUDY

The research context could be discussed at the intersection of sustainability, Indigenous epistemology and place-based learning paradigm. Despite the fact that sustainability education has gained acuity in most regions of the globe, the conceptual and pedagogical foundations of sustainability education have been based largely on the Western academic culture, and sometimes to the detriment of the Indigenous knowing (Ravikumar & Aarthi, 2025b). This has led to educational frames that focus on international frames and standardisation results at the cost of local knowledge systems and cultural particularity. At the same time, Indigenous communities have also carried on with sustainable land stewardship through the lens of experiential learning, oral tradition and shared memory (Jacob et al., 2024). The lack of association between Indigenous knowledge and the formal education systems has restricted the transformative capacity of the sustainability education. Place-based learning has come out as a potential solution to this gap by grounding learning in local places and cultures. Knowledge of this background would be necessary to reframe sustainability education as a relational, ethical, and culturally embedded practice, but not a technical, or even policy-based project.

2.1 Evolution and Limits of Sustainability Education

Sustainability education is the new promotion of environmental education and education on sustainable development, which, in contrast to the former, focuses on competencies in awareness, conservation and policy. The initial models were concerned with scientific literacy and environmental management with the added social and economic aspects. Although this development is taking place, sustainability education is frequently disjointed, discipline-based and abstract. The cognitive indicators are commonly used to evaluate learning outcomes instead of the behavioral change or ethical interaction. Moreover, hegemonic models are prone to generalizing sustainability issues, which do not consider ecological differences and cultural traditions (Housel & Nikoui, 2024). It has resulted in the arrangement of pedagogies that are not really relevant to local communities, especially Indigenous communities, as their systems of knowledge are based in a particular landscape. This marginalization of Indigenous epistemology strengthens the epistemic hierarchies and restricts exposure of learners to other forms of sustainability ways. These restrictions highlight the necessity of pedagogical change that adopts pluralistic knowledge systems and experience and land-based learning. As a way of forming the concept of how sustainability education can be transformed to be of transformative nature, a cyclical approach is suggested. The model incorporates competencies, pluralistic knowledge systems, land-based learning, and behavioral assessment (Ravikumar et al., 2025). The figure 1 shows that sustainability education is consistently a cycle of challenges and solutions, where competences, pluralistic knowledge, cultural respect, land-based learning, and behavioral evaluation are all combined to produce the holistic and transformative learning outcomes.

Figure 1. Cyclical Framework for Sustainability Education

2.2 Indigenous Knowledge, Land, and Culture

The Indian knowledge systems have historical basis on the connections between communities, land, and culture over a long period of time. The imparting of knowledge is done through storytelling, rituals, observation and involvement in the daily life and this strengthens the idea of a worldview land is not a resource but a living being. Indigenous knowledge is epistemologically relational, contextual, and holistic involving the ecological process and moral and spiritual values. This is in comparison to the dominant western epistemologies that focus on abstraction, objectivity and compartmentalization (Sy, 2022). Indigenous knowledge has suffered marginalization in terms of its historical exclusion by the colonizing processes and the formal education systems and has been undermined by the lack of continuity in community-based sustainability practices. Nevertheless, the recent revival of academic and policy concern has raised the question of the usefulness of Indigenous knowledge in solving modern environmental problems. Appreciation of its epistemological underpinnings is critical towards ethical assimilation in the education systems where knowledge is admired as sovereign, situated and inseparable with cultural identity and land. Table 1 Overviews essential principles of Indigenous (relationality, reciprocity, stewardship, intergenerational learning) and how they are applicable to sustainability education.

Table 1. Indigenous Knowledge Principles for Sustainability Education

Indigenous Principle	Description	Relevance to Sustainability Education
Relationality	Emphasizes interconnectedness among humans, land, and non-human beings	Promotes holistic understanding of human–environment systems
Reciprocity	Mutual responsibility between people and nature	Encourages ethical resource use and sustainable behavior
Stewardship	Moral obligation to care for land across generations	Supports long-term ecological responsibility
Intergenerational Learning	Knowledge transmission across generations	Sustains cultural continuity and resilience
Community-Centeredness	Collective knowledge creation and decision-making	Strengthens social cohesion and shared accountability

2.3 Place-Based Learning in Sustainability Education

Place-based learning focuses on local education that is based on local environments, cultures and communities that are experiential. This model of pedagogy is quite close to Indigenous learning models that focus on observation, participation, and relationship to land. Placing learning in ecological and cultural contexts by locating the learning process in context allows place-based education to allow learners to bridge the gap between abstract sustainability ideas and lived experience (Gainsford

& Evans, 2021). It is an interface between Indigenous knowledge and institutional learning providing a platform to allow community knowledge possessors, Elders, and practices into the curricular systems. Place-based learning is also a source of emotional connection to place making learners more responsible and stewarding. It will be able to contribute to knowledge co- creation instead of knowledge extraction when practiced ethically, in observance of cultural guidelines and community jurisdiction. Place-based learning can be described, therefore, as a solution-oriented and pedagogically sound avenue of integrating Indigenous knowledge within sustainability education, which is both academically rigorous and culturally authentic (Datta, 2024). Table 2 compares prevailing models of sustainability education with Indigenous and land-based pedagogies by comparing epistemology, pedagogy, learning context and outcomes.

Table 2. Conventional vs. Indigenous Place-Based Education

Dimension	Conventional Sustainability Education	Indigenous Place-Based Education
Epistemology	Scientific, universalized	Contextual, relational
Learning Context	Classroom-centered	Land- and community-centered
Pedagogy	Theory-driven, instructional	Experiential, participatory
Knowledge Source	Textbooks, experts	Elders, land, lived experience
Learning Outcomes	Awareness and compliance	Stewardship, identity, action

3. RESEARCH OBJECTIVES

The chapter is informed by a clearly established research objectives that attempt to methodically investigate the nature of educational, cultural, and institutional relevance of the integration of Indigenous knowledge and place-based learning in sustainability education. The goals are meant to facilitate a transformative approach to instrumental or additive inclusion to epistemic integration. Going about pedagogical practices, ethically and systemic issues, the objectives give a systematic basis to theoretical examination, empirical research and practical results that will be addressed further. The research objectives of this chapter are the following:

- To examine the role of Indigenous knowledge systems in enabling transformative sustainability education grounded in relational, ethical, and ecological worldviews.

- To analyze the effectiveness of place-based and land-centered pedagogical approaches in fostering experiential, culturally responsive, and action-oriented sustainability learning.
- To explore how Indigenous epistemologies contribute to learners' ecological awareness, cultural identity, and sense of stewardship toward local environments.
- To identify institutional, cultural, and ethical dimensions influencing the integration of Indigenous knowledge into formal education systems, including challenges such as resistance, resource constraints, and risks of cultural appropriation.
- To propose pathways for ethically integrating Indigenous knowledge into sustainability education through policy support, community collaboration, and alignment with global sustainability frameworks such as the UN SDGs.

4. CONCEPTUAL MODEL

The proposed theoretical framework in this chapter is an integrative solution to understanding how the Indigenous knowledge systems can be utilized to bring about transformative sustainability education with regard to the interdependence of land, culture, learning, and the global sustainability requirements. The model in place of Indigenous knowledge offers an adjacent contextual additive would situate it as a structural epistemic framework that restores the cause of education, pedagogy and educational outcomes. It is also the model that is anchored on the Indigenous theoretical principles that emphasize on the relationality of human and nature, the reciprocity of the ecological interaction and the stewardship as the moral duty. Land-based and place-based pedagogies are informed by the principles that encourage experiential learning, continuity of cultures and also environmental responsibility (Opoku & James, 2021). The framework goes a step further to place the Indigenous knowledge in the context of the wider sustainability discourses by contextualizing the local practices with the global sustainability objectives, especially the United Nations Sustainable Development Goals (SDGs) (Aarthi, et al., 2025). The model shows how sustainability education can be locally based but still play a role in global agendas via integrating local epistemologies with global objectives. On the whole, the conceptual model is an analytical tool that allows investigating pedagogical change, moral assimilation, and systemic congruency in sustainability education. Developed based on the theoretical discourse, the chapter suggests a combined conceptual framework that links the Indigenous epistemologies with sustainability learning outcomes. Figure 2 proposes a comprehensive model, which links Indigenous knowledge systems, land-based pedagogies, theoretical concepts and global sustainability goals in order to provide whole body and ethically oriented sustainability education.

Figure 2. Integrated Model for Sustainability Education

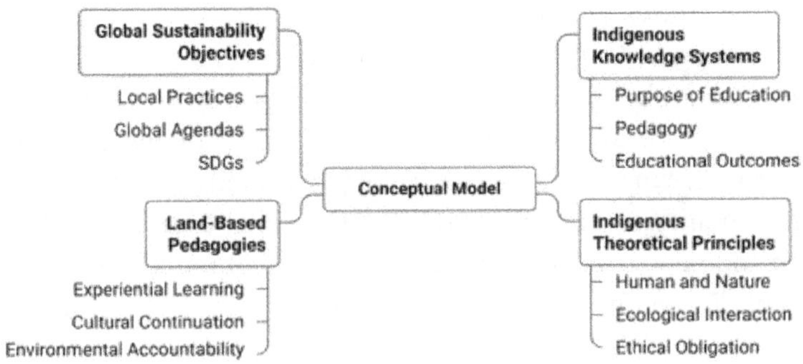

4.1 Foundations of Relational Sustainability

Three concepts, relationality, reciprocity, and stewardship are associated with the three theoretical foundations of the conceptual model because of their centrality in the Indigenous epistemologies. The concept of relationality highlights the interrelatedness of human beings, non-human beings, land and spirituality, which challenges anthropocentric beliefs that divide the society and nature. The concept of knowledge is realized to be a result of relationships but not an abstract observation. Reciprocity is a concept of responsibility between human beings and the environment, in which the responsibility towards the land is returned by sustenance, identity and continuity of cultures (Laver et al., 2024). Ethical decision-making and sustainable use of resources are informed by this principle. Stewardship is a moral responsibility to maintain and preserve ecological systems to serve the current and future generations and the activity is past the individual interest to the collective and intergenerational welfare. The combination of these principles permits a normative framework of sustainability education, which is based on ethical action, long-term orientation, and ecological deference (Aarthi, Ravikumar, Khudayberganov, et al., 2025). By instilling the said pillars within the school systems, students will be able to transform sustainability not only as a technical process, but as a relationship and a moral behavior as well.

4.2 Land–Culture–Sustainability Framework

The overall conceptual map is that there is interdependence of the land, culture, learning, and sustainability as the units of transformative education which support each other. Ecological processes, patterns of seasons and human-environment interaction, are offered by a setting and a place to practice the opportunity of seeing. Culture is an intermediary of learning because it entrenches values, stories and practices that inculcate the sense of sustainability and its practice. It is an intergenerational and participative learning, experiential learning that is executed by land-based activities, storytelling, community mentorship and reflections. An example, in Canada and Australia, other Indigenous education programs are land-based field education, in which the students are taught about the ecological processes, managing biodiversity and environmental stewardship by the direct teaching of the knowledge holders and Elders of the community. Sustainability is the outcome of this kind of interactions in the form of ecological awareness, cultural continuity, and responsible behaviour (Day and McPherson, 2025). The framework is the feedback loops through which learning makes cultural identity and stewardship are stronger which results in improved sustainable relations with the land. A combination of these dimensions will take the model beyond the disintegrated learning experiences in order to provide a comprehensive framework on which to design ecologically informed, culturally conscious, pedagogically radical curriculums. Table 3 Provides the most important elements and relationships of the suggested conceptual framework of Indigenous sustainability education.

Table 3. Model Components of Land–Culture–Learning–Sustainability

Component	Core Role	Educational Function
Land	Living source of knowledge	Site for experiential and ecological learning
Culture	Value system and worldview	Shapes meaning, ethics, and identity
Learning	Relational and experiential process	Enables transformative understanding
Community	Knowledge holders and mentors	Supports co-creation and continuity
Sustainability	Outcome of integrated relationships	Guides responsible action and resilience

4.3 Indigenous Knowledge and the SDGs

The conceptual model introduces the relevance of Indigenous knowledge systems in global sustainability agendas by showing that they can be used to address the United Nations Sustainable Development Goals (SDGs). Practices of indigenous land stewardship, conserving biodiversity, water and community resilience are directly related to climate action, life on land, clean water and sustainable communities goals. These alignments make the point that, global sustainability goals are better attained with local knowledge and practices ingrained within a particular culture (Athayde et al., 2017). Instead of giving universal solutions, the model proposes the SDGs to be adapted to the contexts of Indigenous epistemologies. This will enhance the relevance of policies, local ownership, and sustainability. Also, the alignment promotes educational equity because Indigenous knowledge is accepted as the legitimate participant of the global discourse on sustainability. Through its connection of local wisdom to international outcomes, the framework exemplifies what sustainability education can produce in the form of globally aware, yet locally rooted students who can apply their cultural knowledge to solve complicated environmental issues in culturally relevant ways. Table 4 Aligns Indigenous practices and outcomes of sustainability to applicable UN SDGs.

Table 4. Indigenous Knowledge Practices and the SDGs

Indigenous Practice	Sustainability Focus	Aligned SDGs
Traditional land stewardship	Ecosystem protection	SDG 15, SDG 13
Community water management	Water conservation	SDG 6
Sustainable agriculture	Food security	SDG 2, SDG 12
Intergenerational learning	Knowledge continuity	SDG 4
Collective governance	Social equity	SDG 10, SDG 16

5. RESEARCH METHODOLOGY

The research methodology of this chapter is qualitative and interpretive based as it aims to investigate how Indigenous knowledge systems and place-based pedagogies can be incorporated in sustainability education. With the relational, contextual and culturally embedded characteristics of Indigenous epistemologies, the methodology pays a greater emphasis on depth of knowledge over generalizability. The study plan will be a hybrid of the systematic literature review and the selected community-based case studies that will enable the study to capture both theoretical and practice-based lens. It is an interpretivist orientation that informs the investigation with its focus on meaning-making, lived experience, and the socio-cultural aspects of learning (Dodgson, 2019). The treatment of the subject of the research supports the discourse on the necessity of re-orienting the sustainability education to the ethical participation, ecological responsibility, and cultural relevance. The methodology framework does not depict Indigenous knowledge as an object of analysis in itself, but instead, it recognizes it as a living and situational system of knowledge. The practice enables us to research about the pedagogical change, the forces inside the institutions, and ethical issues involved in the integration of knowledge. Overall, the chosen methodology would be appropriate to the objectives of the chapter since it will enable the analysis of the issue of Indigenous knowledge in sustainability education in a subtle, respectful, and context-sensitive way (Star et al., 2025).

5.1 Research Design and Philosophical Orientation

The underlying research design of the study will be the qualitative, exploratory research design that uses the interpretivist and constructivist philosophical orientations. Such perceptions acknowledge the fact that knowledge is socially constructed and culturally and is created in the interaction between the learners, societies and the environments. This type of orientation may be particularly helpful in the context of probing into the system of Indigenous knowledge (relationality and contextual meaning are more central in this case) (Mbah et al., 2021). The design enables one to venture into different realities and perspectives unlike the effort to seek a single objective truth. The non-positivism approach allows taking into consideration the concept of reflexivity and cultural sensitivity to comprehend the processes of pedagogy and experiences of a specific place in relation to the specific environment. The study adopts an integrative design, which involves the integration of theoretical information with real phenomena and on which theoretical and contextual meaning can be acquired step by step (Gruenewald, 2003). This philosophical basis is also ensuring that the epistemology, methodological and ethical demands of dealing with Indigenous knowledge in educational research are harmonized.

5.2 Literature Analysis and Community Case Studies

To obtain the data, the chapter relies on two primary sources of information, such as the thorough analysis of academic sources and limited number of the community-based case studies. Peer-reviewed journal articles, policy reports, Indigenous authored, and international organizations sustainability education, place-based learning, and Indigenous knowledge systems reports were included in the literature review. This is because this review provides a theoretical background with a bring-out of current themes, gaps and arguments (Castagno & Brayboy, 2008). Community-based case studies are other sources of empirical evidence of the applications of Indigenous and land-based pedagogies in practice to supplement the literature. These cases are selected based on the relevance and the date and the interest that they exhibit towards the Indigenous communities. The materials as described in case studies include written learning programs, curriculum programs and community-institution programs. Taken together, these data sources will allow conducting a process of triangulation, which will enhance the analytical rigor at the cost of the contextual sensitivity (Ram et al., 2024). Integrated methodology is useful in attaining the balanced view of theory and practice as far as transformative sustainability education is concerned.

5.3 Ethics, Indigenous Protocols, and Knowledge Sovereignty

Its research approach is ethically-oriented particularly due to the lack of historical historicity of marginalization and misappropriation of Indigenous knowledge. Under the principles of respect, reciprocity, and responsibility, the research is informed by the principles of addressing the Indigenous epistemologies. The indigenous protocols of response to knowledge sharing, authorship and representation are identified with the special emphasis on the communities rather than the systems of knowledge (Carm, 2014). The approach does not involve extractive research methods since it focuses on documented instances in which communities maintain agency and control. Cultural guidelines, such as recognition of Elders and knowledge holders in the community has been identified as part of ethical engagement. A state of reflexivity is upheld during the research process to reduce the epistemic bias and power sources (Aseron et al., 2013). The study applies methodological rigor and decolonizing research ethics because it pre-empts knowledge sovereignty. This moral inclination makes sure that the assimilation of Indigenous knowledge in the sustainability education is performed with integrity, accountability, and cultural esteem.

6. LITERATURE REVIEW

The sustainability educational literature is becoming more aware of the constraints of the universalized, technocratic models that privilege the scientific knowledge and undervalue the local and Indigenous practices of knowledge. More modern scholarship is underlining the fact that sustainability issues are quite social, cultural, and ethical and should be examined through educational structures with recourse to the lived experience, place, and identity (Manns et al., 2025). In this framework, a novel area of critical pathways to transformative sustainability education has been developed through Indigenous knowledge systems and place-based learning. The reviewed literature in this section is in line with the chapter objectives by discussing the role of the Indigenous epistemology in the practices of sustainability, the role of the land-based pedagogy in the learning process, and the role of decolonizing practices in improving equity and cultural renewal in education systems.

6.1 Indigenous Knowledge and Sustainability Practices

The literature available indicates that Indigenous knowledge systems are entrenched in a long-term ecological process and sustainability on adaptive practices. These systems include a close knowledge of biodiversity, climate variability, soil management, water conservation and food security, which have been acquired by hundreds of years of observation and experience. According to scholars, the sustainability practices in the indigenous community are holistic in nature and they combine the environmental, social, spiritual, and economic aspects of sustainability instead of distinguishing them as distinct domains (Hunt (Nehiyaw), 2025). The ethical aspects of Indigenous knowledge with references to stewardship principles, reciprocity, and collective responsibility are also highlighted in the literature. These values dispute extractive and anthropocentric paradigms of dominant sustainability discourse. Other studies indicate that Indigenous knowledge increases the resilience of communities through the development of adaptive capacity, social cohesion, and intergenerational learning (Jaiswal, 2025). Nonetheless, the literature warns against instrumentalization of the Indigenous knowledge without consideration of the cultural background and sovereignty. All in all, Indigenous knowledge is not only introduced as an alternative source of data but a holistic sustainability model with far-reaching consequences on the education and practice.

6.2 Place-Based and Experiential Learning Models

The educational literature of place-based, experiential, and land-based education highlights the significance of situating education in ecological, and cultural settings of the locality. The character of place-based learning is the focus on real-world, community involvement, and problem solving that operates on the basis of local issues (Miuka & Tufu, 2025). The experiential and land-based models focus on the learning process through physical engagement and contact with the land, observation of the natural processes and involvement in practices that are culturally relevant (Ravikumar & Aarthi, 2025a). The studies have shown that these pedagogies increase the ecological literacy, critical thinking, and emotional sense of place. There is also the association of these approaches with greater motivation to learners and enhanced environmental responsibility (Harris et al., 2025). Notably, researchers mention that place-based learning is consistent with Indigenous pedagogies, which focus on such traditional aspects as observation, participation, and relational learning. In the literature, it is implied that incorporation of these models in formal schooling may help in bridging the divide between theory and practice. However, the difficulties with rigid curricula, test restrictions, and teachers being unprepared are often cited as hindrances to implementation.

6.3 Decolonizing Education and Cultural Revitalization

The idea of decolonization of education has gained a significant place in sustainability and Indigenous studies literature and is considered as a reaction to the necessity to destabilize epistemic hierarchies, which prioritize Western systems of knowledge. According to scholars, sustainability education should deal with the historical injustices and power inequities that discriminated Indigenous cultures and knowledge (Torris-Hedlund et al., 2025). The literature emphasizes the importance of utilizing the Indigenous knowledge integration in promoting cultural revitalization, language retention and identity construction among learners (Ravikumar & Aarthi, 2025c). Educational equity is also associated with decolonizing approaches, which legitimize various forms of knowing and establish universally inclusive learning settings. Science highlights the role of ethical interaction, community partnership, and observation of the Indigenous sovereignty to prevent cultural appropriation (Mdletshe, 2025). Also, the literature identifies institutional resistance, gaps in policies, and resource constraints as some of the ongoing challenges. In general, decolonizing sustainability education is introduced as an ethical agenda and practical approach to attaining the socially fair and culturally competent learning systems.

7. CASE STUDIES

In this section, recent and contextually-specific case studies are offered, which demonstrate the context in which Indigenous land-based learning and collaborative education partnerships are being executed to promote sustainability education. Based on projects in India and the international scene, the cases illustrate how culturally-based pedagogies, community leadership, and school-community partnership can produce significant learning experiences that can enhance ecological knowledge, cultural pride, and sustainability action among the learners (University of Zambia Special Education and Sociology of Education, 2025). Nonetheless, these case studies also reveal the obstacles and opportunities of the operationalization of the Indigenous knowledge in the context of formal and non-formal education. Collectively, they offer empirical support to the discussion of the transformative potential of Indigenous and place-based approaches and also information to inform policy, practice, and future research.

7.1 Case Study I: Indigenous Land-Based Learning in India

A recent study examining Traditional Ecological Knowledge (TEK) in Northeast India in the Indian context offers a good illustration of the Indigenous land-based learning in the formal education frameworks. More than 200 Indigenous communities reside in northeast India and the cultural, linguistic, and ecological knowledge of these communities are deeply rooted in a particular landscape and environmental practice (Sharma, 2025). Examples of research that examines TEK and its incorporation in school-based curricula have demonstrated that the integration of land-based and culture-grounded knowledge into the learning process can greatly boost students learning and retention rates, especially in cases where Indigenous students tend to be disengaged with school-based curricula, which are not relevant to their lived contexts. TEK-based pilot programs have produced reports of greater identification with local landscape, greater ecological sensitivity and better academic performance through the location of learning in the landscapes, seasonal cycles and community activities with which the learners are already acquainted. These include shared curriculum development with the elders of communities, incorporation of traditional agricultural and resource-management methods in the classroom and outdoor experience that links students to land stewardship (Das et al., 2025). The Indian Knowledge Traditions (IKT) system also offers policy-congruent strategies to operationalize land-based and Indigenous pedagogies and connect them to the changes in education policies like the National Education Policy and retain cultural integrity and ecological viability. The figure 3 presents a chronological process of integrating the Traditional Ecological Knowledge into education, through identification of the Indigenous communities and documentation of knowledge to curriculum integration, implementation of programs, and alignment of policies.

Figure 3. Framework for Integrating Indigenous Knowledge in Education

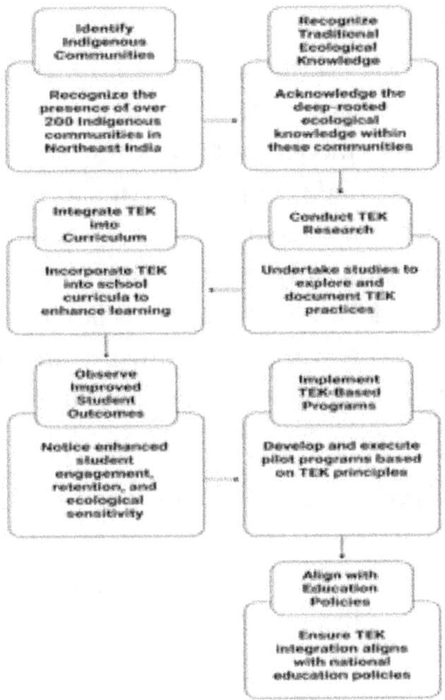

7.2 Case Study II: Community–School Partnerships for Sustainability

A good example of a study on the global scale can be seen through a new model of educational partnership in Canada when Indigenous community leaders and educational institutions engage in the development of sustainability education programs, incorporating Indigenous ecological understanding and Western environmental science. Indigenous knowledge in this model is co-taught by environmental scientists and Indigenous knowledge, forming a curriculum blending and accommodating various epistemologies (Pratiwi et al., 2023). The focus of such partnerships is mutual learning and joint control over the material of the education. Learners practice land through guided field trips, community-based instructions and reflexivity of relationality and stewardship that pre-empts learning (Prabagar et al., 2025). Through these programs there is a development of strong community school connections where schools are considered to be knowledge sharing institutions and not isolated content dispensers. This participatory approach has enhanced the competencies of learners in their ecological awareness, ethical involvement and cultural perceptions, as well as fostering the sense of community pride and intergenerational knowledge sharing. It also considers systemic issues including rigidity of the curriculum and cultural representation by ensuring that the priorities of the community are considered in the design of the formal education (Harushimana, 2022). Besides, community-school relationships support Indigenous procedures of knowledge sharing and positional authority and the educational practices must acknowledge community sovereignty and cultural values. This case underscores how institutional partnership, based on the respect of each other, common goals, and co-creation, can bring sustainability education to relational and culturally-based and grounded paradigms. The figure illustrates a collaborative model where Indigenous peoples and the school mutually design and deliver sustainability curricula, which results in the improved learner competencies, community pride, and the appreciation of cultural sovereignty.

Figure 4. Community–School Sustainability Partnership Model

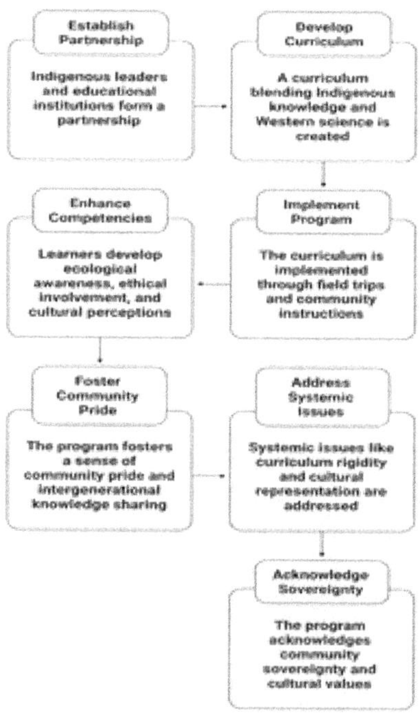

7.3 Comparative Insights from Case Studies

The analysis of the Indian and global cases offers multiple important insights on the approaches to land-based learning and community partners. First, learning trajectories with cultural roots and place-based interactions with the local landscapes and Indigenous knowledge systems are found to make a remarkable contribution to the engagement, place identification, and ecological awareness of the learners. Second, the participation of a community in teaching and curriculum development enhances a sense of reciprocity and shared ownership of the educational outcomes, which increases the relevance and accountability (Sumida Huaman, 2023). Third, the combination of Indigenous and Western approaches with the help of collaborative pedagogies helps to bridge the gaps in the epistemic approach and improve sustainability knowledge without the loss of cultural integrity. Nonetheless, the two cases also indicate systemic issues, such as policy, and resource restrictions, and educator capacity building that should promote culturally responsive practice. All of those cases highlight the value of ethical involvement, the respect towards the Indigenous sovereignty, and the adaptable institutional framework that will be able to sustain the long-term collaboration and land-based learning programs. Table 5 Summarizes the two case studies in terms of context, pedagogical practice, stakeholder engagement and learning realized.

Table 5. Case Study Summary and Outcomes

Case Study	Context	Pedagogical Approach	Key Outcomes
Case Study I	Indigenous communities, India	Land-based, TEK-integrated learning	Ecological awareness, cultural pride
Case Study II	Community–school partnerships	Co-teaching, experiential learning	Stewardship, learner engagement
Cross-Case Insight	Comparative	Community-led education	Identity, ethical sustainability

8. RESULTS AND DISCUSSION

The results that became apparent in the case studies prove that the inclusion of Indigenous systems of knowledge and place-based pedagogies in the sustainability learning process has substantial educational, pedagogical, and socio-cultural effects. In both settings, learners shown had increased ecological knowledge, higher emotional attachment to place and increased cultural identity. These results indicate that land-based and community-based learning strategies can help sustainability education to shift an intellectual cognition to embodied, ethical, and action learning (Mtitu, 2021). As discussed, when Indigenous epistemologies are presented as an essential guideline, they aid in the holistic manner of learning, which incorporates environmental knowledge with the stewardship and collective responsibility values. Pedagogically, the cases demonstrate a transition to the classroom-constrained approach of teaching to the experiential and relational instructional setting, which promotes increased involvement and relatability of the learner. Simultaneously, the findings indicate that there are still deeply-seated issues, such as the inflexibility of the institution, dangers of shallow knowledge assimilation, and the cultural appropriation issues. To overcome these issues, there is a need to focus on deliberate ethical frameworks, capacity building and long-term cooperation with Indigenous people. In general, the discussion confirms that transformative sustainability education is most effective in the case of cultural grounding, ethical questioning, and environmental responsiveness.

8.1 Ecological Awareness, Identity, and Cultural Pride

According to the case studies, there are significant educational results concerning ecological awareness, identity, and cultural pride in learners. The use of land-based activities and Indigenous knowledge practices increased the knowledge of learners about their local ecosystems, their seasonal cycles and sustainable use of resources (Aiello et al., 2021). This environmental consciousness did not just stop at the facts but went on to ethical duty and sentiment of place. Students developed more respect towards the Indigenous cultural practices, stories, and values, which added to their cultural identity and self-worth, especially in Indigenous students. The introduction of the knowledge holders of the community into the learning spaces legitimized the Indigenous epistemologies, which led to the development of pride and confidence in the cultural heritage (Fatima et al., 2022). These results are consistent with the research indicating that culturally responsive education increases the motivation and relevance of the learners. Notably, the examples demonstrate that sustainability education that is based on Indigenous views facilitates environmental literacy and socio-cultural well-being, and ecological and cultural sustainability are two interconnected phenomena.

8.2 Pedagogical Transformation through Indigenous Learning

Pedagogically, the case studies reflect a definite shift in the traditional content-based teaching and learning into the experiential, participatory and relational learning approaches. The indigenous pedagogies also focused on learning by observing, practicing, storytelling and mentorship making land an active location of knowledge creation. These strategies prompted the learners to think and contemplate sustainability problems based on their immediate surroundings. Field-based observations and community-led instruction that comprised of experiential learning activities were able to help students gain a deeper understanding and have long term retention than abstract classroom teaching (Ormiston, 2014). The concurrent operation of Indigenous and formal education practices allowed the co-creation of knowledge, by transcending the borders of the epistemology and preserving the culture at the same time. Teachers have noted the increased engagement of learners and the collaborative learning process and suggested the usefulness of experiential pedagogies to resolve sophisticated issues of sustainability. The results indicate that the concept of incorporating Indigenous and experiential learning methodologies would help to restructure sustainability education with a transformative orientation that helps achieve agency, relevness, and ethical practices.

8.3 Challenges and Mitigation Strategies

The case studies have positive results, but also there are notable challenges and risks in the integration of the Indigenous knowledge into the formal education systems. The problem of institutional resistance became a repetitive pattern, which was usually based on the strictness of curricula, standardized assessment systems and insufficient educator readiness (Dei et al., 2025). The other issue was the danger of cultural appropriation particularly when native knowledge was utilized without consultations to the community or taking into account the cultural processes. The risk of these practices is that they will reduce Indigenous epistemologies to a formal or tokenistic level. To minimise the risks, the cases have brought to the fore the necessity to address these risks by using ethical engagement frameworks that focus on the community consent, knowledge sovereignty and long-term partnerships (Wheeler et al., 2020). The foundational enablers that were identified as being critical were capacity-building efforts among the educators, the flexibility of the policy level and institutional support frameworks. Reflexivity and constant communication between the institution and the Indigenous communities also emerged as a critical action towards ensuring respectful, meaningful and sustainable integration of Indigenous knowledge in sustainability education. Table 6 Identifies the primary institutional, cultural, and ethical problems and fights the pragmatic approaches.

Table 6. Challenges in Integrating Indigenous Knowledge.

Challenge	Risk	Mitigation Strategy
Institutional rigidity	Limited curriculum flexibility	Policy reform and modular curricula
Cultural appropriation	Knowledge misuse	Community consent and protocols
Educator preparedness	Superficial integration	Professional development
Resource constraints	Program discontinuity	Long-term funding mechanisms
Power imbalance	Marginalization of communities	Shared governance models

9. FUTURE RESEARCH DIRECTIONS

It is essential to conduct more research to improve the understanding of how Indigenous knowledge systems and place-based pedagogies can be implemented into sustainability education on different contexts under a systematic and ethically informed research. The current study demonstrates the possible outcomes; however, the field is rather recent and fragmented, and there is not much comparison and longitudinal information (Bodkin-Andrews et al., 2017). The research should not be limited to case studies to draw good transferable knowledge that takes the cultural specificity seriously but educates the broader change. The research agenda should also include methodological innovation, policy alignment, the evolving role of digital technologies in assisting in the transmission of the Indigenous knowledge. High level of priority should be given to the participatory and community-based research methods that will concentrate on other voices and priorities of the Indigenous. In addition, additional studies should identify the viability of the application of culturally-based models of sustainability education in formal education without compromising the epistemic integrity (Tata et al., 2023). Using interdisciplinary and contextual research, the researchers would as well contribute to the process of developing inclusive, resilient, and ethically grounded sustainability education models that have the power to combine the responses to the complicated socio-environmental problems.

9.1 Expanding Empirical Research in Indigenous Contexts

The multiplicity of knowledge systems and practices of sustainability demand the use of empirical studies in a diverse range of Indigenous and regional contexts. The literature available is also likely to be localized and narrow, and might not be in a position to offer general information regarding variation in contexts. More studies should be carried out on the sustainability education programs by Indigenous on the other ecological regions, culture and other systems of governance. The comparative research designs can help in understanding how local knowledge systems respond to the unique environmental issues and socio-political circumstances (Gerlach, 2018). Through such research, common principles and context-specific strategies to inform the effective pedagogical integration would be identified as well. They should focus on collaborative research approaches, which consider the Indigenous communities as co-researchers but not subjects. This method increases validity, ethical soundness, and relevance and knowledge sovereignty. Increasing the empirical coverage will empower the evidence base of culturally based sustainability education and direct policy and practice in other regions.

9.2 Longitudinal Studies on Learning Outcomes

Research is urgently required to determine long-term effects of Indigenous and place-based sustainability education to learners and communities through longitudinal studies. Majority of the current research endeavors revolve around immediate educational success, and they provide insufficient information about long-term behavioral transformation, identity construction, and community resilience. Longitudinal research on the effectiveness of early exposure to land-based and Indigenous pedagogies to determine their effects on the environmental attitudes and civic engagement of learners, and career trajectories in the future, should be conducted. Also, the study ought to delve into the larger community-based effects, such as cultural rejuvenation, intergenerational knowledge transfer and enhanced stewardship practices (Datta, 2018). These studies can also determine the impact of the nature of sustained education engagement on community-institution interactions and policies. Longitudinal designs entail long term cooperation, moral dedication and adaptive designs, which acknowledge customary community processes as well as priorities. Transformative claims and institutional and policy investment in Indigenous sustainability education will require evidence produced through long-term studies to be validated.

9.3 Innovations in Indigenous Sustainability Education

The role of digital technologies, policy, and curriculum innovation in promoting Indigenous sustainability education should also be studied in the future. Such excavations can be used in digital platforms to preserve knowledge, tell stories, and exchange ideas between communities when they are built in accordance with the Indigenous values and governance. Studies on how digital tools can be used to serve the purpose of aiding land-based learning and never replacing relational and experiential practices are required (Kana'iaupuni et al., 2011). The change in education policies, its funding and accreditation systems must be explored in terms of how to establish conditions that will make culturally based sustainability education a possibility. Curriculum research ought to be directed towards the flexible, modular curricular designs which can accommodate the coexistence of Indigenous knowledge and academic rigour. Scalable and ethical integration can be assisted through the investigation of such innovations by use of participative and evaluative research. Digital, policy and curriculum-related research can work together to close the gap between local knowledge systems and institutional structures towards inclusive and future-ready sustainability education.

10. MANAGERIAL IMPLICATIONS

The idea of integrating Indigenous knowledge systems and place-based learning into sustainability education has vast implications in terms of management to educational institutions, policymakers, and other stakeholders of the community. Educational leadership has a vital position of establishing enabling conditions to promote culturally-based pedagogies, ethical practice and flexibility within the institutions (Magni, 2017). Curriculum, faculty development, resource allocation, and community partnerships are all aspects of the work of managerial decision-making, which determine the success of sustainability education efforts. The results of this chapter indicate that sustainability education should be seen as a systemic and relational process and not a curricular inclusion (Cherubini, 2009). Managers and administrators must therefore embrace leadership practices that appreciate pluralistic epistemologies, ensure cross sector cooperation, and long-term commitment to community engagement. The major levers to making Indigenous knowledge embedded in formal education forms are strategic planning, policy alignment, and organizational culture. Learning that Indigenous knowledge is a strategic asset, not a peripheral resource, educational managers can achieve transformative learning outcomes as well as deliver institutional goals in relation to equity, sustainability, and social responsibility.

10.1 Leadership and Curriculum Implications

The key roles played by the educational leaders and curriculum designers in the operationalization of the Indigenous and place-based sustainability education. The leadership practice should focus on the inclusion of curriculum frameworks that support different epistemologies and learning modalities. The curriculum designers are requested to go outside the box of the standardized content to draw up flexible and context-based curricula that integrates land-based learning, community knowledge and experiential pedagogies. As an example, the design of curricula and field instruction using Indigenous Elders have been implemented in universities in countries like Canada and New Zealand in community-guided sustainability curriculum where institutional leadership can be used to operationalize culturally grounded sustainability education. It will imply that the professional development will have to be invested in such a way that trained educators will have the competencies needed to educate the culturally responsive way and to conduct ethical interaction with the Indigenous communities (Zinga & Styres, 2018). This will mean that there must be an investment in professional growth so that educators can possess the skills required to teach in a culturally responsive manner and to interact ethically with Indigenous communities. Leaders should also encourage interdisciplinary cooperation and new evaluation practices that have high regard to experiential and relational learning outcomes. Such initiatives require organizational support systems such as time allocation, recognition systems, and institutional incentives in order to be maintained (Mudaly, 2018). The framework of embedding Indigenous knowledge in the main system of the curriculum as opposed to its elective items can help educational leaders provide the level of coherence, legitimacy, and sustainability. Such steps of managers are essential in aligning institutional missions and societal responsibilities with the sustainability education.

10.2 Policy Recommendations for Inclusive Education

On the policy level, the support of inclusive and culturally based sustainability education needs supportive regulatory frameworks and strategic investments. Indigenous knowledge systems should be accepted by policy makers as valid and powerful national and global sustainability agendas. The policies of education have to permit flexibility in the curriculum to support place-based and community-based learning and the academic standard. The ways to be funded should be the long-term collaborations with Indigenous communities, training of educators, and curriculum development (Malcom, 2024). Coherence and effect can also be increased by policy alignment in the context of education, environment and cultural sectors. Also, accountability structures ought to shift to aspects of cultural relevance, community involvement and ethical actions rather than just performance measures. Indigenous consultation and collaboration are some of the policies that can be used to prevent the phenomenon of tokenistic inclusion and cultural appropriation (Martinez, 2023). Governments can use these principles to establish the enabling conditions to support equitable, sustainable, and culturally responsive education systems by entrenching them in education policy.

10.3 Community–Institution Knowledge Exchange Models

The community-institution collaboration models are key to long-term knowledge exchange and moral incorporation of Indigenous knowledge. The models that are based on mutual respect, shared decision making and long-term commitment should form the basis of such models as opposed to short term project-based engagement. The institutions should consider the communities as equal partners and knowledge communities, and co-designing of curricula, pedagogy, and evaluation processes (Hudson & Vodden, 2020). Accountability and trust can be enhanced by formal methods of governance that institutionalize the community involvement e.g. by the inclusion of the advisory councils or agreements on collaboration. These collaborations facilitate intergenerational knowledge transmission as well since the Elders, practitioners, and youth are all treated as participants of the educational processes. The principles of sustainable collaboration should be grounded on the effective communication, clear formulations of expectations, and conflict resolution tools. The idea of community-institution partnerships can result in increasing the sustainability education through the mutual knowledge exchange and contributing to the cultural continuity and community empowerment. Significant roles of such models in powerful educational systems that can address complex environmental and social problems in concert are notable.

11. CONCLUSION

The chapter has also contributed to the reframing of sustainability education in defining Indigenous ways of knowing as a core aspect of epistemic stance but not peripheral and marginal ones. Indeed, the chapter clearly shows that the Indigenous knowledge systems provide entire, relational, ethically-conversant perspectives that are necessary to tackle the modern environmental and socio-cultural issues through an intensive analysis of the theory, literature, methodology, and the latest case studies. Indigenous epistemologies would shift the image of learning, land and culture within the context of sustainability education by focusing on the concept of relationality, reciprocity and stewardship which would facilitate the transition of abstract and technocratic frames of learning into experience based, place based and community-based ones. These methods produce favorable pedagogic results as it is indicated in the analysis, such as heightened ecological consciousness, sound cultural identity and learner agency. In theory, the chapter provides the conceptual framework, which integrates the aims of land, culture, learning, and global sustainability and in practice, it proposes some tips on the curriculum design, pedagogy, and working with communities. The policy implications illuminate the requirement of flexibility in the institution, moral involvement, and awareness of the Indigenous knowledge sovereignty. Generally, the findings support the idea that indigenization of sustainability education is not merely an inclusion approach, but a radical approach that can build resilience, equity and justice. Education may enable the learner and community to envision and develop sustainable futures based on cultural integrity, environmental responsibility, and community health through the alignment of the knowledge structures surrounding the local with sustainability demands in the global environment.

REFERENCES

Aarthi, S., Ravikumar, R. N., Khudayberganov, S., Ismailov, A., & Khujaniyozova, O. (2025). Using Intelligent Tutoring Systems to enhance student engagement. In *Advances in Computational Intelligence and Robotics* (pp. 155–188). IGI Global. DOI: 10.4018/979-8-3373-1827-1.ch008

Aarthi, S., Ravikumar, R. N., Sobirova, K., & Mamadiyarov, Z. (2025). Empowering a sustainable future through digital learning and the SDGs. In *Harnessing E-Learning to Create a Sustainable Future* (pp. 145–172). IGI Global Scientific Publishing., DOI: 10.4018/979-8-3373-0112-9.ch006

Aiello, S., Aguayo, C., Wilkinson, N., & Govender, K. (2021). Developing culturally responsive practice using mixed reality (XR) simulation in Paramedicine Education. *Pacific Journal of Technology Enhanced Learning*, *3*(1), 15–16. DOI: 10.24135/pjtel.v3i1.89

Aseron, J., Wilde, S., Miller, A., & Kelly, S. (2013). Indigenous student participation in higher education: Emergent themes and linkages. *Contemporary Issues in Education Research*, *6*(4), 417. DOI: 10.19030/cier.v6i4.8110

Athayde, S., Silva-Lugo, J., Schmink, M., Kaiabi, A., & Heckenberger, M. (2017). Reconnecting art and science for sustainability: learning from indigenous knowledge through participatory action-research in the Amazon. *Ecology and Society: A Journal of Integrative Science for Resilience and Sustainability*, *22*(2). DOI: 10.5751/ES-09323-220236

Bodkin-Andrews, G., Whittaker, A., Harrison, N., Craven, R., Parker, P., Trudgett, M., & Page, S. (2017). Exposing the patterns of statistical blindness: Centring Indigenous standpoints on student identity, motivation, and future aspirations. *Australian Journal of Education*, *61*(3), 225–249. DOI: 10.1177/0004944117731360

Carm, E. (2014). Inclusion of indigenous knowledge system (IKS) – A precondition for Sustainable Development and an integral part of environmental studies. *The Journal of Educational Research*, *4*(1), 58–76. DOI: 10.3126/jer.v4i1.10726

Castagno, A. E., & Brayboy, B. M. J. (2008). Culturally responsive schooling for Indigenous youth: A review of the literature. *Review of Educational Research*, *78*(4), 941–993. DOI: 10.3102/0034654308323036

Cherubini, L. (2009). “taking Haig-Brown seriously”: Implications of Indigenous thought on Ontario educators. *Journal of the Canadian Association for Curriculum Studies*, *7*(1), 6–23. DOI: 10.25071/1916-4467.18136

Cruz, A. R. (2020). Land Education. In *Oxford Research Encyclopedia of Education*. Oxford University Press., DOI: 10.1093/acrefore/9780190264093.013.682

Das, T. M., Medhi, M., & Banerjee, D. S. (2025). Indigenous knowledge systems and sustainable learning: A comprehensive regional analysis of Northeast India. *International Journal of Environmental Sciences*, •••, 1999–2012. DOI: 10.64252/vcp0c468

Datta, R. (2018). Decolonizing both researcher and research and its effectiveness in Indigenous research. *Research Ethics*, *14*(2), 1–24. DOI: 10.1177/1747016117733296

Datta, R. (2024). Relationality in Indigenous climate change education research: A learning journey from Indigenous communities in Bangladesh. *Australian Journal of Environmental Education*, *40*(2), 128–142. DOI: 10.1017/aee.2024.13

Day, K., & McPherson, L. (2025). Resilient epistemologies: Integrating Crotty's framework with Indigenous yarning protocols for decolonial research. *International Journal of Qualitative Methods*, *24*(16094069251404324), 16094069251404323. Advance online publication. DOI: 10.1177/16094069251404323

Dei, D.-G. J., Anane-Donkor, L., Dzandza, P. E., Peasah, T., & Puttick, C. P. (2025). Synthesizing the characteristics and applications of indigenous knowledge for sustainable development: A systematic review. *SAGE Open*, *15*(3), 21582440251383843. Advance online publication. DOI: 10.1177/21582440251383843

Dodgson, J. E. (2019). Reflexivity in qualitative research. *Journal of Human Lactation : Official Journal of International Lactation Consultant Association*, *35*(2), 220–222. DOI: 10.1177/0890334419830990 PMID: 30849272

Fatima, Y., Cleary, A., King, S., Solomon, S., McDaid, L., Hasan, M. M., Al Mamun, A., & Baxter, J. (2022). Cultural identity and social and emotional wellbeing in Aboriginal and Torres Strait Islander children. In *Family Dynamics over the Life Course* (pp. 57–70). Springer International Publishing., DOI: 10.1007/978-3-031-12224-8_4

Gainsford, A., & Evans, M. (2021). Integrating andragogical philosophy with Indigenous teaching and learning. *Management Learning*, *52*(5), 559–580. DOI: 10.1177/1350507620972528

Gerlach, A. (2018). Thinking and researching relationally. *International Journal of Qualitative Methods*, *17*(1), 160940691877607. DOI: 10.1177/1609406918776075

Gruenewald, D. A. (2003). The best of both worlds: A critical pedagogy of place. *Educational Researcher (Washington, D.C.: 1972)*, *32*(4), 3–12. DOI: 10.3102/0013189X032004003

Harris, O., Taylor, S., & Mitchell, T. (2025). Telling the land: Aboriginal educational narratives and curriculum integration in Australian schools. *International Journal of Educational Narratives*, *3*(3), 218–229. DOI: 10.70177/ijen.v3i3.2211

Harushimana, I. (2022). Preparing educators who save lives: What can US schools of education do? *Journal for Multicultural Education*, *16*(3), 225–236. DOI: 10.1108/JME-11-2021-0198

Housel, D. A., & Nikoui, M. (2024). Practical recommendations for accommodating students with limited or interrupted formal education in adult education programs in the United States. *Adult Learning*. Advance online publication. DOI: 10.1177/10451595241297660

Hudson, A., & Vodden, K. (2020). Decolonizing pathways to sustainability: Lessons learned from three Inuit communities in NunatuKavut, Canada. *Sustainability (Basel)*, *12*(11), 4419. DOI: 10.3390/su12114419

Hunt (Nehiyaw), D. (2025). Inuit Qaujimajatuqangit (IQ) and the climate of colonialism: pedagogy, intellectual emancipation, and decolonial futures. *AlterNative (Auckland, N.Z. : 2005), 11771801251362692*. DOI: 10.1177/11771801251362691

Jacob, M. M., Sabzalian, L., Anderson, R. N., Muniz, H. R., Simmons, K., & Beavert, V. R. (2024). Indigenous genealogies of relational knowledge: Cedar tree and gray squirrel as important relatives and teachers. *Genealogy*, *8*(1), 19. DOI: 10.3390/genealogy8010019

Jaiswal, D. A.Dr. Anita Jaiswal. (2025). Indigenous pedagogies teaching and learning practices rooted in local contexts. [NIJMS]. *Naveen International Journal of Multidisciplinary Sciences*, *1*(4), 97–104. DOI: 10.71126/nijms.v1i4.40

Kana'iaupuni, S. M., Ledward, B., & Keohokalole, K. (2011). New research on the impact of cultural influences in education on Native Hawaiian student outcomes. *AAPI Nexus Journal: Policy, Practice, and Community*, *9*(1–2), 211–229. DOI: 10.36650/nexus9.1-2_211-229_KanaiaupuniEtAl

Kioupi, V., & Voulvoulis, N. (2019). Education for Sustainable Development: A systemic framework for connecting the SDGs to educational outcomes. *Sustainability (Basel)*, *11*(21), 6104. DOI: 10.3390/su11216104

Laver, Q., Andersen, O., Moore, S., Pottle, C., & Snook, J. (2024). Fostering fisheries futures: The promise of localized stewardship education in Nunatsiavut. *Frontiers in Environmental Science*, *12*(1409358), 1409358. Advance online publication. DOI: 10.3389/fenvs.2024.1409358

Levitan, J., & Johnson, K. M. (2020). salir adelante: Collaboratively developing culturally grounded curriculum with marginalized communities. *American Journal of Education, 126*(2), 195–230. DOI: 10.1086/706921

Magni, G. (2017). Indigenous knowledge and implications for the sustainable development agenda. *European Journal of Education, 52*(4), 437–447. DOI: 10.1111/ejed.12238

Malcom, A. (2024). Integrating indigenous knowledge into Livestock Policy formulation. *International Journal of Livestock Policy, 3*(1), 30–42. DOI: 10.47941/ijlp.1962

Manns, M., Heiss, J., Morohashi, J., & Loupis, L. (2025). Harnessing Indigenous knowledge for education for Sustainable Development. *Journal of Education for Sustainable Development, 19*(2), 157–167. DOI: 10.1177/09734082251360875

Martinez, N. (2023). Indigenous literacies: A look at pedagogies and policy in the Southwest United States. *Journal of Adolescent & Adult Literacy: A Journal From the International Reading Association, 66*(5), 282–289. DOI: 10.1002/jaal.1282

Mbah, M., Johnson, A. T., & Chipindi, F. M. (2021). Institutionalizing the intangible through research and engagement: Indigenous knowledge and higher education for sustainable development in Zambia. *International Journal of Educational Development, 82*(102355), 102355. DOI: 10.1016/j.ijedudev.2021.102355

McCarty, T., & Lee, T. (2014). Critical culturally sustaining/revitalizing pedagogy and Indigenous education sovereignty. *Harvard Educational Review, 84*(1), 101–124. DOI: 10.17763/haer.84.1.q83746nl5pj34216

Mdletshe, L. N. (2025). Towards the application of inclusive indigenous pedagogies in tourism education in South Africa. In *Indigenous Teaching Disciplines and Perspectives for Higher Education* (pp. 171–196). IGI Global Scientific Publishing., DOI: 10.4018/979-8-3693-9296-6.ch009

Miuka, N. K., & Tufu, J. (2025). Decolonizing the curriculum: Policy implications for Indigenous knowledge integration in national education systems. *International Journal of Post Axial: Futuristic Teaching and Learning*, 194–203. DOI: 10.59944/postaxial.v3i3.478

Mtitu, E. A. (2021). *Learner-centred teaching in Tanzania: Geography teachers' perceptions and experiences*. DOI: 10.26686/wgtn.17006401

Mudaly, R. (2018). Towards decolonising a module in the pre-service science teacher education curriculum: The role of indigenous knowledge systems in creating spaces for transforming the curriculum. *Journal of Education, 74*(74). Advance online publication. DOI: 10.17159/2520-9868/i74a04

Nesterova, Y. (2020). Rethinking environmental education with the help of indigenous ways of knowing and traditional ecological knowledge. *Journal of Philosophy of Education, 54*(4), 1047–1052. DOI: 10.1111/1467-9752.12471

Opoku, M. J., & James, A. (2021). Pedagogical model for decolonising, indigenising and transforming science education curricula: A case of South Africa. *Journal of Baltic Science Education, 20*(1), 93–107. DOI: 10.33225/jbse/21.20.93

Ormiston, N. T. (2014). Transforming stories and teachings into social work pedagogies. *Affilia, 29*(3), 368–372. DOI: 10.1177/0886109914522630

Prabagar, S., Mothkur, R., Swetha, C. B., Kurian, A., Pavithra, K., & Aarthi, S. (2025). Generative AI for inclusive special education. In *Advances in Computational Intelligence and Robotics* (pp. 205–236). IGI Global Scientific Publishing. DOI: 10.4018/979-8-3373-3745-6.ch008

Pratiwi, P. H., Dwiningrum, S. I. A., & Sumunar, D. R. S. (2023). Integrated disaster risk management in the education process in schools. *Journal of Integrated Disaster Risk Management, 13*(1). Advance online publication. DOI: 10.5595/001c.91284

Ram, A. T., Netto, S. G., & Mohanan, S. M. (2024). *NEP 2020 vision: Embedding Indian Knowledge System in contemporary Teacher Training and Curriculum Development*. Gurukul International Multidisciplinary Research Journal., DOI: 10.69758/GIMRJ/2412IV02V12P0014

Ravikumar, R. N., & Aarthi, S. (2025b). AI-powered academic libraries. In *Role of Library Resources for Higher Education in the Digital Era* (pp. 19–56). IGI Global Scientific Publishing., DOI: 10.4018/979-8-3373-3005-1.ch002

Ravikumar, R. N., & Aarthi, S. (2025c). The role of educational developers in AI integration at scale. In *Advances in Computational Intelligence and Robotics* (pp. 289–322). IGI Global Scientific Publishing. DOI: 10.4018/979-8-3373-4129-3.ch011

Ravikumar, R. N., Aarthi, S., Ruzimova, F., & Mamadiyarov, Z. (2025). Training and development in AI-driven education scaling solutions. In *Advances in Computational Intelligence and Robotics* (pp. 267–298). IGI Global Scientific Publishing. DOI: 10.4018/979-8-3373-2185-1.ch010

Ravikumar, & Aarthi, S. (2025a). Natural Language Processing in conversational AI tutors. In *Advances in Computational Intelligence and Robotics* (pp. 275–306). IGI Global Scientific Publishing. DOI: 10.4018/979-8-3373-4217-7.ch011

Sharma, S. (2025). Indigenous Knowledge Systems and sustainable Socio-economic development in Northeast India: An analysis through the lens of Astalakshmi and Viksit Bharat. *International Journal For Multidisciplinary Research*, *7*(5), 58933. Advance online publication. DOI: 10.36948/ijfmr.2025.v07i05.58933

Star, J., Ringaert, L., & Larcombe, L. (2025). Qualitative methods case study: Using MAXQDA in Indigenous HIV journey mapping research. *International Journal of Qualitative Methods*, *24*(16094069251356668), 16094069251356667. Advance online publication. DOI: 10.1177/16094069251356667

Sumida Huaman, E. (2023). Earth worlds and Indigenous dream-making: A reflection on teaching for beauty, repair, and balance. *Revista Española de Educación Comparada*, *43*(43), 158–173. DOI: 10.5944/reec.43.2023.36610

Sy, J. M. C. (2022). Till the land, defend the land: reflections on the critical place-based pedagogy of the Alternative Learning Center for Agricultural and Livelihood Development, Surigao del Sur. *AlterNative (Auckland, N.Z. : 2005)*, *18*(3), 402–411. DOI: 10.1177/11771801221115925

Tata, E. F., Yongabi, K. A., Dzelamonyuy, E., Forbang, N. I., & Geh, E. A. (2023). Decolonizing African science: Efforts, challenges, and future directions. *International Journal of Science and Research Archive*, *10*(1), 823–833. DOI: 10.30574/ijsra.2023.10.1.0840

The University of Zambia Special Education and Sociology of Education, D. of E. P., & Kaziya, K. (2025). Harnessing Indigenous Knowledge Systems to transform mathematics instruction in rural Zambian schools: A case study from Kalomo District. *INTERNATIONAL JOURNAL OF SOCIAL SCIENCE AND EDUCATION RESEARCH STUDIES*, *05*(03). DOI: 10.55677/ijssers/V05I03Y2025-13

Torris-Hedlund, M. A., Palakiko, D.-M., & Komene, E. (2025). Asserting Indigenous academic sovereignty through assemblage, Indigenization, and cultural safety. *Creative Nursing*, *31*(3), 243–253. DOI: 10.1177/10784535251357218 PMID: 40636965

Wheeler, H. C., Danielsen, F., Fidel, M., Hausner, V., Horstkotte, T., Johnson, N., Lee, O., Mukherjee, N., Amos, A., Ashthorn, H., Ballari, Ø., Behe, C., Breton-Honeyman, K., Retter, G.-B., Buschman, V., Jakobsen, P., Johnson, F., Lyberth, B., Parrott, J. A., & Vronski, N. (2020). The need for transformative changes in the use of Indigenous knowledge along with science for environmental decision-making in the Arctic. *People and Nature*, *2*(3), 544–556. DOI: 10.1002/pan3.10131

Zinga, D., & Styres, S. (2018). Decolonizing curriculum: Student resistances to anti-oppressive pedagogy. *Power and Education, 175774381881056*. Advance online publication. DOI: 10.1177/1757743818810565

KEY TERMS AND DEFINITIONS

Community–Institution Partnership: Collaborative relationships between educational institutions and Indigenous communities based on mutual respect and shared decision-making.

Cultural Revitalization: Efforts to sustain and strengthen Indigenous languages, traditions, values, and identities through education and community practice.

Decolonizing Education: The process of challenging dominant knowledge systems and recognizing Indigenous epistemologies and histories in education.

Indigenous Knowledge Systems: Holistic, place-based ways of knowing developed by Indigenous communities through long-term interaction with land, culture, and environment.

Knowledge Sovereignty: The right of Indigenous communities to control, protect, and govern the use of their knowledge systems.

Land-Based Learning: Experiential learning that occurs directly on and with the land, emphasizing observation, practice, and ecological relationships.

Place-Based Learning: Learning that is rooted in local environments, cultures, and community experiences to make education relevant and meaningful.

Reciprocity: A value emphasizing mutual care and responsibility between people and the natural world.

Relationality: The principle that humans, nature, and non-human beings are interconnected and exist in mutual relationship.

Stewardship: Ethical responsibility to protect, care for, and sustain land and resources for present and future generations.

Sustainability Education: An educational approach that promotes understanding and action toward environmental protection, social equity, and long-term well-being.

UN Sustainable Development Goals (SDGs): A set of global goals aimed at achieving environmental sustainability, social inclusion, and economic development by 2030.

Chapter 3
Hybrid Sustainability Epistemologies in the UAE Curriculum:
Ethics, Place, and Futurity

Ali Ibrahim
http://orcid.org/0000-0002-1070-6256
United Arab Emirates University, UAE

Rasha Abdallah
United Arab Emirates University, UAE

ABSTRACT

The United Arab Emirates' shift toward a diversified, post-oil economy is reflected in its national curriculum. This chapter argues that the UAE curriculum represents a hybrid sustainability epistemology integrating ethical–religious formation, place-based ecological knowledge, scientific literacy, and technological advancement. Sustainability is embedded across subjects rather than taught independently: Islamic Studies and Moral Education promote ethical responsibility, social and heritage subjects connect learning to local ecological contexts, and science and innovation address climate change, energy transitions, and emerging technologies. The chapter further shows how the curriculum operates as an institutional system that organizes and reproduces sustainability knowledge in alignment with national priorities. While this model demonstrates coherence across school types, it also highlights tensions between global frameworks, local contexts, and development goals. Overall, the chapter frames curriculum as a hybrid knowledge infrastructure in rapidly modernizing societies.

DOI: 10.4018/979-8-3373-7650-9.ch003

1. INTRODUCTION: CURRICULUM AS A SITE OF SUSTAINABILITY KNOWLEDGE FORMATION

Sustainability has emerged as one of the defining global concerns of the twenty-first century, shaping policy discourse, economic planning, and educational reform across diverse national contexts. Yet within education, sustainability is often framed narrowly as environmental awareness or scientific literacy, emphasizing climate science, conservation practices, and technological innovation. While these elements remain essential, scholars increasingly argue that sustainability must be understood more broadly as an epistemological project—one concerned not only with what students know but with how knowledge is organized, transmitted, and mobilized toward future-oriented forms of social life (Bonnett, 2002; Sterling, 2001; Wals, 2011). From this perspective, curriculum functions not merely as a vehicle for environmental content but as a central institutional site where sustainability knowledge is structured, legitimized, and reproduced through formal schooling systems.

Within contemporary policy and academic discourse, sustainability is most commonly conceptualized through environmental, social, and economic dimensions, often referred to as the "three pillars" of sustainable development (Purvis et al., 2019). However, recent scholarship has increasingly problematized this framework as overly reductionist, arguing for a more integrated understanding of sustainability as a complex and systemic form of knowledge rather than a set of discrete domains. Recent work further suggests that sustainability should be understood as a multidimensional and contested concept shaped by epistemological, spatial, and temporal dynamics rather than as a fixed or universally agreed framework (Wijsman et al., 2026). From a systems-oriented perspective, sustainability emerges as a dynamic process shaped by interdependent ecological, social, and economic relations that cannot be fully understood through isolated analytical categories (Stumm, 2025). This shift toward systemic thinking aligns with broader sustainability education debates that emphasize complexity, relationality, and the need to move beyond linear and compartmentalized approaches to environmental knowledge. Accordingly, sustainability is not only a policy objective but also an epistemological formation through which knowledge about human–environment relations is organized, interpreted, and enacted.

Expanding beyond conventional sustainability frameworks, recent scholarship has also emphasized Indigenous and relational epistemologies that foreground interconnectedness, intergenerational responsibility, and place-based knowledge systems. These perspectives challenge dominant techno-scientific approaches by emphasizing that sustainability emerges through lived relationships between communities, environments, and cultural practices rather than through abstract or universal models alone (Escobar, 2001; Williams, 2024). In this context, sustainability knowledge is understood as relational, historically situated, and transmitted across generations through both formal and informal learning processes. Such approaches underscore the importance of integrating ethical, ecological, and cultural dimensions of knowledge, reinforcing the need for educational frameworks that move beyond disciplinary fragmentation toward more holistic and context-sensitive understandings of sustainability.

The United Arab Emirates (UAE) offers a particularly compelling case for examining sustainability as an epistemological formation within curriculum. As a rapidly developing state navigating post-oil economic transition, environmental vulnerability, and ambitious technological modernization agendas, the UAE has positioned sustainability as a central national priority reflected across education and policy frameworks (Abdallah, Ismail, & Alkaabi, 2025; Baroudi & Abi Haidar, 2025; National Committee on Sustainable Development Goals, 2023). Educational policy plays a pivotal role in this transformation, with curriculum serving as a key mechanism for aligning ethical formation, ecological awareness, and future-oriented innovation within broader national development strategies (ADEK, 2025; Baroudi & Abi Haidar, 2025). Unlike many Western models of sustainability education that prioritize environmental science as the primary knowledge base, the UAE curriculum integrates diverse epistemic traditions, reflecting both global policy influences and localized sociocultural foundations (Baroudi & Abi Haidar, 2025; UNESCO, 2017).

This chapter argues that the UAE curriculum institutionalizes a hybrid sustainability epistemology in which ethical–religious formation, place-based ecological knowledge, scientific literacy, and innovation-driven futurity are structurally integrated to produce disciplined, future-oriented citizens prepared for post-oil transformation (Bonnett, 2002; MoE, 2023, 2024; Sterling, 2001; UNESCO, 2017; Wals, 2011). The concept of hybridity here draws on scholarship emphasizing epistemological plurality and the coexistence of diverse knowledge systems within educational contexts. Rather than representing fragmented or competing frameworks, these domains operate relationally, producing a coherent sustainability orientation grounded simultaneously in moral discipline, ecological awareness, scientific understanding, and technological ambition (Alzaim et al., 2024; Baroudi & Abi Haidar, 2025).

Within this hybrid formation, ethical–religious knowledge provides a foundational framework for sustainability understood as moral and behavioral responsibility. Islamic environmental ethics emphasize principles such as stewardship (*khilafah*), trust (*amanah*), moderation, and accountability, framing human–environment relations through ethical obligations derived from religious teachings rather than solely through ecological reasoning (Foltz, 2000). At the same time, sustainability knowledge is also shaped by localized ecological contexts and place-based cultural practices. Scholarship on place and environment highlights how ecological knowledge emerges through people's lived relationships with particular landscapes, where cultural identity, environmental experience, and ecological understanding are mutually constituted (Escobar, 2001). Place-based education further demonstrates that embedding learning within local landscapes and community environments cultivates ecological consciousness and responsibility toward the places people inhabit (Gruenewald, 2003; Smith, 2002).

Clarifying the meaning of "Indigenous" within the UAE context is therefore essential. Rather than referring to sovereignty-based political identities common in settler-colonial scholarship, Indigenous knowledge here is understood as locally-embedded systems of practices, beliefs, and ecological understanding developed over generations in relation to specific environments and cultural traditions (Dorji et al., 2024). In the UAE, such knowledge is reflected in desert survival practices, maritime livelihoods, and ethical orientations shaped by Islamic environmental thought that historically guided adaptation to arid landscapes and scarce natural resources (Foltz, 2000; Heard-Bey, 2005). These traditions continue to influence contemporary sustainability discourse through heritage initiatives, national sustainability agendas, and educational programs that frame environmental stewardship as both a cultural responsibility and a developmental priority (Expo 2020 Dubai, 2020). Viewed through a relational epistemological perspective, Indigenous climate scholarship highlights how such knowledge systems emphasize interdependence, intergenerational responsibility, and ethical stewardship—principles that increasingly intersect with scientific and technological approaches to sustainability governance (Degyansky, 2024; Whyte, 2020). In this sense, the UAE curriculum reflects what Escobar (2001) describes as a pluralistic sustainability epistemology, where diverse knowledge traditions coexist and interact rather than being subordinated to a single modern environmental framework.

This chapter argues that the UAE curriculum represents a hybrid sustainability epistemology in which ethical–religious formation, place-based ecological knowledge, scientific literacy, and technological futurity are structurally integrated. Rather than presenting sustainability as a discrete subject, the curriculum distributes sustainability knowledge across multiple domains, constructing it as a layered and interconnected system of moral responsibility, ecological awareness, scientific understanding, and future-oriented innovation. The sections that follow develop this argument by first outlining a conceptual framework for hybrid sustainability knowledge, then examining ethical foundations, place-based ecological learning, scientific and environmental literacy, and technological futurity, before concluding with an analytical synthesis of the hybrid model and a discussion of its structural tensions within the UAE context.

2. CONCEPTUAL FRAMEWORK: HYBRID SUSTAINABILITY KNOWLEDGE IN EDUCATION

Sustainability education has increasingly been conceptualized not merely as the transmission of environmental knowledge but as the organization of multiple forms of knowledge that shape how societies understand and respond to socio-ecological challenges. Within the field of Education for Sustainable Development (ESD), sustainability is therefore framed as an interdisciplinary learning domain that integrates environmental, social, economic, and ethical dimensions of human development (UNESCO, 2017). Rather than promoting a single body of expert knowledge, contemporary sustainability education emphasizes the cultivation of critical, participatory, and transformative forms of learning that enable learners to engage with complex sustainability challenges across social and ecological systems (Wals, 2011). Scholars have also cautioned against overly instrumental approaches that reduce sustainability education to behavioral modification or technical problem-solving, arguing instead for educational frameworks that recognize the plurality of knowledge systems involved in sustainability thinking (Jickling & Wals, 2008). From a systems-oriented perspective, sustainability can be understood as a dynamic and relational knowledge system in which ecological, social, and economic processes are co-constituted and cannot be meaningfully addressed in isolation (Stumm, 2025).

From this perspective, sustainability education can be understood as a knowledge architecture shaped by the interaction of multiple epistemic systems rather than a fixed or universally agreed framework. Recent scholarship emphasizes that sustainability is inherently multidimensional, encompassing epistemological, spatial, and temporal dynamics that influence how knowledge is produced, organized, and applied across contexts (Wijsman et al., 2026). At the same time, place-based perspectives highlight that knowledge is not abstract or universally transferable but emerges through learners' engagement with their cultural, ecological, and social environments (Nwigwe, 2026). Together, these perspectives reposition sustainability education as a relational and context-dependent process, where meaning is constructed through the interaction between global frameworks and localized knowledge systems. This framing provides a foundation for understanding how national curricula—such as that of the United Arab Emirates—may integrate diverse epistemic traditions, including scientific, cultural, and ethical dimensions, within a unified but non-homogeneous model of sustainability.

Within this chapter, sustainability knowledge is conceptualized through three interrelated domains that together structure how sustainability is understood and taught within educational systems. First, ethical–religious knowledge frames sustainability through moral responsibility and normative principles governing human relationships with the natural world. In many cultural contexts, including Islamic environmental thought, concepts such as stewardship (khilafah), trust (amanah), moderation, and accountability shape sustainability as an ethical obligation rather than solely a technical or scientific concern (Abdullah et al., 2026; Foltz, 2000). Second, place-based ecological knowledge refers to locally-embedded understandings of ecological systems developed through long-term human interaction with specific environments. Scholarship on Indigenous and community-based knowledge highlights how such knowledge systems emphasize relational understandings of land, intergenerational responsibility, and ecological adaptation to local environmental conditions (Battiste, 2013; Dorji et al., 2024; Williams, 2024). Third, scientific and technological knowledge reflects the growing role of sustainability science, climate research, and innovation-driven environmental management in contemporary education systems, particularly as national policies increasingly align educational goals with global sustainability agendas such as the United Nations Sustainable Development Goals (UNESCO, 2017). Taken together, these domains illustrate that sustainability education does not emerge from a single epistemic tradition but rather from the interaction of ethical, ecological, and scientific forms of knowledge that together shape how sustainability is framed within curriculum and educational policy.

Rather than operating as discrete knowledge traditions, these domains intersect within educational systems to produce hybrid sustainability epistemologies that reflect broader cultural, political, and developmental contexts. Contemporary sustainability scholarship increasingly emphasizes that effective sustainability education emerges through the interaction of diverse knowledge systems rather than through the dominance of any single epistemological framework (UNESCO, 2017; Wals, 2011). Decolonial and Indigenous-informed approaches further highlight that sustainability learning is strengthened when ethical traditions, place-based ecological knowledge, and scientific inquiry are integrated within relational and context-sensitive pedagogies (Dorji et al., 2024; Kato & de Carvalho, 2026; Williams, 2024). In rapidly developing contexts such as the UAE, this hybridization is particularly visible, as sustainability discourse reflects the convergence of religious ethics, localized environmental knowledge, and innovation-driven development agendas aligned with global sustainability frameworks. National policy frameworks reinforce this integration by positioning sustainability as both a curricular priority and a broader societal objective embedded across sectors and institutions (ADEK, 2025). Conceptualizing sustainability education through this hybrid lens therefore provides a framework for understanding how curriculum functions not only as a site of knowledge transmission but as an institutional space in which multiple epistemic traditions converge to shape sustainability thinking and practice. This perspective aligns with emerging frameworks that conceptualize sustainability as a multidimensional and contested field shaped by interactions between knowledge systems, values, and power relations rather than as a fixed set of universally agreed principles (Wijsman et al., 2026).

Within the UAE context, this hybrid sustainability epistemology becomes particularly visible through national policy frameworks and curriculum structures that integrate ethical, ecological, and scientific knowledge across educational domains. National sustainability agendas position education as a central mechanism for advancing environmental responsibility and long-term development, aligning schooling with broader policy priorities associated with the Sustainable Development Goals and national climate strategies (The UAE Government, 2018). These frameworks emphasize cross-sectoral collaboration and long-term planning, recognizing sustainability as a societal priority that spans governance, economy, environment, and education. At the implementation level, initiatives such as the UAE Sustainable Schools Initiative illustrate how sustainability education is operationalized through interdisciplinary curricula, applied learning strategies, and whole-school practices designed to foster ecological awareness and responsible behavior among students (Environment Agency - Abu Dhabi, 2026; GEEP, n.d.). Together, these developments reflect an educational vision in which sustainability is not treated as a discrete subject but as an integrative knowledge orientation spanning religious education, social studies, science, and innovation-driven learning.

These knowledge traditions indicate that sustainability education cannot be understood as the transmission of a single disciplinary body of knowledge but rather as the integration of multiple epistemic traditions that shape how humans understand and act within socio-ecological systems. Sustainability education scholarship increasingly emphasizes the necessity of such epistemic plurality, arguing that meaningful sustainability learning requires engagement with ethical values, ecological relationships, and scientific knowledge simultaneously (Jickling & Wals, 2008; Wals, 2011). Systems-oriented and relational approaches further reinforce that sustainability challenges are inherently complex and require integrative forms of knowledge capable of addressing interdependent environmental and social processes (Stumm, 2025; Whyte, 2020). Within this perspective, the hybrid sustainability framework proposed in this chapter conceptualizes curriculum as a space where ethical–religious traditions, place-based ecological knowledge, and scientific–technological knowledge interact to shape sustainability learning in complementary and mutually reinforcing ways.

Building on this conceptualization, the chapter adopts a hybrid sustainability framework that understands curriculum as an institutional and epistemic space where diverse knowledge traditions converge to shape sustainability learning. Rather than treating ethical–religious, ecological, and technoscientific knowledge as discrete domains, the framework emphasizes their interaction within curricular structures, pedagogical practices, and school-based initiatives. This perspective enables a more nuanced analysis of how sustainability is constructed across subject areas and educational policies in the UAE context. The sections that follow apply this framework by examining ethical foundations in Islamic Studies and Moral Education, place-based ecological learning across curricular domains, scientific and environmental literacy, and technological futurity within innovation-driven education policy, ultimately synthesizing these strands into a hybrid sustainability model grounded in the UAE's sociocultural and developmental context.

3. ETHICAL FOUNDATIONS: MORAL DISCIPLINE AND CIVIC RESPONSIBILITY IN CURRICULUM

The first dimension of this hybrid framework is ethical formation, through which sustainability is initially constructed as moral and behavioral responsibility. Within sustainability education scholarship, ethical values and moral worldviews are increasingly recognized in shaping how individuals understand their responsibilities toward the environment and society. Rather than emerging solely from scientific knowledge about ecological systems, sustainability orientations are often shaped through ethical frameworks that regulate human behavior, discipline, and responsibility within broader social and spiritual relationships (Jickling & Wals, 2008; Wals, 2011). In many cultural contexts, religious traditions play a central role in articulating these ethical orientations, emphasizing stewardship, accountability, and moral discipline as guiding principles for human interaction with the natural world (Foltz, 2000). Contemporary scholarship on Islamic approaches to sustainability similarly highlights how environmental responsibility is grounded in moral accountability before God and in ethical obligations toward both present and future generations (Abdullah et al., 2026; Ferdinan et al., 2026). From this perspective, sustainability is not framed primarily as ecological knowledge but as a form of disciplined moral conduct embedded within broader systems of belief and social responsibility. This ethical dimension establishes the foundational layer of the hybrid sustainability framework, shaping how sustainability-related attitudes are formed prior to engagement with ecological or scientific knowledge.

Within the UAE curriculum, Islamic Studies plays a central role in shaping sustainability-related dispositions through frameworks of moral discipline, restraint, and accountability rather than through explicit ecological doctrine. Core ethical principles embedded in Islamic teachings—including the avoidance of waste (*israf*), personal accountability before God, and the cultivation of moral self-regulation—frame responsible behavior as an expression of ethical obligation grounded in faith. Islamic environmental ethics scholarship similarly emphasizes that human responsibility toward the natural world is articulated less through environmental abstraction than through disciplined conduct guided by stewardship, moderation, and accountability (Abdullah et al., 2026; Foltz, 2000). Within educational contexts, such principles position sustainability-related behaviors not as technical competencies but as moral practices cultivated through internalized ethical norms and spiritual responsibility. In this way, sustainability is introduced to learners as a matter of conduct before it is approached as a domain of knowledge.

Beyond individual moral discipline, Islamic ethical teachings within the curriculum also cultivate sustainability-related orientations through relational principles that emphasize justice, responsibility toward others, and intergenerational accountability. Concepts such as fairness (*adl*), responsibility for one's actions, and respect for parents and elders situate ethical conduct within networks of social and familial relationships that extend across generations. Islamic ethical traditions frequently frame stewardship of resources and responsible conduct as obligations not only to God but also to future generations and the wider community, thereby reinforcing the moral duty to act with restraint and foresight (Ferdinan et al., 2026; Foltz, 2000). Within the educational context, these principles shape students' understanding of responsibility as a relational practice grounded in justice, care, and social continuity. Rather than presenting sustainability primarily as an environmental concern, this ethical framework situates responsible behavior—including moderation in consumption and intergenerational consideration—within broader systems of moral obligation embedded in religious and social life.

While Islamic Studies establishes the ethical foundations of responsible conduct through faith-based moral discipline, the UAE Moral Education curriculum extends these ethical orientations into the domain of civic responsibility and social participation. Introduced as part of a national effort to cultivate responsible and engaged citizens, Moral Education focuses on values such as tolerance, coexistence, community engagement, volunteerism, respect for laws and social institutions, and commitment to national cohesion, positioning students as active participants in sustaining social and civic life. Rather than framing sustainability primarily through environmental content, Moral Education encourages learners to understand responsibility in terms of collective well-being, social harmony, and engagement with community challenges. In contrast to Islamic Studies, where ethical responsibility is framed through spiritual accountability and personal moral regulation, Moral Education emphasizes the social enactment of responsibility within pluralistic communities and national life. This distinction reflects complementary educational roles: Islamic Studies cultivates internalized moral discipline grounded in religious principles, while Moral Education translates these dispositions into civic practices that sustain social cohesion and collective well-being. Together, these subjects establish a progression from moral formation to civic responsibility within the broader sustainability framework.

Sustainability education scholarship similarly highlights the importance of civic responsibility and participatory citizenship as essential dimensions of sustainability learning, emphasizing that sustainable societies depend not only on environmental awareness but also on socially responsible behaviors and collaborative action (Jickling & Wals, 2008; UNESCO, 2017; Wals, 2011). Within this framework, Moral Education extends the ethical foundations established through religious instruction into civic contexts, enabling students to translate moral values into forms of responsible participation that support long-term societal sustainability.

Taken together, the ethical orientations cultivated through Islamic Studies and the civic values advanced in Moral Education position sustainability within the UAE curriculum primarily as a form of moral and relational responsibility. Rather than emerging initially through ecological or scientific knowledge, sustainability is first constructed through disciplined behavior, accountability, justice, and responsible participation in social life. Within Islamic Studies, these dispositions are grounded in spiritual accountability and moral self-regulation, while Moral Education extends them into civic contexts through responsible citizenship and social cohesion. This configuration establishes an ethical foundation in which sustainability is understood as a pattern of conduct before it is approached as a domain of knowledge. It is upon this moral and relational grounding that subsequent forms of sustainability learning—particularly ecological, scientific, and technological—are built, thereby reinforcing the integrated and sequential nature of the hybrid sustainability framework advanced in this chapter.

4. PLACE-BASED ECOLOGICAL LEARNING IN THE UAE CURRICULUM

The second dimension of this framework situates sustainability within place-based ecological knowledge grounded in the environmental realities of the UAE. Building on the ethical foundations through which sustainability is first introduced as moral and relational responsibility, the UAE curriculum further develops sustainability understanding through ecological learning rooted in lived environmental contexts. Sustainability education scholarship highlights that ecological knowledge is most meaningfully acquired through engagement with specific landscapes, ecosystems, and material environmental realities, rather than through abstract environmental concepts alone (Lotz-Sisitka et al., 2016; Wals, 2011). Such approaches are further supported by research demonstrating that ecological knowledge grounded in community practices and lived experience can be integrated alongside scientific concepts to enhance place-based learning (Singh-Pillay & Madlala, 2026). This perspective is further reinforced by scholarship emphasizing the integration of multiple knowledge systems, where locally grounded and experiential ways of knowing complement scientific approaches in understanding environmental change (Leonard et al., 2022). Place-based education research similarly demonstrates that place-based learning is most effective when it connects learners to the ecological, cultural, and social contexts in which they live, fostering experiential and community-based engagement with environmental systems (Yemini et al., 2025). Decolonial and relational perspectives further suggest that such knowledge emerges through historically grounded interactions between communities and their environments, shaped by local ecological constraints and cultural practices (Escobar, 2001; Kato & de Carvalho, 2026). Recent place-based education research reinforces this position by highlighting how environmental learning becomes transformative when it is anchored in local socio-ecological systems that students directly encounter in their everyday lives (Williams, 2024). Within this framework, place-based learning becomes a process of understanding how environmental conditions—such as aridity, coastal ecosystems, and resource limitations—structure both human adaptation and ecological awareness. In the UAE context, this place-based dimension complements the ethical domain by embedding sustainability within the lived ecological realities that define the region's environmental and cultural landscape.

In the UAE context, place-based ecological learning is closely tied to desert environments, where ecological limits have historically shaped patterns of settlement, resource management, and social organization. Life in arid landscapes required communities to develop adaptive strategies grounded in restraint in resource use, mobility across landscapes, and careful management of scarce water and food resources. Research on the region demonstrates how desert livelihoods—including pastoral movement, oasis agriculture, and seasonal coastal migration—were structured around intimate knowledge of climate variability and ecological constraints (Heard-Bey, 2005). These adaptive practices cultivated forms of environmental awareness rooted in understanding the limits of fragile ecosystems and maintaining balance between human activity and natural resources. Within educational contexts, desert ecology therefore provides a critical entry point for place-based learning, enabling students to understand environmental responsibility as historically situated adaptation rather than abstract environmental knowledge. By grounding ecological learning in the lived conditions of arid landscapes, the curriculum positions sustainability as emerging from local environmental realities rather than from universalized ecological discourse.

Beyond desert landscapes, coastal and marine ecosystems constitute another key ecological domain through which sustainability knowledge is embedded within the UAE's environmental and educational landscape. Mangrove forests and marine habitats play vital roles in biodiversity conservation, shoreline stabilization, and climate resilience across the Arabian Gulf. National environmental initiatives increasingly emphasize mangrove restoration and marine conservation as central components of broader sustainability strategies. These efforts are reinforced through educational programs that introduce students to marine ecosystems, biodiversity protection, and coastal stewardship through experiential and field-based learning. Initiatives such as the Sustainable Schools Initiative promote applied environmental engagement, enabling students to encounter ecological systems directly while developing awareness of coastal vulnerability and conservation priorities (GEEP, n.d.; National Committee on Sustainable Development Goals, 2023). Through this ecological framing, place-based learning becomes grounded in the protection of ecosystems that sustain both biodiversity and human livelihoods.

Water scarcity represents one of the most defining environmental conditions shaping sustainability awareness in the United Arab Emirates. Situated within an arid climatic zone characterized by extremely limited freshwater resources, the country has historically relied on careful water management practices while increasingly turning to technological solutions such as desalination to meet growing demand. Contemporary national sustainability strategies emphasize both conservation and innovation in addressing water security, recognizing water as central to long-term environmental and economic sustainability (MOCCAE, 2017; National Committee on Sustainable Development Goals, 2023). Within educational contexts, water scarcity provides a concrete entry point for sustainability learning by linking ecological constraints with technological adaptation. Students engage not only with environmental challenges associated with freshwater scarcity but also with scientific and engineering responses that enable large-scale water production and management. This intersection between environmental limits and technological innovation reflects the hybrid sustainability model in which ecological awareness and scientific advancement operate in tandem to shape place-based learning.

Beyond ecological systems and technological adaptation, sustainability knowledge in the UAE is also embedded within cultural heritage practices that historically enabled communities to survive and adapt to challenging environmental conditions. Traditional livelihoods such as oasis agriculture, pearl diving, and desert pastoralism required detailed ecological knowledge of seasonal cycles, marine systems, and fragile resource environments. Research highlights how these practices cultivated practical environmental knowledge grounded in observation, experience, and adaptive resource management (Escobar, 2001; Heard-Bey, 2005). Contemporary cultural continuity initiatives extend this ecological legacy through educational programs designed to preserve and transmit environmental knowledge across generations. Programs such as SANAA reinforce the connection between cultural identity and environmental awareness by demonstrating how traditional ways of life embodied principles of conservation, adaptation, and ecological balance. In this way, place-based learning becomes linked not only to scientific knowledge but also to cultural memory and intergenerational transmission of environmental understanding, thereby further reinforcing the relational dimensions of the hybrid sustainability framework.

Collectively, these ecological contexts demonstrate how sustainability knowledge within the UAE curriculum is embedded in the environmental conditions and lived realities shaping human–environment relations. Desert ecosystems, coastal environments, water scarcity, and heritage survival practices provide concrete contexts through which learners encounter sustainability as a function of ecological limits, adaptation, and resource awareness. This place-based grounding extends the ethical foundations established earlier by situating responsibility within specific environmental contexts and linking moral conduct to ecological understanding. At the same time, these experiences prepare learners to engage with more formalized scientific interpretations of environmental systems, creating a progression from lived ecological awareness to analytical environmental knowledge. In this way, place-based ecological learning operates as a critical bridge between ethical formation and scientific literacy within the broader hybrid sustainability framework developed in this chapter.

5. SCIENTIFIC AND ENVIRONMENTAL LITERACY

The third dimension views sustainability as a domain of scientific understanding through which environmental challenges are analyzed and addressed. Extending the ethical and place-based ecological dimensions of the framework, the UAE curriculum further advances sustainability understanding through scientific and environmental literacy. While ethical education establishes responsibility and ecological place-based learning grounds sustainability in lived environmental contexts, scientific knowledge provides the analytical frameworks necessary to examine, explain, and respond to complex environmental challenges. Sustainability education scholarship emphasizes that scientific literacy enables learners to understand systemic relationships, interpret environmental data, and evaluate evidence-based solutions to issues such as climate change, biodiversity loss, and resource management (Jickling & Wals, 2008; UNESCO, 2017; Wals, 2011). From a systems-oriented perspective, such literacy is essential for understanding the interdependence of environmental, social, and economic processes that characterize contemporary sustainability challenges (Stumm, 2025). Within the UAE context, this scientific dimension is closely aligned with national priorities related to environmental innovation, climate resilience, and sustainable development (MOCCAE, 2017; National Committee on Sustainable Development Goals, 2023). Positioned within the hybrid sustainability framework advanced in this chapter, scientific literacy complements ethical and ecological knowledge by equipping learners with the conceptual tools required to understand and address sustainability challenges in technologically complex and rapidly developing contexts.

Climate science forms a central component of scientific and environmental literacy within sustainability education by enabling learners to understand the systemic drivers and consequences of global environmental change. Educational approaches to climate literacy emphasize understanding atmospheric systems, global warming processes, and the interconnected nature of human and natural systems, equipping students with the knowledge required to interpret environmental risks and evaluate mitigation strategies (UNESCO, 2017, 2022; Wals, 2011). In the UAE context, climate science holds particular relevance given the country's exposure to extreme heat, water stress, and climate-related vulnerabilities associated with arid environments. National sustainability strategies highlight the importance of climate resilience and adaptation across sectors, emphasizing scientific monitoring, environmental planning, and technological innovation as key responses to climate change (MOCCAE, 2017; National Committee on Sustainable Development Goals, 2023). Within curricular contexts, engagement with climate science enables students to connect global environmental processes with local environmental realities, fostering sustainability understanding grounded in scientific analysis while remaining responsive to regional ecological conditions.

Biodiversity education represents another key dimension of scientific and environmental literacy, enabling learners to understand the complexity of ecological systems and the importance of conserving biological diversity across terrestrial and marine environments. Sustainability education frameworks emphasize that knowledge of ecosystems, species interactions, and habitat preservation is essential for developing informed environmental decision-making and long-term conservation practices (UNESCO, 2017, 2022; Wals, 2011). In the United Arab Emirates, biodiversity conservation has become an important component of national sustainability initiatives, particularly in relation to desert ecosystems, coastal habitats, and marine biodiversity within the Arabian Gulf. National sustainability efforts include the expansion of protected areas and the restoration of sensitive ecosystems such as mangrove forests and coastal wetlands as part of broader strategies to safeguard ecological resilience (Expo 2020 Dubai, 2020; National Committee on Sustainable Development Goals, 2023). Within educational contexts, engagement with biodiversity science allows students to explore the interdependence of species and ecosystems while developing awareness of the environmental consequences of habitat degradation and resource overuse.

Renewable energy education constitutes a critical component of scientific sustainability literacy by introducing learners to the technological and environmental dimensions of energy production and consumption. Sustainability education literature emphasizes that understanding energy systems, carbon emissions, and alternative energy technologies is essential for addressing climate change and supporting long-term environmental sustainability (UNESCO, 2017; Wals, 2011). In the United Arab Emirates, renewable energy development has become a central national priority, reflecting broader efforts to diversify energy sources and reduce dependence on fossil fuels while advancing sustainable economic growth. National sustainability strategies highlight large-scale investments in solar energy, clean energy infrastructure, and technological innovation as key components of long-term environmental planning (MOCCAE, 2017; National Committee on Sustainable Development Goals, 2023). Within educational contexts, engagement with renewable energy systems enables students to understand how scientific research and engineering solutions contribute to energy transitions, reinforcing the role of technological innovation within the broader sustainability framework.

Urban sustainability represents a key domain within scientific and environmental literacy, focusing on how cities can be designed and managed to balance environmental, economic, and social demands. Sustainability education emphasizes the importance of understanding urban systems—including energy use, water management, transportation, and waste reduction—as interconnected components of sustainable development (UNESCO, 2017; Wals, 2011). In the United Arab Emirates, rapid urbanization has positioned cities at the center of sustainability planning, with national strategies promoting resource efficiency, sustainable infrastructure, and environmentally responsive urban design. These priorities are further operationalized through planning frameworks such as the Dubai Development Authority Master Planning Guidelines, which promote integrated approaches to land use, mobility, infrastructure, and resource efficiency within urban development (Dubai Development Authority, 2019). Initiatives aimed at developing smart and sustainable cities further highlight the integration of clean energy systems, green building practices, and efficient resource management as essential to long-term sustainability (MOCCAE, 2017). Within educational contexts, engagement with urban sustainability enables students to examine how scientific knowledge is applied to real-world challenges associated with urban growth and environmental impact.

The scientific domains explored in this section illustrate how sustainability within the UAE curriculum is grounded in systematic, evidence-based understanding of environmental processes. Through engagement with climate science, biodiversity, energy systems, urban sustainability, and earth systems knowledge, learners develop the capacity to analyze environmental challenges and evaluate potential solutions using scientific reasoning. Sustainability education research highlights that addressing such challenges requires not only disciplinary knowledge but also systems thinking and the ability to integrate scientific understanding with social and ethical considerations (Jickling & Wals, 2008; Lotz-Sisitka et al., 2016; Stumm, 2025; UNESCO, 2017; Wals, 2011). Within the hybrid framework presented in this chapter, scientific literacy does not replace ethical or ecological knowledge but operates alongside them, forming a complementary domain that enables informed and responsible engagement with sustainability issues. This integrated configuration positions the UAE curriculum as a distinct model of sustainability education that moves beyond conventional disciplinary approaches by embedding sustainability across ethical, ecological, and scientific domains.

6. TECHNOLOGICAL FUTURITY AND POST-OIL PREPARATION

Beyond ethical, ecological, and scientific dimensions, sustainability within the UAE curriculum is further extended through a forward-looking orientation toward technological futurity. In this context, sustainability is not limited to environmental preservation but is reconfigured as the capacity to anticipate, design, and adapt to emerging socio-technical realities. This reflects broader shifts in sustainability discourse, where innovation, digital transformation, and technological advancement are increasingly positioned as central to long-term resilience and development, particularly within national policy frameworks that emphasize innovation-driven growth and economic diversification (National Committee on Sustainable Development Goals, 2023).

Within the UAE curriculum, this dimension is reflected through the integration of artificial intelligence, coding, robotics, space science, and entrepreneurship as core areas of learning. These domains are not treated as isolated technical competencies but as future-oriented knowledge systems through which learners engage with complex global challenges and envision alternative sustainable futures. In this sense, technological learning extends sustainability beyond present environmental concerns into future possibilities, positioning learners as active agents in shaping socio-economic transformation.

Within the hybrid sustainability framework, this dimension introduces a future-oriented epistemology in which knowledge is not only used to understand systems but also to anticipate, design, and transform them. This orientation aligns with national development agendas that prioritize knowledge economies and technological innovation as key drivers of post-oil transition and long-term sustainability. However, technological futurity does not operate independently; rather, it is embedded within ethical, ecological, and scientific frameworks that shape how technological solutions are understood, evaluated, and applied. This integration ensures that innovation is not pursued as an end in itself but is guided by considerations of responsibility, environmental impact, and societal well-being.

Consequently, technological futurity expands the epistemic scope of sustainability education by introducing a future-oriented dimension in which knowledge is not only used to understand existing systems but also to design and transform them. It reflects a shift from sustainability as conservation toward sustainability as innovation, where the capacity to create adaptive, resilient, and context-responsive solutions becomes central to educational outcomes. Within the UAE context, this positions technology as a key mediator between global sustainability challenges and localized development strategies, reinforcing the role of education in preparing learners for uncertain and rapidly evolving futures.

7. INSTITUTIONAL COHERENCE ACROSS SCHOOL TYPES

The fifth dimension of the hybrid sustainability framework concerns institutional coherence across school types, through which sustainability becomes structurally embedded within the UAE education system. Extending the ethical, ecological, scientific, and technological domains examined earlier, this section demonstrates how sustainability learning is not only conceptually integrated across subject areas but also systematically institutionalized through national curriculum structures and governance mechanisms. In contrast to education systems where sustainability remains fragmented across subjects or dependent on localized initiatives, the UAE model reflects centralized coordination that aligns curriculum frameworks, national priorities, and school-level implementation across both public and private sectors (National Committee on Sustainable Development Goals, 2023). Within this institutional configuration, sustainability operates not as an optional curricular theme but as a structurally embedded educational orientation reinforced through policy mandates, regulatory frameworks, and cross-sector coordination. From this perspective, the hybrid sustainability model advanced in this chapter is sustained not only through curricular design but also through institutional architecture.

A central mechanism supporting institutional coherence is the requirement that both public and private schools implement core national subjects aligned with federal educational priorities. Across school types, mandatory subjects such as Islamic Studies, Moral Education, Social Studies, and Arabic language ensure that key ethical, civic, and cultural dimensions of sustainability are consistently integrated within the curriculum regardless of institutional affiliation or curricular model. These nationally mandated subjects provide a shared educational foundation through which students encounter themes related to responsibility, environmental awareness, social cohesion, and national development. By embedding these domains across diverse school systems—including national, private, and international schools—the UAE ensures that sustainability-related knowledge and values remain consistently reinforced across institutional contexts (UNESCO, 2017). This structural consistency reflects broader policy objectives aimed at cultivating shared civic and developmental orientations among students while maintaining curricular diversity across school types.

Institutional coherence is further reinforced through centralized curriculum frameworks that guide content development, learning outcomes, and assessment structures across the education system. Federal and emirate-level authorities provide overarching curriculum standards that define educational priorities while allowing contextual flexibility in implementation. Within these frameworks, sustainability-related themes—particularly those connected to national development, environmental stewardship, and innovation—are embedded across subject areas rather than confined to a single discipline. Centralized planning ensures vertical alignment across grade levels and horizontal alignment across subject domains, enabling sustainability learning to develop progressively through the curriculum (UNESCO, 2017). This governance structure supports continuity between policy goals and classroom practice, reinforcing the institutional stability of the hybrid sustainability model across educational levels (The UAE Government, 2018).

A distinctive feature of the UAE education system is the integration of national educational priorities within diverse international curricular models. Private schools implementing British, American, International Baccalaureate, and other global curricula remain required to incorporate national subjects and align with regulatory standards established by educational authorities (National Committee on Sustainable Development Goals, 2023). This regulatory framework enables schools to maintain curricular diversity while ensuring alignment with national educational goals. Within this system, sustainability learning is reinforced through dual curricular structures: international curricula contribute global perspectives on sustainability, while national subjects embed locally grounded ethical, ecological, and civic dimensions. This layered configuration supports institutional coherence by integrating global sustainability frameworks with national priorities, ensuring that sustainability learning remains consistent across heterogeneous school environments.

Multiple institutional mechanisms further sustain coherence across school types by monitoring curriculum implementation and ensuring alignment with national priorities. Regulatory frameworks governing curriculum approval, inspection systems evaluating educational quality, and professional development initiatives for educators collectively reinforce consistent curricular delivery. Inspection frameworks assess schools' alignment with national standards, including cross-curricular priorities such as innovation, citizenship, and environmental responsibility, through a unified system of performance standards and indicators that ensures consistency in evaluation across school types (ADEK, 2025; UAE School Inspection Framework, 2015). At the same time, teacher training programs and professional learning communities at schools support educators in implementing sustainability-related content across subject areas. These institutional processes operate together to maintain consistency while allowing contextual flexibility in pedagogical practice. Through such governance structures, sustainability education becomes embedded within routine institutional operations rather than dependent on isolated initiatives.

These institutional arrangements demonstrate how the hybrid sustainability epistemology within the UAE curriculum is structurally embedded across school types through coordinated governance and regulatory systems. Mandatory national subjects establish shared ethical and civic foundations, centralized curriculum frameworks ensure continuity and alignment, international curricula provide complementary global perspectives, and regulatory mechanisms sustain implementation across diverse educational settings. Within this institutional configuration, sustainability operates not as a discrete curricular objective but as a systemic educational orientation embedded across subjects, policies, and school structures and governed through inspection, accountability, and system-wide quality assurance mechanisms (National Committee on Sustainable Development Goals, 2023; UAE School Inspection Framework, 2015).

8. ANALYTICAL SYNTHESIS: THE HYBRID SUSTAINABILITY MODEL

The analysis reveals that sustainability within the UAE curriculum is not constructed as a singular or purely environmental domain, but rather as a multidimensional and relational knowledge system emerging from the interaction of diverse epistemic traditions. This reflects broader sustainability debates that conceptualize sustainability as a complex and contested construct shaped by epistemological, spatial, and temporal dimensions (Nwigwe, 2026). Within this framework, knowledge is not treated as universally transferable; instead, it is situated and context-dependent, developing through the interaction between learners and their cultural, ecological, and social environments (Nwigwe, 2026). Such a perspective aligns with scholarship emphasizing that sustainability is strengthened through the integration of diverse knowledge systems—including scientific, experiential, and culturally grounded forms of knowledge—rather than the dominance of a single epistemological framework (Leonard et al., 2022). It further resonates with research demonstrating that sustainability education is shaped by the dynamic interaction of multiple pedagogical, institutional, and contextual factors rather than isolated curricular elements (Hu et al., 2025). Consequently, the UAE curriculum can be understood as embodying a hybrid sustainability epistemology in which scientific knowledge, technological innovation, ethical-religious values, and place-based cultural understandings are structurally integrated.

Research highlights a persistent disjunction between institutional sustainability policies and their practical implementation within curricula, resulting in fragmented and uneven integration across disciplines (Poon, 2017). Empirical studies further demonstrate that, despite policy recognition, the integration of diverse knowledge systems—particularly Indigenous and locally grounded forms of knowledge—often remains superficial or tokenistic, with limited pedagogical frameworks to support meaningful implementation (Singh-Pillay & Madlala, 2026). At the same time, scholarship emphasizes that sustainability understanding is strengthened when multiple knowledge systems are brought into dialogue, enabling more holistic, context-sensitive, and epistemologically inclusive approaches to environmental learning (Leonard et al., 2022). Within this context, sustainability concepts become distributed across subjects and learning experiences, where environmental awareness, national identity, technological advancement, and ethical responsibility intersect. Such interaction reflects the inherently relational nature of sustainability, where different knowledge systems both complement and, at times, challenge one another. In this sense, the curriculum operates as a dynamic epistemic architecture that organizes sustainability through interconnected layers of meaning shaped by both global policy discourses and localized sociocultural foundations.

At the foundation of the hybrid sustainability model lies ethical grounding, through which sustainability is initially constituted as moral and behavioral responsibility. As demonstrated in earlier sections, Islamic Studies and Moral Education position sustainability not primarily as a technical or environmental issue but as a matter of disciplined conduct, accountability, and relational responsibility toward others and future generations. Ethical principles such as stewardship, moderation, justice, and self-regulation shape sustainability attitudes at the level of individual behavior and social interaction, establishing a normative framework within which environmental awareness becomes meaningful. This ethical grounding is critical, as it frames sustainability as a question of how individuals ought to act before it becomes a question of what they ought to know, thereby anchoring sustainability learning in moral dispositions that guide subsequent forms of knowledge acquisition (Abdullah et al., 2026; Ferdinan et al., 2026; Jickling & Wals, 2008).

Building on this ethical foundation, sustainability is further developed through place-based ecological learning grounded in the environmental realities of the UAE. Desert ecosystems, coastal environments, water scarcity, and heritage-based survival practices provide concrete contexts through which learners encounter ecological limits, adaptation strategies, and human–environment interdependence. Unlike abstract environmental education models, this dimension situates sustainability within lived ecological conditions, allowing students to understand how environmental knowledge emerges through interaction with specific landscapes and resource constraints. In this way, place-based learning translates ethical responsibility into ecological awareness, linking moral conduct with environmental understanding and reinforcing the relational nature of sustainability within local contexts (Escobar, 2001; Kato & de Carvalho, 2026; Lotz-Sisitka et al., 2016; Yemini et al., 2025).

The third dimension of the model introduces sustainability as a domain of scientific understanding, enabling learners to analyze environmental systems and evaluate evidence-based solutions to complex sustainability challenges. Through engagement with climate science, biodiversity, renewable energy, and urban sustainability, students develop the analytical tools required to interpret environmental processes and assess technological and policy responses. Scientific literacy complements ethical and ecological knowledge by providing explanatory frameworks that deepen understanding of systemic relationships within natural and human systems. Importantly, within the hybrid model, scientific knowledge does not replace ethical or place-based learning but operates alongside them, contributing a distinct yet interdependent mode of understanding that supports informed decision-making and critical engagement with sustainability issues (UNESCO, 2017; Wals, 2011).

Extending beyond scientific understanding, the model incorporates technological futurity as a critical dimension of sustainability within the UAE context. Educational initiatives focused on artificial intelligence, coding, robotics, space science, and entrepreneurship reflect national priorities associated with economic diversification and innovation-driven development. Within this framework, sustainability is linked not only to environmental protection but also to the capacity to design, innovate, and adapt within rapidly evolving technological landscapes. This dimension positions learners as future-oriented agents capable of contributing to post-oil transformation, emphasizing the role of technological knowledge in shaping sustainable economic and societal futures. As such, technological futurity expands the scope of sustainability beyond environmental concerns to include the development of skills and competencies necessary for long-term national resilience and global competitiveness (National Committee on Sustainable Development Goals, 2023).

Crucially, these four dimensions do not operate in isolation but interact dynamically within the curriculum to produce a coherent sustainability orientation. Ethical grounding shapes how sustainability is valued and enacted, place-based ecological learning contextualizes environmental awareness within lived realities, scientific literacy provides analytical depth, and technological futurity extends sustainability into the domain of innovation and future development. The interaction among these domains reflects a systems-based understanding of sustainability in which knowledge is relational, interconnected, and context-dependent. This aligns with research demonstrating that sustainability education outcomes emerge from the interaction of multiple interdependent factors—including teacher practices, institutional structures, and contextual conditions—rather than from isolated curricular inputs (Hu et al., 2025). Rather than privileging one form of knowledge over others, the hybrid model demonstrates how sustainability education emerges through the integration of multiple epistemic traditions that together shape how learners interpret and respond to socio-ecological challenges (Stumm, 2025; Wals, 2011).

This synthesis positions the UAE curriculum as a distinct model of sustainability education that moves beyond conventional frameworks centered primarily on environmental science. By integrating ethical, ecological, scientific, and technological domains within a unified curricular structure, the hybrid sustainability model offers a more comprehensive approach to sustainability learning that reflects both global sustainability agendas and localized sociocultural contexts. The model contributes to broader debates in sustainability education by demonstrating how curriculum can function as an epistemic infrastructure that organizes diverse forms of knowledge into a coherent and actionable framework. In doing so, it highlights the importance of epistemological plurality in addressing complex sustainability challenges, suggesting that effective sustainability education requires not only interdisciplinary integration but also recognition of multiple ways of knowing (Jickling & Wals, 2008; UNESCO, 2017).

9. STRUCTURAL TENSIONS AND CONTRADICTIONS

While the hybrid sustainability model presented in this chapter highlights the integrative strengths of the UAE curriculum, it also reveals a set of inherent structural tensions that reflect broader complexities within sustainability education. Rather than representing shortcomings, these tensions can be understood as characteristic of sustainability itself, which is widely conceptualized as a multidimensional and contested construct shaped by competing priorities, knowledge systems, and temporal orientations (Wijsman et al., 2026).

One key tension emerges between global sustainability frameworks and localized sociocultural foundations. On the one hand, the UAE curriculum aligns with international sustainability agendas, incorporating themes such as climate change, innovation, and global citizenship. On the other hand, it simultaneously embeds sustainability within national identity, cultural heritage, and ethical-religious values. This dual orientation creates a productive but complex interplay between universalized sustainability discourses and context-specific interpretations, where global models are adapted rather than adopted wholesale. Such dynamics reflect place-based perspectives, which emphasize that sustainability knowledge is always mediated through local contexts, cultural meanings, and lived experiences (Nwigwe, 2026).

A second tension arises between environmental sustainability and economic development, particularly within the context of post-oil transformation. National strategies emphasize economic diversification, technological innovation, and large-scale urban development, all of which are framed as components of sustainable futures. However, these priorities may coexist with environmental challenges associated with rapid urbanization, resource consumption, and ecological impact. This reflects a broader contradiction within sustainability discourse, where economic growth and environmental protection are often pursued simultaneously despite their potential incompatibilities. In the UAE context, this tension is not resolved but rather managed through a model that seeks to balance development ambitions with sustainability commitments.

A third tension concerns the relationship between centralized curriculum structures and the contextual flexibility required for meaningful sustainability learning. While the curriculum integrates multiple domains of sustainability across subjects, its implementation remains shaped by national standards, prescribed learning outcomes, and structured curricular frameworks. This may limit the extent to which place-based, experiential, and community-linked learning can be fully realized in practice. Similar patterns have been observed in other contexts, where centralized educational systems constrain the integration of locally grounded knowledge despite policy-level commitments to contextual relevance (Nwigwe, 2026).

Finally, a tension can be identified between technological futurity and ethical-ecological grounding. While technological innovation is positioned as a key driver of sustainable futures, it also raises questions about the extent to which technological solutions remain aligned with ethical principles and environmental limits. Within the hybrid model, this tension is addressed through the integration of ethical and scientific perspectives that guide technological development; however, the balance between innovation and responsibility remains an ongoing challenge.

The above tensions do not undermine the coherence of the hybrid sustainability model but instead highlight its dynamic and evolving nature. They demonstrate that sustainability education within the UAE curriculum operates as a negotiated space in which multiple priorities, knowledge systems, and developmental trajectories intersect. This reinforces the argument that sustainability is not a fixed or unified concept but a relational and context-dependent construct, shaped by continuous interaction between global discourses and local realities. In this sense, the presence of tension is not a weakness but a defining feature of sustainability itself, reflecting the complexity of addressing socio-ecological challenges within rapidly transforming societies.

10. CONCLUSION: CURRICULUM AS A HYBRID SUSTAINABILITY EPISTEMOLOGY INFRASTRUCTURE

This chapter has argued that sustainability within the United Arab Emirates curriculum is best understood not as a discrete subject or a narrowly defined environmental domain, but as a hybrid epistemological formation structured through the integration of multiple knowledge traditions. Moving beyond conventional models of sustainability education that privilege scientific or environmental knowledge alone, the analysis has demonstrated how the UAE curriculum institutionalizes sustainability through the interaction of ethical–religious formation, place-based ecological learning, scientific literacy, and technological futurity. Together, these domains form a coherent yet dynamic framework through which sustainability is constructed as both a way of knowing and a way of acting within educational contexts.

Central to this argument is the recognition that curriculum operates as a key institutional site where sustainability knowledge is organized, legitimized, and transmitted. Within the UAE context, sustainability is not confined to a single subject area but is distributed across multiple curricular domains, embedded within national policy frameworks, and reinforced through institutional structures that ensure coherence across school types. Ethical education establishes sustainability as moral responsibility, place-based learning grounds it in ecological realities, scientific literacy provides analytical depth, and technological education extends it into future-oriented innovation and economic transformation. This layered configuration positions curriculum as a form of sustainability infrastructure—an organized system through which diverse epistemic traditions converge to shape how learners understand and engage with sustainability challenges.

At the same time, the chapter has highlighted that this hybrid model operates within a set of structural tensions associated with economic development, environmental pressures, and global policy alignment. These tensions do not diminish the coherence of the framework but instead reveal the complex and negotiated nature of sustainability within rapidly transforming societies. By integrating global sustainability agendas with localized ethical and cultural knowledge systems, the UAE curriculum reflects a context-sensitive approach that navigates these tensions through adaptation and synthesis rather than through rigid or singular models of sustainability education.

The findings of this chapter contribute to broader debates in sustainability education by illustrating the importance of epistemological plurality in addressing complex socio-ecological challenges. They suggest that effective sustainability education requires moving beyond disciplinary silos toward integrative frameworks that combine ethical values, ecological understanding, scientific reasoning, and technological innovation. In this sense, the UAE curriculum offers a distinctive model that challenges dominant paradigms of sustainability education and demonstrates how curriculum can function as a site of epistemic integration within diverse cultural and developmental contexts. Conceptualizing curriculum as a form of hybrid sustainability epistemology infrastructure provides a useful lens for rethinking the role of education in shaping sustainable futures. It foregrounds the capacity of educational systems to organize knowledge in ways that reflect both global challenges and local realities, while also preparing learners to navigate the uncertainties and complexities of a rapidly changing world.

REFERENCES

Abdallah, A. K., Ismail, L. S., & Alkaabi, A. M. (2025). Green careers: Educating for the future of sustainability. In *Legal frameworks and educational strategies for sustainable development* (pp. 337–366). IGI Global Scientific Publishing. https://doi.org/10.4018/979-8-3693-2987-0.ch017

Abdullah, W., Tarihoran, N., Supardi, S., Saenong, F., Wasehudin, W., Syarifuddin, E., Fauzi, A., & Mamur, I. (2026). Advancing sustainable Islamic management education through education for sustainable development. *Discover Sustainability, 7*(1), 257. Advance online publication. DOI: 10.1007/s43621-025-02571-1

Abu Dhabi Department of Education and Knowledge (ADEK). (2025). *School policy on sustainability* (Version 1.2). https://adek.gov.ae/-/media/Project/TAMM/ADEK/Policies/School-Policies/Health-safety-and-wellbeing/ADEK_S_Sustainability-Policy_EN-12.pdf

Alzaim, M. A., AlAli, M., Mattar, Y., & Samara, F. (2024). Integrated framework for enhancing liveability and ecological sustainability in UAE communities. *Sustainability (Basel), 16*(22), 9872. Advance online publication. DOI: 10.3390/su16229872

Baroudi, S., & Abi Haidar, H. (2025). Sustainability in the United Arab Emirates secondary schools: A policy practice analysis. *Sustainability (Basel), 17*(7), 3129. Advance online publication. DOI: 10.3390/su17073129

Battiste, M. (2013). Indigenous knowledge and indigenous peoples' education. In United Nations (Ed.), *Traditional knowledge in policy and practice* (pp. 31–51)., DOI: 10.18356/aa0ced95-

Bonnett, M. (2002). Education for sustainability as a frame of mind. *Environmental Education Research, 8*(1), 9–20. DOI: 10.1080/13504620120109619

Degyansky, M. (2024). Relational, ecological cooperation with and as part of more-than-human world(s). In Warren, J. N., Biggiero, L., Hübner, J., & Ogunyemi, K. (Eds.), *The Routledge handbook of cooperative economics and management* (pp. 433–442). Routledge., DOI: 10.4324/9781003449850-34

Dorji, T., Rinchen, K., Morrison-Saunders, A., Blake, D., Banham, V., & Pelden, S. (2024). Understanding how indigenous knowledge contributes to climate change adaptation and resilience: A systematic literature review. *Environmental Management, 74*(6), 1101–1123. DOI: 10.1007/s00267-024-02032-x PMID: 39215837

Dubai Development Authority. (2019). *Master planning guidelines*. https://dda.gov.ae/-/media/5F89FD14770B491BB843C84BC8031C40.pdf

Environment Agency—Abu Dhabi. (2026). *Sustainable Schools.* https://sustainableschools.ead.ae/sis/ssiWebsite/

Escobar, A. (2001). Culture sits in places: Reflections on globalism and subaltern strategies of localization. *Political Geography, 20*(2), 139–174. DOI: 10.1016/S0962-6298(00)00064-0

Expo 2020 Dubai. (2020). *Sustainability report 2020: Connecting minds, creating the future.* https://www.expo2020dubai.com/-/media/expo2020/sustainability/expo-designed-report-digital-version-v12.pdf

Ferdinan, A., Abdillah, , Rahman, A., & Bahtiar, A. R. (2026). Eco Islamic education and its contribution to sustainable development goals in Muhammadiyah boarding schools of South Sulawesi. *Discover Sustainability, 7*(1), 310. DOI: 10.1007/s43621-026-02647-6

Foltz, R. (2000). Is there an Islamic environmentalism? *Environmental Ethics, 22*(1), 63–72. DOI: 10.5840/enviroethics200022149

Global Environmental Education Partnership (GEEP). (n.d.). *Sustainable schools initiative* (Case study). https://thegeep.org/resources/case-studies/sustainable-schools-initiative

Gruenewald, D. A. (2003). The best of both worlds: A critical pedagogy of place. *Educational Researcher, 32*(4), 3–12. DOI: 10.3102/0013189X032004003

Heard-Bey, F. (2005). The United Arab Emirates: Statehood and nation-building in a traditional society. *The Middle East Journal, 59*(3), 357–375. https://www.jstor.org/stable/4330153. DOI: 10.3751/59.3.11

Hu, L., Cheng, J., Wu, C., Liu, Z., & Zheng, L. (2025). Influencing factors for sustainable high-quality implementation of comprehensive practical activity courses in rural areas. *Humanities & Social Sciences Communications, 12*(1), 1937. Advance online publication. DOI: 10.1057/s41599-025-06226-4

Jickling, B., & Wals, A. E. J. (2008). Globalization and environmental education: Looking beyond sustainable development. *Journal of Curriculum Studies, 40*(1), 1–21. DOI: 10.1080/00220270701684667

Kato, D. S., & de Carvalho, L. M. (2026). Afro-Amerindian cosmopolitics in environmental education: A decolonial analysis of academic discourses involving the epistemological disputes around the term sustainability. *British Educational Research Journal, 52*(1), 25–46. DOI: 10.1002/berj.70104

Leonard, K., Buttigieg, P. L., Hudson, M., Paul, K., Pearlman, J., & Juniper, S. K. (2022). Two-eyed seeing: Embracing the power of indigenous knowledge for a healthy and sustainable Ocean. *PLoS Biology*, *20*(10), e3001876. Advance online publication. DOI: 10.1371/journal.pbio.3001876 PMID: 36269777

Lotz-Sisitka, H., Ali, M. B., Mphepo, G., Chaves, M., Macintyre, T., Pesanayi, T., Wals, A., Mukute, M., Kronlid, D., Tran, D. T., Joon, D., & McGarry, D. (2016). Co-designing research on transgressive learning in times of climate change. *Current Opinion in Environmental Sustainability*, *20*, 50–55. DOI: 10.1016/j.cosust.2016.04.004

Ministry of Climate Change and Environment (MOCCAE). (2017). *National climate change plan*. https://www.moccae.gov.ae/assets/1d66432d/Infosheet-%20NCCP.pdf.aspx

Ministry of Education (MoE). (2024). *Ministry of Education launches a national campaign to instil the values of "We the UAE 2031" vision among students*. https://www.moe.gov.ae/En/MediaCenter/News/Pages/Ministry-of-Education-launches-a-national-campaign-to-instil-the-values-of-We-the-UAE-2031-vision-among-students.aspx

MoE. (2023). *UAE Ministry of Education launches Greening Education Hub themed 'Legacy from the Land of Zayed' ahead of COP28*. https://www.moe.gov.ae/En/MediaCenter/News/Pages/MOECOP28.aspx?utm_source=chatgpt.com

National Committee on Sustainable Development Goals. (2023). *UAE and the 2030 Agenda for Sustainable Development*. https://mohap.gov.ae/documents/20117/1212145/the+2030+agenda+for+sdg.pdf/6e96b55c-73b6-d1d1-36cf-f1277da306f1?t=1739164957965

Nwigwe, E. O. (2026). Understanding place-based education in Nigeria: A structured narrative review of its status, barriers, and African comparative insights. *Social Sciences & Humanities Open*, *13*, 102409. Advance online publication. DOI: 10.1016/j.ssaho.2025.102409

Poon, J. (2017). Engaging sustainability good practice within the curriculum design and property portfolio in the Australian higher education sector. *International Journal of Sustainability in Higher Education*, *18*(1), 146–162. DOI: 10.1108/IJSHE-09-2015-0149

Purvis, B., Mao, Y., & Robinson, D. (2019). Three pillars of sustainability: In search of conceptual origins. *Sustainability Science*, *14*(3), 681–695. DOI: 10.1007/s11625-018-0627-5

Singh-Pillay, A., & Madlala, A. (2026). Legitimising Indigenous knowledge systems as a pathway to sustainable development. *Discover Education*, *5*(1), 87. Advance online publication. DOI: 10.1007/s44217-025-01062-9

Smith, G. A. (2002). Place-based education: Learning to be where we are. *Phi Delta Kappan*, *83*(8), 584–594. https://www.jstor.org/stable/20440205. DOI: 10.1177/003172170208300806

Sterling, S. (2001). *Sustainable education: Re-visioning learning and change*. Green Books.

Stumm, S. (2025). Epistemological implications of a system—Theoretical understanding for sustainability models. *Systems Research and Behavioral Science*. Advance online publication. DOI: 10.1002/sres.3149

UNESCO. (2017). *Education for sustainable development goals : Learning objectives*. UNESCO. https://unesdoc.unesco.org/images/0024/002474/247444e.pdf

UNESCO. (2022). *United Arab Emirates: Climate change communication and education, Education Profiles*. https://education-profiles.org/northern-africa-and-western-asia/united-arab-emirates/~climate-change-communication-and-education?utm_source=chatgpt.com

Wals, A. E. J. (2011). Learning our way to sustainability. *Journal of Education for Sustainable Development*, *5*(2), 177–186. DOI: 10.1177/097340821100500208

Whyte, K. (2020). Too late for indigenous climate justice: Ecological and relational tipping points. *Wiley Interdisciplinary Reviews: Climate Change*, *11*(1), e603. Advance online publication. DOI: 10.1002/wcc.603

Wijsman, K., Bogner, K., Avelino, F., Kalfagianni, A., & Weis, C. U. (2026). Towards a lexicon for just sustainability transitions: Thinking with capabilities, epistemology, intersectionality, spatiality, and temporality. *Environmental Innovation and Societal Transitions*, *59*, 101104. Advance online publication. DOI: 10.1016/j.eist.2026.101104

Williams, L. (2024). Indigenous intergenerational resilience and lifelong learning. *Sustainability (Basel)*, *16*(11), 4494. Advance online publication. DOI: 10.3390/su16114494

Yemini, M., Engel, L., & Ben Simon, A. (2025). Place-based education: A systematic review of literature. *Educational Review*, *77*(2), 640–660. DOI: 10.1080/00131911.2023.2177260

Chapter 4
Place-Based Education and the Ecology of Learning

Zahid Ahmad Thoker
http://orcid.org/0009-0002-6897-5835
Lovely Professional University, India

Lovleen Marwaha
http://orcid.org/0000-0002-5175-2590
Lovely Professional University, India

ABSTRACT

Place-Based Education (PBE) relates education to the local settings, cultures and communities to enhance sustainability and environmental consciousness. Based on the Indigenous knowledge systems, PBE promotes experiential and land-based learning that can develop environmental stewardship, critical thinking, and social responsibility. The chapter examines the theoretical perspectives, pedagogical approaches and policy models supporting PBE in the world and its contribution to decolonizing education and the promotion of Education for Sustainable Development (ESD). It brings forth the successful school-community partnerships and new teaching strategies through case studies, both global and local. The chapter wraps up by discerning the important measures in incorporating PBE into the curriculum that can help develop climate resilience and sustainability literacy.

DOI: 10.4018/979-8-3373-7650-9.ch004

1. INTRODUCTION

Over the past few years, education systems worldwide have experienced increasing pressure to develop meaningful responses to complex problems such as climate change, biodiversity loss, and rapid transformations in social-ecological systems. The conventional conception of schooling, with many of its tendencies endorsing the privilege of decontextualised knowledge and standardised forms, is becoming apparent as ill-prepared to engage learners with the realities of the world around them, which are increasingly interconnected and intertwined. Place-based education (PBE) can be described as a pedagogical response to this changing context, which regrounds the learning process in the proximate ecological, cultural, and spatial environments of learners' lives (Yemini et al., 2025). PBE provides a pathway toward more meaningful, generative, and sustainable learning, as curriculum, pedagogy, and community engagement are grounded in local landscapes, histories, and systems.

1.1 Introduction and Background of Place-Based Education

PBE has the concept that place is a defining aspect of learning because the environments around students, local ecologies, community networks, and cultural heritage are not peripheral to education but full of learning and inquiry (Hernandez Gonzalez, 2023). PBE in a globalised, rapidly urbanised, and disconnected world is a form of learning that relocalises, inviting learners to be active participants in their communities and to discover their local environments rather than passively receiving them as observers (Yemini et al., 2025). The arguments in support of PBE are manifold: it promotes the more engaged approach, as it utilises the familiarity and relevance of the surrounding where learners live; it promotes the ecological literacy and awareness as it draws attention to the interrelations of place, culture, and nature; and it fits into the demands of more community-centred, experiential, and contextualised learning (Dartevelle, 2012). In addition, PBE is closely connected to Indigenous and local knowledge systems, which have always had learning embedded in place, time, and relationality (Gerelkhuu et al., 2026). In this regard, PBE is not merely a pedagogical innovation but a philosophic restructuring that situates learners within the living systems of their places, rather than treating place as an objective, a priori knowledge container.

1.2 Correlation Between Ecology, Sustainability, and Learning

The main idea of the connection between ecology, sustainability, and learning is at the centre of PBE. Ecology provides a metaphor and a scientific perspective of interdependence, systems dynamics, resilience, and change. The ecological concept of learning leads to less discrete and more relationship-based, flow-based, and coevolutionary relationships among human beings, nonhuman systems, and place (Hernandez Gonzalez, 2023). Sustainability, on the other hand, offers a normative orientation to education that goes further to promote sustainability, which must not only develop knowledge but also system thinking, agency, and responsible action as capabilities (Bascopé et al., 2019). PBE realises such a connection: putting learners into contact with their local ecosystems, communities, and cultures, PBE facilitates the sustainability-related cognitive, socio-emotional, and behavioural aspects of teaching (Hernandez Gonzalez, 2023). For example, young people who participated in place-based stewardship programmes reported greater agency, a stronger sense of place, and a greater likelihood of taking action to promote ecological well-being (Gallay et al., 2016). In this way, PBE is more a medium through which education, environment, and culture interact, in which learning is no longer about the world but rather in and with the world.

1.3. Conceptualizing the Ecology of Learning

Ecology of learning is a critical theoretical perspective that allows us to address the construction of knowledge based on the dynamic forces of interactions between the learner, the environment, and the socio-cultural context. In contrast to the traditional educational models where learning is viewed as a linear and classroom-based process, the concept of ecology of learning views learning as a complex and adaptive system that is developed in relation to the connections between individuals, communities, and the environment (Barron, 2006; Sterling, 2011). Learning in this perspective does not simply mean the acquisition of knowledge but an event that comes up as a result of interaction with place, culture, and lived experience.

Although place-based education (PBE) focuses on the significance of a local context and experience, the ecology of learning builds upon this practice, making the experience of learning multiple dimensions, including cognitive, affective, social, and environmental, into one. It points to the fact that the learning environments are network systems in which knowledge moves through both the formal and informal contexts, such as schools, communities, and the natural ecosystems. This facilitates the holistic approach to sustainability learning processes, especially within the context of the implementation of Indigenous knowledge systems and community-based practices.

The theoretical input of the chapter is placing the ecology of learning as a theoretical transition between place-based education and sustainability education. Through the synthesis of the information provided by the Indigenous epistemologies, systems thinking, and experiential pedagogy, the chapter proceeds to develop a cohesive framework according to which learning is perceived as relational, context-responsive, and ecologically based. Such a reconceptualization not only enriches the theoretical context of PBE but also offers grounds for how to structure educational practices that would be more responsive to the realities of 21 st -century sustainability challenges.

1.4. Objectives and Scope of This Chapter

The main idea of this chapter is to critically examine how place-based education (PBE) can serve as a powerful learning process focused on sustainability, grounded in ecological knowledge, community interactions, and culturally oriented pedagogies. In particular, the chapter seeks to elucidate the pedagogical processes through which learners acquire ecological literacy and a sense of place, while simultaneously nurturing environmental stewardship and social responsibility. PBE shows students that education can be understood not only as a place but also as a living laboratory that represents intricate interdependencies between humans and nature by situating education within local ecosystems and cultural landscapes (Yemini et al., 2025).

Another role of this chapter is to examine the theoretical connections among ecology, sustainability, and education through an interdisciplinary approach. It emphasises how systems theory, interdependence, and resilience-based ecological thinking can be used to design curricula and pedagogical approaches that will develop sustainability competencies. Through that, it fills the gap between environmental science and practice in education and proves that ecological awareness can influence cognitive, affective, and ethical development of learners. It is also discussed in terms of how Indigenous epistemologies and local systems of knowledge can enhance the conceptual basis of PBE by drawing on relationally oriented concepts of knowing and living that oppose anthropocentric and extractive conceptualizations (Shrivastava & Kennelly, 2013).

The chapter has a broad scope that addresses the theoretical and practical aspects of place-based education. In theory, it integrates existing models that bridge ecological systems thinking with experiential and community-based learning models. In practice, empirical case studies from various settings (rural schools, Indigenous communities, and urban sustainability projects) are needed to demonstrate how place-based practices have been applied to yield ecological and educational outcomes, as in Figure 1. The policy and institutional aspects of PBE are also considered in the chapter, where the authors discuss the possibility of reorganising education systems to promote place-based curricula in accordance with the United Nations Sustainable Development Goals (Yemini et al., 2025).

Finally, the chapter is intended to offer conceptual and practical guidance to educators, policymakers, and researchers who may wish to repurpose learning spaces as ecologically situated, socially responsive systems. By articulating the educational possibilities of reconnecting learners to their local settings, the research establishes

PBE as a pillar of sustainability education in the 21st century, a methodology that not only develops knowledge but also fosters ecological awareness, resilience, and the welfare of the group.

Figure 1. Place-Based education for sustainability and its synergy

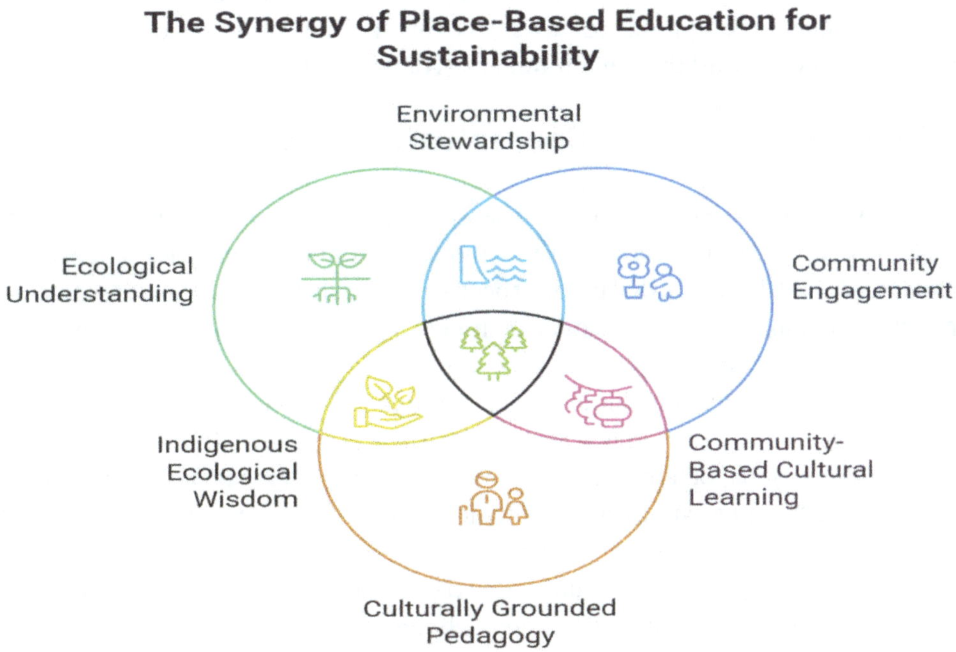

2. THEORETICAL FOUNDATIONS

2.1 Indigenous Knowledge Systems and Place-Based Learning

Indigenous knowledge systems (IKS) are networks of ecological, cultural, spiritual, and relational understandings upheld across generations, providing a strong foundation for place-based learning. These systems emphasize place attachment, local ecosystems, seasonality, and community memory, situating learning within relationships rather than treating it as abstract and decontextualized. For example, Peltier (2021) draws on Anishinaabe pedagogy to illustrate how land-, community-, and culturally centered ways of knowing position learners as active participants within ecosystems rather than detached observers. Through place-based learning grounded in Indigenous epistemologies, education becomes a process of reconnecting with local environments and cultural heritage. Studies indicate that learners value the integration of Indigenous knowledge with scientific perspectives in experiential learning contexts (Jones, Williams-Jones, & Nyce Jr., 2025). Aligning PBE with Indigenous knowledge systems supports the development of learner identity, cultural continuity, ecological stewardship, and resilience. By situating learning within local ecological and cultural contexts, students develop stronger connections and responsibilities toward their environments (Hlalele, 2019).

2.2 Learning Ecologies: Conceptual Frameworks

Learning ecologies conceptualize learning as a dynamic and interconnected system involving learners, communities, tools, environments, and knowledge. Hecht and Crowley (2019) argue that this perspective emphasizes adaptive processes, complexity, and situated change. Rather than viewing learning as linear transmission, the ecological metaphor frames it as emergent, relational, and context-dependent. In sustainability education, this approach highlights the co-evolution of knowledge, communities, and environmental action. Hernandez Gonzalez (2023) demonstrates how place-based education engages cognitive, socio-emotional, and behavioral dimensions of learning within such ecological systems. From this perspective, PBE represents not merely a pedagogical method but a reconfiguration of learning environments, where schools, communities, and ecosystems function as interconnected spaces for knowledge creation. This approach supports responsive, context-sensitive, and relational curriculum design, reflecting an integrated ecology of learning (Hecht & Crowley, 2019).

2.3 Decolonizing Education Through Local Contexts

Decolonizing education involves challenging dominant Eurocentric epistemologies while affirming Indigenous and local ways of knowing, being, and learning. Place-based education provides a pathway for this transformation by grounding learning in local contexts, valuing community knowledge, and promoting learner agency. For instance, Skerrett and Vlach (2022) demonstrate how decolonized educational practices in the Caribbean engage learners with local histories, languages, and identities. Similarly, Quilaqueo and Torres (2024) emphasize Aboriginal Country-centered learning as a foundational element of curriculum, highlighting relational, cultural, and ecological dimensions of place. This decolonial perspective shifts PBE beyond the use of local context toward critical engagement with issues of knowledge, power, and representation in education. It supports Indigenous self-determination and ecological justice, positioning learning as a transformative process that connects sustainability with social and cultural responsibility.

2.4 Cognitive and Affective Aspects of Place-Based Learning

Sustainability education should not be limited to cognitive (facts, concepts) but should also involve affective (values, attitudes, emotions) and behavioural (actions, dispositions) aspects. PBE supports this multidimensional learning by exposing learners to their locations and by fostering reflection, involvement, and agency. Hernandez Gonzale z (2023) offers the empirical results of how PBE promotes cognitive, socio-emotional, and behavioural learning: cognitive: becoming knowledge of interdependency; socio-emotional: being a part of and empathising; behavioural: acting effectively and responsibly (Hernandez Gonzalez, 2023). PBE facilitates connection, responsibility, and agency by engaging learners with their place through an emotional, experiential approach that is often difficult to cultivate through classroom instruction. To illustrate, in a study of youth rural sustainability in Taiwan, learners acquired not only knowledge of local systems but also a greater sense of purpose and identity (Chang and Kuo, 2021). Place-based learning, therefore, has the capability of fostering all three dimensions of learning, namely cognitive, affective, and behavioural, which are critical to sustainability.

2.5 Sustainability Literacy and Core Competencies

Sustainability literacy is the ability of learners to comprehend the complicated socio-ecological systems and take action. ESD competency frameworks define systems thinking, anticipatory thinking, normative competency, strategic competency, collaboration, critical thinking, self-awareness, and integrated problem-solving as the primary competencies (National Academies of Sciences, Engineering & Medicine, 2020). The scoping review of sustainability competences in secondary education by Sposab and Rieckmann (2024) attests to an increase in empirical research on the frameworks (Sposab and Rieckmann, 2024). Within the framework of PBE, sustenance literacy is cultivated by connecting local ecological and cultural environments to global issues: students explore their environment, learn about interdependencies, interact with the community, and effect change. Bascopé et al (2019) illustrate place-based STEM programmes and show how students can acquire attitudes, competencies, and community connections to build socio-ecological resilience. Place-based pedagogy can be used as a powerful tool to empower learners to navigate and create sustainable futures by integrating sustainability competency frameworks into place-based pedagogy.

3. PEDAGOGICAL FRAMEWORKS AND INNOVATIONS

3.1 Experiential, Outdoor, and Land-Based Learning

Outdoor and land-based experiential learning places learners in direct physical and mental contact with their local environments through hands-on activities, physical immersion, and thinking and acting on site. This form of pedagogy also makes learning not confined to the classroom walls but to the landscapes, ecosystems, and community spaces where learners can observe, ask questions, engage, and reflect on the connection between human and more-than-human worlds. For example, Lloyd and Grey (2014) explain how a one-year place-based outdoor learning programme in a primary school in Australia led to sustained engagement with the same outdoor environment, the development of place connections, and the establishment of environmental sustainability behaviours. Equally, Thompson and Chapman (2025) demonstrate the support for a green corridor in an urban Australian environment as an outdoor classroom where transdisciplinary experiential learning took place on planetary health and built-environment problems.

The methods are particularly applicable to sustainability education, as they help contextualise learning within real ecological-social situations rather than purely theoretical frameworks. Land-based pedagogy relies on the idea of relationality to land, culture, community, and time, and invites learners to become stewards (rather than simply learners) of nature. The difficulty lies in planning curricula that combine outdoor activities and reflection, community work, and follow-through in ways that ensure experience does not exist in isolation but is embedded in the developing ecology of learners' knowledge.

3.2 Incorporating Cultural and Ecological Knowledge

The presence of cultural (including Indigenous) and ecological knowledge systems is a crucial component of the pedagogy of place-based education. It implies that educational design recognises the cultural, historical, spiritual, and ecological significance of specific local landscapes, and the place knowledge that is plural, layered, and embedded. For example, in their study of eight rural primary schools in Taiwan, Chang and Kuo (2021) employed a place-based pedagogical action study that combined rural tacit knowledge with scientific inquiry in their Knowledge-Ties Youth Rural Sustainability (KYRS) model, thereby adding cultural-landscape wisdom to ecological learning.

Within the cultural-ecological integration paradigm, learners interact with storeys, place practices and histories, and official ecological science or concepts of sustainability. The hybridised method contributes to local relevance, learner engagement, and relationality to place. It also endorses decolonising policies by valuing non-Western epistemologies and grounding sustainability education in local contexts rather than abstract global frameworks.

3.3 Systems Thinking and Transformative Pedagogy

Transformative place-based pedagogy is grounded in systems thinking, the capacity to view and engage with intricate, dynamic, interrelated socio-ecological systems rather than discrete entities. The concept of systems thinking in education also enables learners to comprehend feedback, interdependencies, scale, emergence, resilience, and adaptability, all of which are essential to sustainability. As an illustration, Goode and MacGillivray (2023) studied a systems-thinking pedagogy developed by teachers at a professional development institute, which can be used to understand how schools could create a curriculum to support systems literacies.

Within the PBE framework, systems thinking is applied by asking learners to explore, for example, the interactions among local watershed dynamics, community land-use behaviours, and cultural practices, or among local biodiversity, the economy, and community identity as a living system. Transformative pedagogy is also a process of reversing the roles of learners and teachers, shifting them from passive receivers of information to active co-creators of place-based inquiry, reflection, and action. The location itself is transformed into a laboratory of transformation. This pedagogical orientation is compatible with sustainability education, which is not merely about imparting knowledge but rather about facilitating agency, participation, and change.

3.4 Digital Storytelling in PBE and Technological Mediation

While place-based learning focuses on locality and close engagement with the surrounding world and community, digital technologies and storytelling can augment, record, and magnify these experiences in meaningful ways. Digital storytelling: In digital storytelling, learners construct narratives of their place-based learning experiences using multimedia technologies to facilitate reflection and voice, cultural expression, and community engagement. To illustrate, Staley and Freeman (2017) found that rural U.S. high school students who engaged in digital storytelling produced more authentic and engaging work and initiated a re-evaluation of their identities and futures. On the same note, Meier (2022) defines the roles of digital storytelling in higher learning: multimodal coherence, identity work, and cultural-situated learning. Frontiers

Technology-mediated storytelling, which is part of the PBE, enables learners to work together, produce work for others in the community to read, and, further, write about ecological and cultural observations and reflect on their experiences. It also facilitates translation into global networks and broader audiences, enabling local place-based insights to engage in global discussions. Although caution should be exercised to ensure that technologies are not isolated but are relevant to the local context, they should not merely be superficial imitations of tech use. They should not obscure the relational, embodied, and place-based aspects of learning.

Table 1. Pedagogical Frameworks and Innovations in Place-Based Education

Pedagogical Dimension	Core Principles and Approaches	Learning Processes and Strategies	Expected Sustainability Outcomes
Experiential, Outdoor, and Land-Based Learning	Learning through direct interaction with natural and community environments, sensory engagement, reflection, and action.	Field-based inquiry, observation, ecological mapping, reflective journaling, stewardship projects	Ecological literacy, environmental awareness, personal connection to place, and action-oriented learning
Cultural and Ecological Knowledge Integration	Integration of Indigenous, local, and scientific knowledge systems; plural epistemologies; respect for cultural landscapes	Storytelling, intergenerational dialogue, cultural mapping, blending traditional and scientific interpretations	Cultural continuity, ethical engagement with land, and decolonized environmental understanding
Systems Thinking and Transformative Pedagogy	Seeing and understanding interconnected systems; promoting adaptive and critical thinking; fostering agency and reflection.	Systems mapping, feedback loop analysis, scenario modeling, collaborative problem-solving	Systems literacy, resilience thinking, collective responsibility, and community problem-solving capacity
Technology-Mediated and Digital Storytelling	Connecting physical place-based experience with digital reflection and sharing, multimodal and collaborative expression	Creation of multimedia narratives, virtual field blogs, community archives, and social media storytelling for awareness	Expanded audience reach, digital ecological citizenship, narrative-based reflection, and cultural-technological equity

4. KNOWLEDGE IN PRACTICE: GLOBAL AND LOCAL CASE STUDIES

Figure 2. Interconnected dimensions of transformative sustainability education through place-based learning

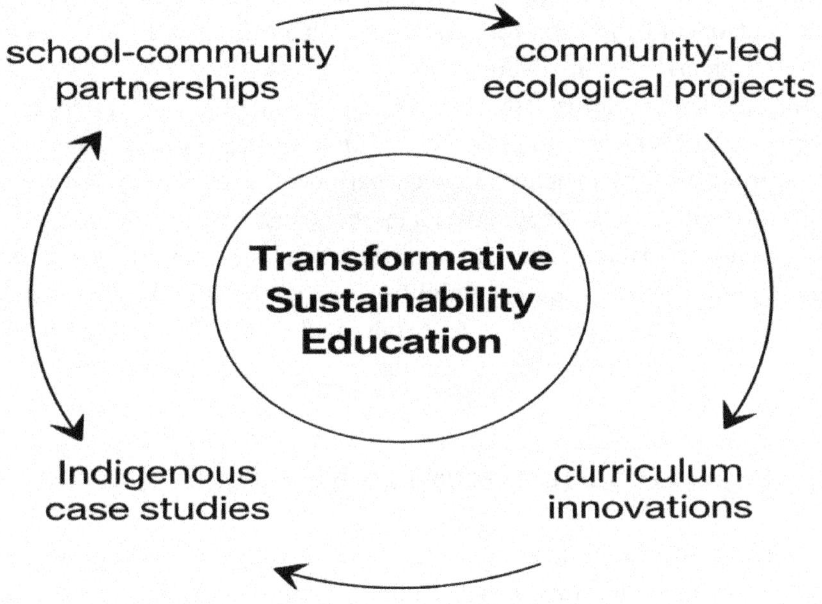

4.1 School–Community Partnerships

School-community partnerships provide powerful platforms on which the theory of place-based education can be translated into practical use. For example, South African research shows that although schools and communities often engage in collaboration aimed at sustainability, the long-term effectiveness of such collaboration requires goal alignment, shared resources, and sustained communication (Myende, 2019). The three school-community partnerships examined in this research were analysed through the qualitative approach, and results indicated that overlapping spheres of influence, as espoused by Epstein, are necessary to facilitate decision-making, sharing, and sustaining resources (Myende, 2019). In other locations, the case studies of urban environmental education in Canada, Australia, China, and Spain all recognise schools that combine local green spaces with local actors (parents, local NGOs, municipal agencies) to make the school a participant in a broader socio-ecological network of learning, stewardship, and resilience. The examples given above show that as schools incorporate place-based pedagogy and actively involve the community in these goals - garden programmes, local heritage research, or neighbourhood clean-up projects, students become more deeply relational to their place, and community bonds become stronger. Nevertheless, such partnerships must be supported by institutional backing, defined channels of communication, and substantive contributions from all partners to ensure that they are not based on one-time or superfluous involvement.

In this case, it is emphasized that place-based education can best be effective when learning is shaped within the community relations and the local ecological systems. In terms of an ecology of learning, such partnerships explain how knowledge is produced in terms of exchanges between learners, institutions, and the environment and support the interrelatedness of cognitive, social, and environmental processes of learning. Nonetheless, the sustainability of such programs lies in the institutional support, organized communication, and a fair involvement of the stakeholders. In the absence of these conditions, partnerships are prone to being short-term or symbolic as opposed to transformative.

4.2 Ecological Projects Headed by the People

In addition to formal education, place-based learning is a rich experience in lived practice in the form of community-based ecological restoration and sustainability projects. It is noteworthy that the community-based ecological mangrove rehabilitation initiative in Indonesia involved local fishers and aquaculture pond owners collaborating to restore mangrove habitats through co-management. Among the main issues, land tenure problems and stakeholder alignment were identified, but the project demonstrated that ecological restoration, community welfare, and local knowledge may overlap. Although this example is not directly related to a school curriculum, it offers a useful paradigm for how community-led ecological action coalesces into an ecological living classroom, where culture and ecology intersect. The other example is the Sustainable Schools Initiative in the United Arab Emirates that gathers more than 150 schools together with stakeholders (government, non-government, and private) to cut down on resource consumption, to incorporate the concept of sustainability in school activities, and to spread the effect into the neighbouring communities (GEEP, n.d.). These instances highlight that ecological projects embedded in the local cultural, spatial, and temporal contexts and in which stakeholders actively participate are effective platforms for transformative sustainability education.

The examples given demonstrate that community-based ecological initiatives are effective extensions of the ecology of learning, whereby knowledge is produced collaboratively by engaging in the process of environmental problem-solving in reality. They support the importance of experiential and situated learning in the acquisition of sustainability competencies. Nevertheless, the challenges associated with these initiatives are also based on scalability, governance, and long-term institutional integration, implying that such initiatives can only have an effect when there is sustained policy support and sectoral coordination.

4.3 Case Study in the Context of the Indigenous and the Global South

Place-based learning is closely connected to local epistemologies, land-based practices, and relational worldviews in most contexts in the Global South, as well as among Indigenous communities. In Nepal, there is an empirical example of how schools in rural areas introduced so-called sustainability clubs, in which students participated in community activities aligned with the SDGs (Brett et al., 2023). Students developed as global citizens through local sustainability practices and relationships with sister schools worldwide. The Green School initiative is another Surinamese-based project that introduced biodiversity, waste management, and water- and energy-education into primary schools nationwide, with a focus on local ecologies and cultural settings. Though these instances reveal the promise of place-based education in underserved settings, they also reveal structural problems, including resource limitations, teacher capacity, and infrastructure that should be overcome in the implementation of place-based education. Generally, these Indigenous and Global South lenses broaden our perspective of place-influenced pedagogy by preempting equity, cultural revitalisation, and ecological justice.

Concerning ecological learning, such instances draw the significance of contextual and culture-based approaches to sustainability learning. They demonstrate the influence of the local environment and social construct on knowledge, which proves that learning is a relational process and is place-specific. The existence of structural inequalities, however indicates that such practices would only be effective when broader socio-economic and institutional challenges are involved.

4.4 The Examples of Curriculum Innovation

Curriculum innovation provides a bridge between pedagogical theories and practice. In one example, a programme in Ł ódź, Poland (Schools for Climate) engaged schools, teachers, families, and businesses in a structured curriculum that included garden-based learning, teacher-cascade professional development, and community participation (AIPH, n.d.). Students designed their vegetable and herb gardens, attended family- and community-based learning events, and collaboratively designed learning tasks related to community ecological solutions. In a similar manner, one of the cases reported by the Education for Sustainability initiative concerns youth-led initiatives in which students developed refillable water bottles, organised beach-cleaning activities, and participated in campaigns around the idea of Refill not Landfill (Education for Sustainability and Localise, n.d.). These innovations illustrate how the curriculum can be transformed from a more textbook- and lecture-based model to a more dynamic, communal, action-based learning sequence. More so, they demonstrate how teachers can scaffold the experience in which students envision, design, execute, and reflect on the action of sustainability- thus, constituting the ecology of learning in the curriculum.

These illustrations demonstrate the application of curriculum innovation in the ecology of learning by combining knowledge, action, and community incurred in the teaching profession. They demonstrate that learning can be experiential, reflective, and transformative in the case when learners actively participate in the planning of sustainability initiatives as well as in their implementation. Nevertheless, these innovations need curriculum flexibility, teacher training, and institutional commitment in order to scale up, or otherwise they can be an isolated practice, as opposed to a systematized practice. The interconnected dimensions of transformative sustainability education through place-based learning are illustrated in Figure 4.1, highlighting the relationships among school-community partnerships, ecological projects, Indigenous perspectives, and curriculum innovation.

5. POLICY, LEADERSHIP, AND INSTITUTIONAL SUPPORT

5.1 Policy Frameworks Underpinning Place-Based Education (PBE) and Education Sustainable Development (ESD)

The insights derived from the case studies highlight the critical role of policy, leadership, and institutional support in sustaining place-based and ecological learning approaches. Without enabling frameworks, many of the successful practices observed remain localized and difficult to scale. Representing one of the key factors in institutionalising Place-Based Education (PBE) and Education for Sustainable Development (ESD) in the education system, national and international policy frameworks play a crucial role. The ESD Programme 2030 of UNESCO provides detailed policy guidance and technical advice, and encourages international cooperation to make sustainability part of education systems worldwide (UNESCO, 2015). This project highlights the significance of situating learning in local environments, cultures, and communities, which aligns with PBE.

The National Education Policy (NEP) 2020 in India highlights experiential learning, critical thinking, and community involvement, which are the focus of PBE. The policy proposes a shift from memorization to a holistic learner-centered approach and encourages the development of real-life skills (National Education Policy, 2020; Times of India, 2023a). Moreover, the NEP promotes the use of local context and resources in curriculum development, thereby enabling the implementation of PBE practices. According to the ecological learning approach, these leadership practices emphasize the need to match the institutional structures to the local knowledge systems and the environmental setting. Nevertheless, they can be effective only in the case of continued commitments, investment, and transformation of localized innovations into systemic change.

5.2 Decolonizing Education Leadership Models

Decolonizing education needs to be led by individuals who question colonial, outdated systems and adopt Aboriginal knowledge systems and Indigenous realities. A decolonizing leadership paradigm will help thinkers reflect on power hierarchies, facilitate the inclusion of diverse epistemological perspectives in education, and incorporate other epistemologies into the educational process (Woldegiorgis, 2025). The Uttar Pradesh State Council of Educational Research and Training (SCERT) is underway with the creation of open-air theatres in District Institutes of Education and Training (DIETs) in the Indian context. Located in the greenest village, Khonoma, in Nagaland, India, these theatres will form green, open, circular learning environments by using natural resources. The spaces will serve as meditation centres, venues for discussion, yoga, and reflection, and as sites for cultural events, thereby promoting creative expression and interaction among trainee teachers (Times of India, 2023b). These examples show that teacher training acts as an important mediator of theory to practice in the ecology of learning. Nonetheless, access to professional growth and lack of institutional support are still a major barrier especially in resource-strained environments.

5.3 Teacher Training and Professional Learning

PBE requires adequate teacher training and continuous professional development to be implemented effectively. Programmes such as those offered by Teton Science Schools provide teachers with opportunities for field experience between classroom and field instruction. These curricula enable educators to develop competencies in practice through a coherent pedagogical model (Teton Science Schools, n.d.). The Shell NXplorers programme, funded by the Atal Innovation Mission of NITI Aayog, is a professional development programme in India that teaches teachers sustainability, innovation, and critical thinking. It has been introduced in Atal Tinkering Lab (ATL) schools, where STEM education has improved, and students' problem-solving attitudes have developed (Times of India, 2023a). Such instances reveal that teacher training is an important mediator in the ecology of learning theory and practice. Nevertheless, unequal opportunities to pursue professional development and a lack of institutional support are major obstacles, especially in resource-bound environments.

5.4 Hardships, Obstacles, and Success Stories

Although the importance of PBE continues to be recognized, several obstacles impede its proliferation. Place-based approaches are often constrained by institutional resistance, fixed curricula, and standardised testing, which require greater flexibility. Moreover, insufficient resources and poor teacher preparation may impede effective implementation (UNESCO, n.d.). Nonetheless, the positive examples of PBE's use demonstrate its potential under supportive policies and leadership. For example, the Green School project in Suriname incorporated biodiversity and waste-management education in primary schools, emphasizing local ecologies and cultural backgrounds. This strategy was successful in increasing environmental awareness, thereby fostering a sense of community and responsibility among students (Woldegiorgis, 2025).

Finally, to implement PBE in the educational system, it is necessary to secure support from the policymaking process, transformational leadership, and effective teacher training. A more inclusive, context-oriented, and sustainable framework for education may be achieved by overcoming current challenges and drawing on successful models (UNESCO, 2015; National Education Policy, 2020; Times of India, 2023). Policymaking, leadership, and institutional support discussions prove that place-based education is impossible to operate without more systemic systems. Whereas theoretical frameworks focus on relational and context-dependent learning and case studies demonstrate useful applications, policy frameworks are the final determinant of how far such methods can be maintained and extended. This interrelatedness brings forth how the ecology of learning is not just a pedagogical notion, but also an institutional and policy issue that necessitates coordinated actions at more than just one level of the education system.

Table 2. Policy, Leadership, and Institutional Support in Place-Based Education

Dimension	Description	Example Initiatives and Programs	Challenges and Success Factors
Policy Frameworks Supporting PBE & ESD	National and global policies guiding the integration of sustainability and place-based learning.	UNESCO ESD for 2030 roadmap; India's NEP 2020 emphasizing experiential, community-based learning and local contexts	Need for contextualization, alignment with SDGs, and supportive governance.
Leadership Models for Decolonizing Education	Leadership that promotes Indigenous epistemologies, equitable decision-making, and power reflection	Uttar Pradesh SCERT's open-air theatres are inspired by Khonoma village, fostering circular, nature-based learning environments and cultural engagement.	Navigating colonial legacies, encouraging inclusivity, and building reflective leadership
Teacher Training and Professional Learning	Immersive, experiential teacher training linking field and classroom; continuous professional development	Teton Science Schools workshops; Shell NXplorers program with NITI Aayog's Atal Innovation Mission, enhancing STEM and innovation skills in India	Scaling teacher skills, ongoing support, and bridging theory and practice
Challenges, Barriers, and Success Stories	Institutional constraints, rigid curricula, resource gaps, versus models of successful implementation	Green School Suriname is integrating local ecology and culture; challenges include resistance, resource constraints, and rigid testing frameworks.	Institutional buy-in, policy backing, sustained leadership, and community involvement

6. GLOBAL PERSPECTIVES AND FUTURE DIRECTIONS

6.1 Comparative Studies Across Regions

Comparative research on Place-Based Education (PBE) reveals varied practices and outcomes across contexts. The PBE is used at institutions such as Appalachian State University in the United States, where it is implemented through comparative regional studies that encourage students to interact with local communities to develop a better appreciation of their locality and past (Yemini et al., 2025). Likewise, place-based knowledge has been shown to integrate into the pedagogical learning process in Nepal and to have the potential to facilitate contextualised learning and sustainability, and to emphasise the significance of local ecological and cultural contexts in education (Maharjan et al., 2024). PBE has been applied internationally to address environmental problems by incorporating local knowledge into curricula. For example, in the Pacific Islands, community-based learning supported by PBE helps students relate to their local settings, deepen their knowledge of ecological systems, and develop a sense of responsibility for environmental stewardship (Institute of Education Sciences, 2020). Such comparative studies emphasise the flexibility of PBE across diverse cultural and ecological contexts, demonstrating its ability to address region-specific challenges and promote sustainability.

6.2 Climate Resilience and Adaptive Learning

Education is essential for enhancing climate resilience, as it equips individuals with the knowledge and skills to adapt to changing environmental conditions. The Global Center on Adaptation highlights that education can enhance the adaptive capacity of individuals, particularly children and youth, by increasing awareness and readiness regarding climate-related issues (Global Center on Adaptation, 2024). PBE in this context is an effective strategy, as it accounts for local environmental issues in the learning process and makes learning more pertinent and active. There is also the use of innovative technologies to enhance climate-resilience education. The Future Atmospheric Conditions Training System (FACTS) integrates current atmospheric data from Internet of Things (IoT) sensors and, using Generative AI, provides educational content to generate place-based, personalized learning. This system has proven effective in capturing learners' attention and improving their knowledge of climate resilience (Bibri & Huang, 2025). These technological developments underscore the potential of applying digital technologies in conjunction with PBE to establish adaptive learning systems that respond to environmental changes in real time.

6.3 Future Research Directions

The future development of PBE aims to address several aspects to improve its efficiency and applicability. To begin with, longitudinal studies are required to determine the long-term effects of PBE on student performance, academic success, environmentally responsible attitude, and community involvement. Such research would be insightful into the long-term advantages of PBE and the best practices for implementing it. Second, the studies investigate how PBE can be combined with other teaching methods, including STEM education and digital learning, to develop interdisciplinary programmes that address multidimensional global problems. As an example, the integration of PBE and STEM may develop critical thinking and problem-solving skills in students as they will be able to address such problems as climate change and sustainability. Finally, research is needed on the scalability and flexibility of PBE across different educational organisations, such as urban and rural schools, to identify how the principles can be implemented in these settings. Insight into the issues and opportunities associated with scaling PBE can inform policies and measures to expand its reach.

7. CONCLUSION

Place-Based Education (PBE) has become one of the most radical methods for connecting local ecological, cultural, and social conditions with formal learning settings. The chapter has also explained how PBE plays a primary role in promoting sustainability education through the incorporation of Indigenous knowledge systems, community involvement, and place-based learning (UNESCO, 2015; Maharjan et al., 2024). Regional comparative studies indicate that PBE is highly versatile, enabling teachers to design learning experiences that suit local needs while addressing global sustainability issues (Yemini et al., 2025). Placing learning in a real-life context enables students to acquire knowledge and build a sense of belonging, civic responsibility, and ecological literacy, which makes them have the capacity to tackle complex environmental and social problems (Institute of Education Sciences, 2020). This is where leadership and policy frameworks come in to help PBE, and an example of this is the National Education Policy (NEP) 2020 in India, which prioritizes experiential, learner-centered approaches (National Education Policy, 2020; Times of India, 2023a). The key role of the policy in effective classroom practices is institutional support, teacher professional development, and strategic leadership, which would make PBE not merely an ideal in pedagogy but a viable and sustainable model of education (Woldegiorgis, 2025). The potential scope of PBE can be expanded by incorporating technologies, such as IoT, to create adaptive learning systems that provide students with access to real-time environmental data and enable personalised learning (Bibri & Huang, 2025). The innovations increase the engagement, offer real-world problem-solving, and contribute to an education of climate resilience, mediating the local and global views of sustainability.

7.1 Implications for Education, Policy, and Sustainability

Discussed in this chapter, the findings highlight a number of important implications as far as education, policy, and sustainability are concerned. First, education systems are to focus on locally contextualised learning to improve relevance, engagement, and environmental stewardship among the students (UNESCO, 2015). Installing Indigenous knowledge and community-based learning in the curriculum allows schools to build a set of holistic competencies, such as critical thinking, problem-solving, and social-emotional skills (Maharjan et al., 2024). Politically, supportive systems are necessary to institutionalise PBE in different regions. The policies, such as NEP 2020 and international guidelines on ESD, give a framework for integrating sustainability principles in the curricula of nations, and also fostering teacher training, local participation, and inclusive leadership (National Education Policy, 2020; UNESCO, n.d.). Local knowledge, cultural identity, and ecological stewardship have an interactional effect on policymakers in order to establish equitable and contextually acceptable educational environments. PBE provides a model that relates education directly to environmental action, which is a notion that guarantees sustainability. Schools can be facilitators of sustainability efforts at the community level, enhancing resilience and adaptability to climate change and other environmental issues by involving students in local ecological projects (Global Center on Adaptation, 2024; Institute of Education Sciences, 2020). Moreover, technology-mediated solutions, including the AI-based climate learning tools, could be used to supplement standard PBE strategies that would allow learners to operate within complex socio-ecological systems (Bibri & Huang, 2025).

7.2 Pathways for Action

A number of strategic pathways are suggested in order to make the most of PBE on education and sustainability. To start with, place-based pedagogical models must be integrated into all levels of education, and learning goals must be aligned with the environmental and cultural specificities at the local level (Yemini et al., 2025). Second, teachers must be sustained in the process of professional development to ensure they possess competencies and confidence to support the learning activities, such as experience and community-based learning (Teton Science Schools, n.d.). Third, the policy and institutional support should be able to offer adequate resources and flexibility, so that schools can effectively apply the PBE practices, but use resources and flexibility to meet the local needs and constraints (Woldegiorgis, 2025). Lastly, PBE initiatives should incorporate research and evaluation to track the results, determine the best practices, and facilitate scaling in different settings (Maharjan et al., 2024). The focus on community partnerships between teachers, households, policy-makers, and scholars will help to make place-based strategies more effective and sustainable, and eventually make a generation of more ecologically literate, socially responsible, and resilient learners.

To sum up, Place-Based Education is an effective mechanism to support local knowledge and world sustainability objectives. PBE helps learners gain skills and attitudes that enable them to navigate and deal with complex environmental and social challenges through nurturing ecological literacy, cultural awareness, and adaptive learning. PBE can help become a source of transformative, contextually based, and sustainable education in the world with its supportive policies and leadership and new approaches to pedagogy. This chapter helps us realize that place-based education, as comprehended through the perspective of the ecology of learning, entails a logical synthesis of theoretical concepts, pedagogical methods, and policy assistance. Although the theoretical views define learning as being relational and context-specific, the case studies that are empirically done show how such learning occurs via the community and the interaction with the environment. Simultaneously, the policy and leadership systems dictate how far these approaches can be institutionalized and scaled. This inter-relational view supports the fact that sustainability education relies on convergence on the various facets of the education system and not on single-subject interventions. Here, place-based education becomes a strong methodology towards the reconstruction of education in the period of ecological and social insecurity. Through its ability to bridge between local knowledge and global sustainability objectives and its potential to combine theory, practice, and policy, PBE creates ecologically literate, socially responsible, and adaptable learners. The future should be aimed at reinforcing the institutional support, increasing the number of teachers, and securing equitable access to place-based learning, thus allowing the education systems to address complex socio-ecological issues in the Anthropocene in the most effective way.

REFERENCES

Barron, B. (2006). Interest and self-sustained learning as catalysts of development: A learning ecology perspective. *Human Development*, *49*(4), 193–224. DOI: 10.1159/000094368

Bascopé, M., Perasso, P., & Reiss, K. (2019). Systematic Review of Education for Sustainable Development at an Early Stage: Cornerstones and Pedagogical Approaches for Teacher Professional Development. *Sustainability (Basel)*, *11*(3), 719. DOI: 10.3390/su11030719

Bibri, S. E., & Huang, J. (2025). Generative AI of things for sustainable smart cities: Synergizing cognitive augmentation, resource efficiency, network traffic, cybersecurity, and anomaly detection for environmental performance. *Sustainable Cities and Society*, *133*, 106826. Advance online publication. DOI: 10.1016/j.scs.2025.106826

Brett, P., Gurung, G., Shah, R., Yogi, K. K., & Dhungana, J. (2023). Enacting the Sustainable Development Goals in Nepalese Schools and Teacher Education. In Beasy, K., Smith, C., & Watson, J. (Eds.), *Education and the UN Sustainable Development Goals. Education for Sustainability* (Vol. 7). Springer., DOI: 10.1007/978-981-99-3802-5_10

Chang, S. E., & Kuo, M.-Y. (2021). A Place-Based Pedagogical Action Study to Enrich Rural Sustainability: Knowledge Ties of National Taiwan University's 10-Year Partnership with Pinglin. *Sustainability (Basel)*, *13*(5), 2916. DOI: 10.3390/su13052916

Dartevelle, O. (2012). State of the Art of Environmental Sustainability in Academic Curricula and Conditions for Registration.

Gallay, E., Marckini-Polk, L., Schroeder, B., & Flanagan, C. (2016). Place-based stewardship education: Nurturing aspirations to protect the rural commons. *Peabody Journal of Education*, *91*(2), 155–175. DOI: 10.1080/0161956X.2016.1151736

Gerelkhuu, S., Fiel'ardh, K., Fujii, H., Yembuu, B., & Dembereldorj, U. (2026). Towards place-responsive climate change education: Mongolian primary teachers' pedagogical judgement across urban and rural contexts. *Environmental Education Research*, •••, 1–22. DOI: 10.1080/13504622.2026.2635431

Global Center on Adaptation. (2024). *Educate to adapt: The interconnected paths of learning and climate.* https://gca.org/educate-to-adapt-the-interconnected-paths-of-learning-and-climate/

Goode, G. S., & MacGillivray, L. (2023). The construction of systems thinking pedagogy during a professional development institute. *Journal of Pedagogical Research*, *7*(4), 275–302. DOI: 10.33902/JPR.202318879

Hecht, M., & Crowley, K. (2019). Unpacking the learning ecosystems framework. *Journal of the Learning Sciences*, *28*(2), 264–284. DOI: 10.1080/10508406.2019.1693381

Hernandez Gonzalez, F. (2023). Exploring the Affordances of Place-Based Education for Advancing Sustainability Education: The Role of Cognitive, Socio-Emotional and Behavioural Learning. *Education Sciences*, *13*(7), 676. DOI: 10.3390/educsci13070676

Hlalele, D. J. (2019). Indigenous knowledge systems and sustainable learning in rural South Africa. *Australian and International Journal of Rural Education*, *29*(1), 88–100. DOI: 10.47381/aijre.v29i1.187

Institute of Education Sciences. (2020). *Supporting Pacific Island communities through place-based education.* U.S. Department of Education. https://ies.ed.gov/learn/blog/supporting-pacific-island-communities-through-place-based-education

Jones, T. J., Williams-Jones, G., & Nyce, H.Jr. (2025). Braiding Indigenous knowledge systems and Western science through co-creation and co-teaching. *Frontiers in Earth Science (Lausanne)*, *13*, 1587092. DOI: 10.3389/feart.2025.1587092

Lloyd, A., & Gray, T. (2014). Place-based outdoor learning and environmental sustainability within Australian primary schools. Journal of Sustainability Education, 41883.

Maharjan, B., Pant, B. P., & Dahal, N. (2024). Participatory learning: Exploring place-based pedagogy for future teachers. *Pedagogical Research*, *9*(4), em0222. Advance online publication. DOI: 10.29333/pr/15155

Meier, S. (2022). Digital Storytelling: A Didactic Approach to Multimodal Coherence. *Frontiers in Communication*, *7*, 906268. DOI: 10.3389/fcomm.2022.906268

Myende, P. E.P. E. (2019). Creating functional and sustainable School–Community Partnerships: Lessozns from three South African cases. *Educational Management Administration & Leadership*, *47*(6), 1001–1019. DOI: 10.1177/1741143218781070

National Education Policy. 2020. (2020). *Ministry of Education, Government of India.* https://www.education.gov.in/sites/upload_files/mhrd/files/NEP_Final_English_0.pdf\

Peltier, S. M. (2021). The Child Is Capable: Anishinaabe Pedagogy of Land and Community. *Frontiers in Education*, *6*, 689445. DOI: 10.3389/feduc.2021.689445

Quilaqueo, D., & Torres, H. (2024). Aboriginal Country-centred learning. *Curriculum Perspectives*, *44*(2), 205–216. Advance online publication. DOI: 10.1007/s41297-024-00233-2

Shrivastava, P., & Kennelly, J. J. (2013). Sustainability and place-based enterprise. *Organization & Environment*, *26*(1), 83–101. DOI: 10.1177/1086026612475068

Skerrett, A., & Vlach, S. K. (2022). Advancing a sociocultural approach to decolonizing literacy education: Lessons from a youth in a "postcolonial" Caribbean geography. *Research in the Teaching of English*, *57*(1), 67–88. DOI: 10.58680/rte202232002

Sposab, K., & Rieckmann, M. (2024). Development of Sustainability Competencies in Secondary School Education: A Scoping Literature Review. *Sustainability (Basel)*, *16*(23), 10228. DOI: 10.3390/su162310228

Staley, B., & Freeman, L. A. (2017). Digital storytelling as student-centred pedagogy: Empowering high school students to frame their futures. *Research and Practice in Technology Enhanced Learning*, *12*(1), 21. DOI: 10.1186/s41039-017-0061-9 PMID: 30595726

Sterling, S. (2011). Transformative learning and sustainability: Sketching the conceptual ground. *Learning and Teaching in Higher Education*, *5*(1), 17–33. DOI: 10.1177/1469787411429184

Teton Science Schools. (n.d.). *Preparing educators for the place-based education classroom.* https://www.tetonscience.org/preparing-educators-for-the-place-based-education-classroom/

Thompson, S. M., & Chapman, N. (2025). Learning Along the GreenWay: An Experiential, Transdisciplinary Outdoor Classroom for Planetary Health Education. *Sustainability (Basel)*, *17*(9), 4143. DOI: 10.3390/su17094143

Times of India. (2023a, June 24). NEP 2020: Bringing experiential learning to the heart of education. *The Times of India.* https://timesofindia.indiatimes.com/education/news/nep-2020-bringing-experiential-learning-to-the-heart-of-education/articleshow/10124564.cms

United Nations Educational, Scientific and Cultural Organization. (2015). Education 2030: Incheon Declaration and Framework for Action for the implementation of Sustainable Development Goal 4. UNESCO. https://unesdoc.unesco.org/images/0024/002456/245656e.pdf

Woldegiorgis, E. T. (2025). Navigating the decolonial frontier: Leadership perspectives on epistemic justice in South African higher education. *International Journal of Educational Research*, *134*, 102816. DOI: 10.1016/j.ijer.2025.102816

Yemini, M., Engel, L., & Ben Simon, A. (2025). Place-based education – a systematic review of literature. *Educational Review*, *77*(2), 640–660. DOI: 10.1080/00131911.2023.2177260

Chapter 5
Strategic Governance and Sustainability in the United Arab Emirates Education System

Ali Ibrahim
http://orcid.org/0000-0002-1070-6256
United Arab Emirates University, UAE

Noha Zoelfakar
http://orcid.org/0009-0003-7525-5577
United Arab Emirates University, UAE

ABSTRACT

This chapter examines how sustainability is institutionalized within the governance architecture of the UAE education system, defined as the interaction of national strategic visions, centralized policy design, regulatory oversight, and leadership practices shaping education planning and implementation. National frameworks UAE Centennial 2071 and Net Zero 2050 place education within long-term economic and environmental planning supporting post-oil diversification and national development. In this context, sustainability extends beyond environmental protection to include economic transformation, technological modernization, and institutional resilience. While global agendas such as SDG 4 and 16 influence the policy environment, the UAE model reflects selective alignment rather than direct transfer. Centralized authorities translate national priorities into regulatory frameworks and planning requirements aligning schools with national goals. The chapter highlights how strategic vision and regulatory coordination support sustainability governance in a rapidly evolving education system.

DOI: 10.4018/979-8-3373-7650-9.ch005

1. INTRODUCTION: SUSTAINABILITY AS GOVERNANCE PROJECT

1.1 Sustainability Beyond Environmental Reform

Sustainability has traditionally been understood and conceptualized primarily from the perspective of environmental preservation. While this framing remains important, contemporary governance scholarship has expanded the concept beyond ecological protection. Sustainability has increasingly been framed as the integration of economic development, social inclusion, and environmental protection within long-term development strategies (Sachs, 2015). This broader framing aligns with research on environmental policy integration and institutional coherence. Jordan and Lenschow (2010) argue that achieving environmental policy integration requires administrative coordination mechanisms across sectors, supported by policy instruments (e.g., appraisal and assessment) and stronger monitoring and evaluation capacity. Biermann et al. (2012) extend this argument within the context of Earth system governance, suggesting that sustainable development in rapidly transforming societies requires multi-level coordination and long-horizon planning. In this wider sense, sustainability is more than a sectoral policy concern; it is an organizing principle that shapes long-term national planning, institutional design, and public sector coordination. The education sector is integral to this broader conceptualization of sustainability, functioning not only as a site of environmental learning but also as a driver of human capital formation within knowledge-based economies (Areepattamannil, 2024; Powell & Snellman, 2004) and as a strategic mechanism for aligning workforce development with national economic transformation.

1.2 Post-Oil Transformation

In the United Arab Emirates (UAE), sustainability must be understood in the context of post-oil transformation. Since the founding of the UAE federation in 1971, rapid modernization has been underwritten by hydrocarbon revenues, a trajectory consistent with what the literature describes as Gulf rentier states (Hertog, 2011; Krane, 2018). Sustainability in this context extends beyond environmental protection to encompass economic transformation, technological development, and long-term institutional resilience (Badawy et al., 2025; Ibrahim et al., 2023). Diversification, thus, becomes a sustainability imperative rather than solely an economic strategy. Education has been framed in this context, not as a social welfare function, but as a key instrument of national transformation.

1.3 SDG 4 and SDG 16

At the international level, the 2030 Agenda for Sustainable Development constitutes an essential normative reference context for this positioning. In this regard, Goal 4 of the SDGs stresses the importance of inclusive and equitable quality education and lifelong learning for all, while Goal 16 highlights the need for effective and accountable institutions for peaceful and just societies (United Nations, 2015). In this manner, these two goals indicate that sustainable development requires the simultaneous development of and interdependence between human capital and strong institutions.

Comparative education research warns against the direct transplantation of policies from global contexts to national contexts (Rizvi & Lingard, 2010; Steiner-Khamsi, 2016). Global documents such as the SDGs serve as discursive points of reference that intersect but do not determine the strategic ambitions of the domestic context. In this manner, the UAE's sustainability governance strategy does not appear to be driven by a compliance strategy with these internationally set goals. The UAE's strategy is one of selective alignment, where the principles of the SDGs are recognized within the wider policy discourse, but the architecture of governance is nationally determined and centrally coordinated through long-horizon strategic planning. This is evident in initiatives such as Centennial 2071 and Net Zero 2050, which embed sustainability within multi-decadal strategic planning rather than short-term reporting cycles.

1.4 Governance Architecture

This chapter employs the concept of governance architecture to analyze how sustainability is institutionalized. Governance architecture is defined in this chapter as the organized interaction between long-term national strategies, centralized policy design, regulatory oversight mechanisms, inspection and evaluation systems, and leadership practices that shape education planning and implementation. Sustainability is operationalized through this architecture, not via isolated initiatives, but through systemic alignment. As Meuleman (2021) contends, sustainability governance requires high-quality public administration capable of fostering policy and institutional coherence across levels and sectors. Without long-term planning and centralized alignment mechanisms, sustainability can become a patchwork of unconnected initiatives. Empirical analyses of UAE education reform illustrate how centralized governance structures have shaped inspection frameworks, performance reporting, and leadership roles that translate national strategic visions into institutional practice (Rai, 2025).

It is, however, important to note that the governance system in the UAE cannot be understood solely through a technocratic institutional lens. Sustainability governance often requires combining hierarchical, market, and network governance styles in ways that are adapted to specific cultural and institutional contexts (Meuleman, 2018). In this sense, cultural dimensions of governance are not merely symbolic residues but active components of governance architecture. They reinforce policy coherence during rapid modernization, contributing to a hybrid configuration in which political legitimacy and administrative rationality operate in tandem rather than in tension.

1.5 Thesis and Scope

This chapter advances the thesis that the UAE institutionalizes sustainability through centralized, long-horizon leadership embedded within a coherent governance architecture. Sustainability is conceptualized not as a thematic enhancement to education policy but as a strategic logic that informs the planning, regulation, and leadership of the education system as a whole. School leaders are strategic intermediaries in this architecture, translating national strategic agendas into institutional planning, decision-making, and program implementation, while navigating accountability pressures and global benchmarking expectations.

The discussion that follows will concentrate on governance and leadership architecture rather than curriculum-level sustainability initiatives or student performance metrics. By focusing on architecture as expressed through strategic foresight, centralized coordination, regulatory alignment, and leadership culture, the chapter situates the UAE within broader debates on sustainability governance in transitional economies. The UAE case illustrates how sustainability governance in education is positioned within broader national transformation agendas, where strategic alignment between long-horizon visions, regulatory frameworks, and institutional practice has become central to system coordination (Ahmed et al., 2024).

2. LONG-HORIZON STRATEGIC VISION: POST-OIL TRANSFORMATION

Long-horizon strategic planning has become a defining feature of states navigating structural economic transition. Strategy-oriented sustainability governance emphasizes iterative, long-term cycles of planning, monitoring, and revision that can embed learning within administrative routines (Meadowcroft, 2007). This approach is particularly consequential for economies that have been primarily built on revenues from natural resources. The same revenues that have historically financed rapid modernization are also the revenues whose long-term viability is increasingly uncertain.

In the Gulf region, the awareness that oil wealth will not last forever has not arrived suddenly; it has been accumulating quietly for decades, gradually reshaping how states think about their futures (Hvidt, 2013). What makes the UAE particularly notable is not simply that it has recognized this challenge, but the scale and temporal reach of its strategic instruments. Sustainability, in this context, operates as a governing logic for organizing transition beyond hydrocarbons. Strategic frameworks such as UAE Centennial 2071 and UAE Net Zero 2050 function not as isolated policy documents but as temporal anchors that structure governance priorities across decades, integrating economic diversification, technological modernization, environmental transformation, and human capital development within a unified long-horizon national project.

2.1 UAE Centennial 2071: Planning Across Generations

Announced in 2017, UAE Centennial 2071 establishes a fifty-year national roadmap aimed at positioning the UAE among the world's leading nations by 2071, marking the 100th anniversary of the country's founding (UAE Government, 2017). The framework articulates four interconnected pillars: government excellence, a diversified knowledge economy, a cohesive society, and a sustainable environment. Most national development strategies operate within short planning cycles, bounded by political transitions or budget reviews. Centennial 2071 extends governance horizons across multiple leadership generations. This aligns with developmental state scholarship emphasizing strategic coordination and forward planning in guiding structural transformation (Evans, 1995). In such models, state institutions actively shape economic transition through long-term policy coherence rather than reactive regulation.

This temporal extension reshapes sustainability from reactive environmental management into a principle of long-term structural continuity. Sustainability is situated as the capacity to maintain institutional stability, human capital, and governance systems across decades of change, while reconfiguring economic foundations. Education is embedded in this logic as the central enabling mechanism through which all four pillars become operational. A diversified knowledge economy requires educated workers; a cohesive society depends on shared values transmitted through schooling; effective government needs a skilled civil service. UAE Centennial 2071, thus, frames education as a prerequisite for everything else.

2.2 UAE Net Zero 2050 - Environmental Commitment as Economic Strategy

The UAE's Net Zero 2050 strategic initiative, announced in 2021, commits to carbon neutrality by mid-century (UAE Government, 2021). On the surface, this appears to be primarily an environmental pledge. Analytically, however, it is more accurately understood as a comprehensive economic and statecraft strategy. Jordan and Lenschow (2010) argue that sustainability governance requires integrating environmental objectives into broader policy domains rather than treating them as discrete regulatory silos. Net Zero 2050 reflects this integration logic, systematically embedding environmental ambition within industrial diversification, renewable energy expansion, and innovation policy.

Education, therefore, becomes central to climate governance, though not always in the ways that sustainability scholarship typically emphasizes. In Gulf economies, energy reform and economic diversification are structurally intertwined (Krane, 2018). The concern is not primarily about environmental awareness in curricula; it is about preparing graduates capable of leading energy transition, digital innovation, and sustainable finance at a professional level. In this sense, Net Zero 2050 positions education as a form of anticipatory governance: an investment in human capability that must be made decades before the economic transformation.

Importantly, effective SDG implementation scholarship emphasizes the importance of vertical policy coherence in translating national ambition into sectoral practice (Mulholland & Berger, 2017; OECD, 2016, 2019, 2023). In the UAE, this vertical alignment is facilitated through centralized planning structures that translate national strategic goals into ministerial directives, regulatory requirements, inspection systems, and school-level planning mandates. Without such coherence, the knowledge economy agenda would remain largely aspirational rather than operationalized in practice.

2.3 Economic Diversification and the Knowledge Economy Agenda

Both UAE Centennial 2071 and UAE Net Zero 2050 are embedded within a broader diversification strategy that has been unfolding across several decades. Economic diversification strategies aim to reduce the relative contribution of hydrocarbons to GDP by expanding sectors such as finance, tourism, logistics, renewable energy, and advanced services (Hvidt, 2013; Krane, 2018). This structural reorientation requires not only sectoral investment but also deep institutional transformation.

Central to this diversification agenda is the pivot toward what scholars define as the knowledge economy, characterized by intellectual capabilities and knowledge-intensive production rather than natural resource extraction (Powell & Snellman, 2004). Education, thus, becomes central to producing the knowledge capital necessary for sustaining diversified growth. This transition demands reliable regulatory environments, high-quality education systems, research infrastructure, and innovation ecosystems, none of which can be constructed quickly or through episodic reform. Such capacities must be built over time across planning cycles. This is where long-horizon planning becomes necessary, as knowledge-based transformation unfolds over decades rather than years.

2.4 Education as Infrastructure for National Transition

Viewing education as infrastructure is analytically consequential because it shifts attention from curriculum content to governance architecture. Infrastructure is designed, maintained, aligned, and evaluated against system-level performance criteria. When the UAE positions education as infrastructure for post-oil transition, it signals that schools and universities are to be managed with the same strategic seriousness as energy grids or transport networks, that is, as essential systems whose performance underpins economic resilience and long-term stability.

This framing carries significant administrative implications. Central authorities deploy performance indicators, strategic planning mandates, inspection systems, and accountability frameworks designed to ensure that the education system remains aligned with long-term national objectives. Policy coherence thus becomes a stabilizing mechanism across a diverse educational landscape of public and private schools, operating across multiple emirates and serving highly varied student populations. The emphasis of UAE Centennial 2071 on STEM, entrepreneurship, digital literacy, and research capacity reflects recognition of human capital as a primary input into economic resilience. Strategic planning mandates, curriculum modernization initiatives, and performance benchmarking frameworks ensure that the education system's outputs align with the labor market demands of diversification sectors.

2.5 The Relationship With Global Sustainability Frameworks

It would be analytically incomplete to discuss the UAE's long-horizon planning without considering its relationship to global sustainability frameworks that form the broader international policy environment. SDG 4 emphasizes equitable quality education and lifelong learning, while SDG 16 underscores the importance of peace, justice, and strong institutions (United Nations, 2015). These goals substantively resonate with the UAE's emphasis on system performance and governance coherence.

Yet the relationship is one of selective resonance, not compliance. Comparative education scholarship suggests that states do not passively adopt global policy frameworks; they interpret them through domestic strategic priorities (Rizvi & Lingard, 2010). The UAE's governance model reflects this dynamic clearly. The UAE's quality assurance systems, inspection mechanisms, and institutional performance metrics align substantively with SDG 4's emphasis on measurable outcomes. Similarly, centralized regulatory coordination and long-term planning align with SDG 16's focus on institutional effectiveness. But these alignments emerge from domestically generated ambitions rather than from external reporting requirements.

The governance of the UAE education system reflects what Meuleman (2018) describes as sustainability meta-governance: the deliberate coordination of governance instruments and institutional actors to foster policy coherence and effective SDG implementation. National visions provide direction; regulatory systems translate that direction into binding mandates; school leaders operationalize national priorities at the institutional level, and inspection bodies evaluate alignment. Global frameworks such as the SDGs may legitimize aspects of this trajectory or provide useful comparative reference points, but strategic direction remains domestically anchored.

2.6 Sustainability as Statecraft, Not Rhetoric

Taken together, UAE Centennial 2071, UAE Net Zero 2050, and the knowledge economy agenda represent a coherent approach to long-horizon governance for generational transition. The UAE is not simply declaring environmental commitment or setting development targets. It is constructing a governance architecture designed to maintain national stability, economic competitiveness, and human capital formation across a period of fundamental structural change.

Education sits at the center of this architecture because it is the primary mechanism through which each successive generation is prepared for the economic and social conditions that long-term planning anticipates. School leaders, curriculum designers, inspection bodies, and regulatory institutions are all, in different ways, participants in a project of national resilience. Sustainability, understood in this sense, is not an agenda item within education policy. It is the frame within which education policy is conceived, organized, and evaluated. This distinction, between sustainability as a thematic concern and sustainability as a governing logic, is what the following sections aim to examine.

3. CENTRALIZED GOVERNANCE ARCHITECTURE IN EDUCATION

The institutionalization of sustainability within the UAE education system is inseparable from its centralized governance architecture. While the UAE maintains a federal structure that distributes administrative responsibilities across emirates, education governance is organized through a layered system of ministerial oversight, regulatory bodies, standardized accountability mechanisms, and inspection systems. This architecture does not function merely as bureaucratic administration; it operates as a strategic alignment mechanism designed to translate long-horizon national visions into operational coherence across a highly diverse and internationally oriented educational landscape. In this context, centralization is not synonymous with rigidity; rather, it functions as a structural instrument for scalability, policy continuity, and system-wide synchronization.

Comparative education scholarship has long debated centralized and decentralized governance arrangements. Early decentralization discourse frequently linked local autonomy to improved responsiveness (Bray, 1996). However, subsequent research suggests that governance outcomes depend less on formal structural labels and more on how authority, accountability, and institutional capacity are configured and enacted in practice (Honig & Hatch, 2004; Verger et al., 2019). In systems pursuing large-scale transformation, centralized steering mechanisms may serve as instruments for coordinating reform across diverse institutional settings. Without robust coordination mechanisms, sustainability risks becoming dispersed across isolated initiatives rather than embedded within systemic institutional design. The UAE case illustrates this dynamic: central coordination operates as a stabilizing force within a highly heterogeneous ecosystem comprising public schools, private schools, international institutions, and multiple curricular models operating across several emirates.

3.1 Ministry and Emirate-Level Oversight

At the federal level, the Ministry of Education (MoE) establishes system-wide national policy frameworks, curriculum standards, teacher qualification requirements, and performance benchmarks that define directional expectations across the federation. Yet the UAE governance model does not rely exclusively on a single regulatory authority. Emirate-level authorities play central roles, particularly in regulating private education and applying quality assurance mechanisms aligned with each emirate's strategic priorities. In Dubai, the Knowledge and Human Development Authority (KDHA) regulates a competitive private education sector through quality assurance and accountability instruments that shape institutional behavior through comparability, transparency, and reputational signaling (Ben Jaafar et al., 2022; UAE Government, 2025). In Abu Dhabi, the Department of Education and Knowledge (ADEK) conducts inspections and manages performance reporting systems that operate alongside licensing and governance requirements that structure how private schools are run and held accountable (UAE Government,

2025). Comparable regulatory authorities operate in other emirates, including the Sharjah Private Education Authority (SPEA) and the Ras Al Khaimah Department of Knowledge (RAK DOK), reflecting aligned oversight within the broader federal governance framework.

This multi-level structure is often described in policy terms as "coordination," but analytically it reflects centralized steering combined with bounded institutional autonomy. Earlier research on decentralization in UAE schools suggested that the success of transferring responsibilities from the Ministry to schools would depend on multiple factors, with staff willingness to implement the changes identified as a major condition (Al Taneiji & McLeod, 2008). That insight remains relevant because it reframes centralization not as a static condition, but as an ongoing governance configuration. The UAE system centralizes the "ends" of education policy, priorities, standards, and performance definitions, while selectively distributing certain "means" to schools within tightly monitored parameters.

3.2 Inspection and Evaluation as Governance Instruments

Inspection and evaluation are not merely technical activities in the UAE system; they are governing instruments. They translate national aspirations into measurable descriptors, rendering quality visible to regulators, institutions, and stakeholders. Ehren and Visscher (2006) argue that inspection systems operationalize standards of educational quality and, through their indicators and reporting mechanisms, influence how schools prioritize and respond to performance expectations.

Empirical evidence from the UAE context illustrates both the effectiveness and the administrative demands of this approach. In Abu Dhabi's public schools, the school self-evaluation process linked to Irtiqaa was experienced by school administrators as central to reform, yet demanding in terms of evidence production, interpretation of standards, and internal alignment work (Blaik Hourani & Litz, 2016). Similarly, by coupling self-evaluation with performance standards and inspection expectations, Irtiqaa makes school leaders responsible for translating reform requirements into daily evidence, documentation, and improvement routines (Blaik Hourani & Litz, 2016).

In Dubai's regulated private education sector, inspection visibility is amplified by competitive market logic, as recurring inspection cycles and public ratings generate shared performance language across diverse schools that reinforce alignment through reputational incentives (Ben Jaafar et al., 2022). Inspection can thus embed sustainability and reform goals within institutional routines. This dynamic corresponds with what accountability scholars describe as "steering at a distance," whereby central authorities influence local actors through standards and evaluation rather than direct managerial control (Ozga, 2009).

3.3 Strategic Planning Mandates and Alignment

Strategic planning mandates constitute a further instrument of centralized alignment. Schools are required to develop multi-year improvement plans that demonstrate responsiveness to inspection findings and alignment with national objectives. These plans typically include measurable targets related to student achievement, innovation initiatives, sustainability practices, and leadership development. The requirement for documented strategic planning reflects broader governance trends in which performance measurement and goal-setting operate as steering instruments (Verger et al., 2019).

What distinguishes the UAE model is that these mandates are embedded within long-horizon strategic frameworks. School improvement planning does not reset with each inspection; it accumulates, building institutional memory and aligning schools' development trajectories with national strategies extending to 2050 and 2071. Planning, thus, becomes an alignment practice, linking school trajectories to long-term national strategies rather than short-term compliance cycles. This corresponds to what Honig and Hatch (2004) describe as "crafting coherence": an ongoing process in which schools negotiate external policy demands with internal conditions and institutional capacity to produce aligned institutional practice.

3.4 Accountability Frameworks and KPIs

Accountability in the UAE education system operates vertically and horizontally. Schools are accountable upward to regulators through compliance, inspection evidence, and improvement responses; and outward through publicly accessible ratings that influence parental choice, particularly in private education where parental choice functions as a governance mechanism. Key performance indicators (KPIs) translate strategic ambitions into quantifiable metrics, enabling system-level monitoring and benchmarking. By standardizing measurement, KPIs enhance comparability and facilitate scalability. At the same time, measurement necessarily compresses institutional complexity into a bounded set of indicators. This simplification shapes which aspects of performance become visible within the system. Yet the same standardization enables reforms to be monitored consistently across schools, supporting coordinated implementation at scale.

Accountability mechanisms in the UAE, including inspection processes, professional standards, and licensure expectations, frame what effective leadership looks like in practice by specifying observable performance domains and compliance expectations (Rai & Beresford-Dey, 2023). Research indicates that leadership selection and development are increasingly shaped by standards, licensure expectations, and formal evaluation (David & Abukari, 2020). Schools may concentrate resources on activities that map clearly onto inspection descriptors, prioritizing what is measured over what is not, a pattern consistent with indicator effects such as tunnel vision and measure fixation in inspection systems (Ehren & Visscher, 2006). Framed from a governance perspective, this narrowing can be understood as a structural trade-off: system coherence and comparability are often achieved by privileging a bounded set of observable quality signals.

3.5 Policy Coherence Across a Diverse School Ecosystem

The UAE's school ecosystem is structurally diverse, encompassing public national curriculum schools alongside private schools offering multiple international curricula and serving a highly multicultural student population, a complexity that further heightens coordination demands. Policy coherence is not automatically produced across such a landscape; it must be actively constructed and continuously maintained. Centralized governance achieves this through three interlocking mechanisms: standard-setting that establishes shared quality definitions applicable across models and curricula; strategic planning mandates that align institutional trajectories with national priorities; and inspection systems that create recurring performance signals reinforcing those definitions through public accountability.

Nevertheless, centralized governance does not imply the absence of institutional agency. School leaders retain responsibility for contextual adaptation, pedagogical innovation, and operational decision-making within the parameters set by system-level expectations. Empirical studies from the UAE illustrate this governance mediation at the emirate level: in Abu Dhabi through Irtiqaa and leadership standards integration (Blaik Hourani & Litz, 2016); in Dubai through KHDA's regulated accountability instruments (Ben Jaafar et al., 2022). However, their autonomy operates within structured parameters defined by inspection frameworks, strategic mandates, and KPI benchmarks. Such a model balances flexibility with alignment, reinforcing sustainability as systemic coherence rather than fragmented experimentation across a heterogeneous educational landscape.

3.6 Alignment at Scale: Centralization and Structural Trade-offs

Centralization in the UAE education system functions as an alignment and scalability strategy. It does so by standardizing what quality means, routinizing evaluation, requiring schools to plan against identified expectations, and circulating performance signals through public reporting and accountability frameworks.

Through these mechanisms, governance is rendered operable at scale across a highly diverse educational landscape.

Yet this architecture may also generate predictable tensions. Self-evaluation and inspection processes can strengthen improvement discipline while introducing administrative load and interpretive pressure on leaders (Blaik Hourani & Litz, 2016). Accountability in a regulated private education sector can drive improvement, while contributing to reputational competition (Ben Jaafar et al., 2022). Policy mandates, such as inclusive education, can clarify expectations while revealing variations in capacity across schools, particularly where resourcing, staffing, and training vary (Massouti et al., 2024).

Coherence, therefore, is not automatically achieved. It is produced through governance instruments that shape school priorities, redistribute institutional attention, and channel institutional effort toward measurable domains. Centralization secures scalability while structuring how autonomy is exercised within defined constraints.

4. LEADERSHIP CULTURE AND HISTORICAL LEGITIMACY

The governance architecture of the UAE education system cannot be fully understood without examining the cultural and historical foundations of political leadership that underpin it. While previous sections have analyzed structural mechanisms of centralization and strategic alignment, those mechanisms operate within a broader legitimacy framework shaped by consultative authority, nation-building narratives, and a long-term orientation toward stability. Governance in this context is not solely technocratic; it is culturally embedded. The result is a hybrid model in which administrative instruments operate alongside historically grounded leadership legitimacy, and in which sustainability functions as both a strategic priority and a normatively anchored state project.

4.1 Consultative Leadership Traditions

Political sociologists examining Gulf governance frequently emphasize the role of consultative leadership traditions as a historical foundation of legitimacy. Although the UAE is a federal monarchy, political authority has historically drawn legitimacy from consultative traditions commonly associated with the principle of *Shura*, which emphasize dialogue between leadership and community, responsiveness to social needs, and collective national purpose (Herb, 1999).

In governance theory, legitimacy is not secured solely through procedural rationality. Weber's typology of authority distinguishes between traditional, charismatic, and legal-rational authority as ideal types; however, empirical governance arrangements often combine elements of multiple forms (Weber, 1978). Gulf governance, including that of the UAE, exhibits a blend of modern bureaucratic administration and institutionalized hereditary authority structures (Herb, 1999). This matters analytically because reform initiatives, including educational transformation, are more likely to be sustained when framed as extensions of national stewardship rather than as externally imposed managerial directives. Administrative modernization is presented not as a rupture with tradition but as its evolution.

In the education sector, this consultative ethos is often expressed through structured stakeholder engagement, advisory councils, and policy dialogues that accompany major reforms. Although decision-making authority remains centralized, consultation functions as a legitimizing practice that reinforces public confidence in strategic direction. Institutional scholarship distinguishes between pragmatic, moral, and cognitive forms of organizational legitimacy. Legitimacy, in this tradition, refers to a "generalized perception or assumption" that the actions of an entity are desirable, proper, or appropriate within a socially constructed system of norms, values, and beliefs (Suchman, 1995). Governance structures, therefore, sustain public acceptance when they align with shared normative expectations rather than relying solely on procedural authority. In the UAE, the consultative framing of educational reform aligns governance decisions with culturally embedded expectations of responsible leadership, reinforcing acceptance of centralized oversight while maintaining policy coherence.

4.2 Nation-Building Ethos and Developmental Legitimacy

Leadership culture in the UAE is inseparable from its nation-building trajectory. Since the federation's formation in 1971, state legitimacy has been closely tied to visible modernization, rapid development, and the delivery of social services as markers of governing capacity (Davidson, 2009). Education has occupied a central symbolic and practical position within this project, functioning as a vehicle for cultivating national identity and social cohesion while also serving as a measurable indicator of national progress.

Developmental state scholarship provides a useful interpretive lens here. Evans (1995) argues that successful developmental states combine bureaucratic capacity with close coordination between state agencies and industry to guide structural transformation. Although the UAE differs institutionally from the developmental state cases Evans analyzes, coordinated long-term planning and state-led capacity building remain analytically useful as a comparative heuristic. Leadership is culturally expected to guide transformation, not merely regulate markets. Within this developmental framing, long-horizon planning frameworks such as UAE Centennial 2071 are not simply policy roadmaps; they are narrative instruments linking contemporary governance to intergenerational responsibility, projecting an arc of continuity that signals both foresight and commitment to future generations.

4.3 Stability and Long-Term Planning Culture

Rapid modernization can destabilize societies if institutional continuity is not maintained alongside structural change. Governance literature on complex adaptive systems emphasizes that resilience depends on balancing adaptive transformation with institutional stability (Duit & Galaz, 2008). The UAE leadership model prioritizes this equilibrium. Reform initiatives, including curriculum modernization, inspection intensification, and accountability expansion, have typically been framed within broader continuity narratives emphasizing gradual evolution, national identity preservation, and stability rather than abrupt disruption.

Political economy analyses of Gulf states suggest that centralized authority can function as a stabilizing mechanism in contexts of rapid demographic and economic change (Hertog, 2011). As the UAE's public–private education ecosystem expanded and its economic diversification accelerated, governance coherence became increasingly important for maintaining strategic direction across institutional plurality. Leadership culture emphasizing stability provides the normative grounding for this coherence, reducing resistance to reform by framing centralized oversight not as bureaucratic imposition but as protective coordination during structural transition.

This dynamic has direct consequences for how sustainability is institutionalized within the education system. Long-term planning horizons reduce policy volatility by stabilizing strategic direction across cycles. When inspection systems, accountability frameworks, and strategic mandates are introduced, they are embedded within a broader narrative of national progress and collective advancement. Institutional theory suggests that durable reform depends on embedding new practices within established norms and routines, thereby institutionalizing change in ways that reinforce continuity and reduce systemic volatility across reform cycles (March & Olsen, 1989). The UAE's long-term strategic planning frameworks and coordinated oversight mechanisms exemplify this dynamic, as reforms are framed as evolutionary extensions of existing institutional commitments rather than abrupt systemic shifts.

4.4 Leadership as Stewardship and Continuity

A defining characteristic of UAE governance culture is the framing of leadership as stewardship, a guardianship of long-term collective welfare and particularly intergenerational prosperity. Governance systems pursuing long-term sustainability require mechanisms that balance adaptive capacity with institutional stability, alongside forward-looking institutional steering (Duit & Galaz, 2008; Meadowcroft, 2007).

Leaders must justify present constraints and administrative demands as necessary investments in future resilience. The stewardship narrative performs this justificatory function, framing centralized oversight as protective rather than punitive, and long-horizon planning as an expression of responsibility rather than ambition.

This framing filters downward through the governance architecture. School leaders are positioned not merely as institutional managers but as intermediaries who translate strategic vision into institutional practice while maintaining alignment with regulatory expectations. Their authority within schools is therefore shaped not only by instructional expertise but by their demonstrated capacity to embody national priorities and navigate accountability frameworks in alignment with system-level strategic objectives.

This stewardship framing operates within a governance structure characterized by centralized authority and established hereditary leadership institutions (Herb, 1999). Consultative mechanisms function within this broader framework, shaping participation and deliberation while preserving institutional coherence. Sustained legitimacy, therefore, depends on maintaining credible consultation mechanisms alongside technocratic management, ensuring that the stewardship narrative is reinforced by governance practices that are perceived as responsive to institutional feedback. Legitimacy is not a static possession but an ongoing accomplishment. In institutional terms, legitimacy rests on alignment with broader cultural norms and belief systems rather than on formal authority alone and must be continually reproduced through practice (Suchman, 1995).

4.5 Cultural Legitimacy and Administrative Modernity

The central claim of this section is that the UAE governance architecture is culturally legitimized while administratively modern. Inspection systems, KPIs, and regulatory mandates operate through legal-rational authority. Yet they operate within a leadership culture that draws on nation-building narratives, consultative framing, and stewardship ethics. This hybridization strengthens institutional durability.

Rather than interpreting tradition and modernity as opposites, the UAE case suggests a mutually reinforcing relationship. UAE modernization intercalates modern institutions into preexisting authority frameworks, with centrally steered development framed as continuity rather than rupture. This hybrid approach helps explain why large-scale educational reforms can be implemented within relatively stable political conditions. Governance instruments are not perceived solely as managerial impositions; they are embedded within a broader cultural logic of collective advancement and intergenerational commitment. The analytical question is not whether legitimacy exists, but how it is sustained under centralized steering. Legitimacy, as a socially constructed perception of appropriateness, is not a fixed attribute, but depends on continued alignment with shared normative expectations (Suchman, 1995). The durability of the UAE's education governance architecture thus rests on its capacity to balance stability with adaptive modernization in a complex and evolving policy environment (Duit & Galaz, 2008).

5. TRANSLATING VISION INTO SCHOOL-LEVEL PRACTICE

National sustainability visions acquire operational meaning only when institutional actors translate them into daily practice. In the UAE education system, school leaders function as the critical hinge between centralized governance architecture and classroom reality. Strategic frameworks such as UAE Centennial 2071 and UAE Net Zero 2050 do not reach students directly; they travel through school development plans, performance targets, innovation initiatives, operational decisions, and professional learning structures. Sustainability governance, therefore, becomes actionable through leadership practice. This translation is not automatic but interpretive and contextually mediated.

Policy does not travel from the center to schools as a simple linear instruction chain. Instead, it is translated into practice through interpretation and negotiation within situated contexts, shaped by histories, cultures, and available resources (Ball et al., 2012; Spillane et al., 2002). In the UAE context, school leaders operate within clearly defined strategic mandates shaped by national visions and inspection systems. However, these mandates require contextualization within diverse school communities that vary in curriculum model, student demographics, and organizational culture. The translation of sustainability into practice, therefore, involves strategic sense-making and adaptive alignment rather than mechanical compliance.

5.1 School Strategic Plans as Translation Mechanisms

Strategic planning at the school level is the primary instrument through which national priorities are localized and institutionalized. Governance frameworks require schools to develop multi-year improvement plans demonstrating responsiveness to inspection findings and alignment with national objectives, typically encompassing measurable targets related to student achievement, students' personal and social development, innovation skills, digital integration, environmental initiatives, and leadership development.

In the UAE context, inspection systems and performance indicators provide the parameters within which schools construct strategic plans. In Abu Dhabi, the Irtiqaa self-evaluation process ties school improvement planning directly to inspection criteria and performance standards, requiring leaders to identify strengths, diagnose gaps, and articulate improvement actions that are both internally credible and externally demonstrable (Blaik Hourani & Litz, 2016). Sustainability, whether expressed through environmental initiatives or digital transformation targets, enters school planning not as an abstract aspiration but as a structured organizational commitment embedded within accountability cycles and aligned with inspection performance standards.

5.2 Embedding Sustainability in Operations

Operationalizing sustainability extends beyond documentation. It enters school practice through curriculum integration, environmental programming, digital transformation, and innovation initiatives. Sustainability governance scholarship emphasizes that institutional change requires embedding norms into organizational routines and resource allocation decisions, not merely declaring them in policy texts (Meadowcroft, 2007). For schools, this means revising curriculum to incorporate environmental literacy and STEM competencies, investing in energy-efficient infrastructure, and aligning instructional practices with knowledge economy objectives articulated within national development strategies.

Innovation integration is particularly salient. Knowledge economy literature defines economic transformation as dependent on intellectual capital and knowledge-intensive activities (Powell & Snellman, 2004). UAE school leaders operate within policy frameworks that prioritize digital capacity, STEM orientation, and entrepreneurial competencies while maintaining alignment with inspection systems and quality standards. This dual expectation, innovate and comply, requires sophisticated organizational coordination. Leaders must build institutional cultures that support experimentation and adaptive thinking, while ensuring that innovation initiatives are embedded within formally recognized performance domains. When sustainability priorities are codified within UAE inspection frameworks and integrated into mandatory school improvement planning cycles, they can become embedded within institutional routines rather than dependent solely on individual initiative or discretionary leadership preference.

5.3 Leadership Mediation Between Policy and Classroom

Leadership research consistently emphasizes that principals act as policy mediators rather than passive conduits (Leithwood, Harris, & Hopkins, 2020). In the UAE, this mediation occurs within reform environments characterized by both opportunity and clearly defined structural expectations. Self-evaluation processes can strengthen strategic clarity while intensifying administrative expectations (Blaik Hourani, & Litz, 2016). Leadership development approaches aligned with professional standards further reinforce expectations that school leaders enact system-defined priorities in their practice (David & Abukari, 2020). Principals are therefore positioned simultaneously as instructional leaders, accountability stewards, and innovation catalysts within nationally defined reform parameters.

Ball et al. (2012) argue that schools do not simply "implement" policy; they enact it within situated contexts shaped by resources, histories, and cultures. In the UAE's diverse school ecosystem, leaders may interpret sustainability and innovation mandates differently depending on curriculum model, ownership structure, and community expectations. Yet centralized governance architecture shapes the parameters of acceptable interpretation through defined performance expectations and accountability thresholds embedded in inspection and regulatory frameworks.

Leadership mediation also involves organizational buffering. Sustained reform demands can generate fatigue among teachers when accountability pressures intensify alongside innovation expectations. Effective leaders manage this by framing sustainability initiatives not as additional demands but as integrated dimensions of instructional purpose aligned with school improvement goals. Professional development frameworks aligned with national priorities support this capacity-building function, helping teachers understand how their daily work connects to the broader national transformation project.

5.4 Navigating Accountability Pressures

Accountability structures shape school-level translation of national priorities. Inspection ratings, public reporting, and KPI benchmarks shape institutional incentives and leadership priorities in ways that are both enabling and constraining. Research on Dubai's regulated private education sector illustrates how accountability and quality assurance mechanisms can shape school leadership priorities and institutional strategies in response to performance comparisons and reputational signals (Ben Jaafar et al., 2022). Accountability does not merely measure schools; it shapes attention allocation, resource prioritization, and how improvement is narratively framed for internal and external audiences.

For school leaders, navigating this environment involves balancing alignment with sustainability and innovation mandates against the demands of measurable performance. On the one hand, leaders must demonstrate credible responsiveness to inspection criteria and performance benchmarks. On the other hand, they must maintain strategic coherence and protect instructional priorities from becoming fragmented by short-term performance pressures. Navigation therefore requires deliberate prioritization, pacing of initiatives, and the integration of sustainability goals into existing improvement structures rather than layering them as additional programs. Within the UAE's centralized inspection and accountability architecture, performance pressures can direct school attention toward visible and measurable indicators, potentially narrowing institutional focus and shaping how schools balance immediate performance priorities with longer-term developmental objectives aligned with national sustainability agendas (Ball et al., 2012).

5.5 Operationalizing National Sustainability Goals

The cumulative effect of planning mandates, inspection cycles, innovation expectations, and accountability frameworks is the institutionalization of national sustainability goals at the school level. Leaders translate and embed these goals into improvement plans, resource allocation decisions, staff development priorities, and communication strategies.

The core argument of this section is that school leaders are the hinge between macro-vision and micro-practice. They render sustainability actionable within institutional contexts. Without leadership mediation, a long-horizon strategy remains abstract. Through strategic planning, instructional supervision, curriculum alignment, and innovation management, school leaders translate national ambitions into daily organizational routines. Yet this process unfolds within governance architecture that defines standards, evaluation cycles, and performance metrics. Leadership agency is real but exercised within structured parameters of quality assurance, evaluation cycles, and performance metrics. Sustainability governance at the school level, therefore, reflects negotiated alignment rather than autonomous transformation.

The UAE case illustrates how a centralized long-horizon strategy can permeate and structure institutional practice while preserving contextual variation across schools. Sustainability becomes operational not through rhetoric alone, but through structured leadership practice embedded within accountability architecture and reinforced through inspection and planning cycles.

6. HYBRID GOVERNANCE MODEL: ALIGNMENT AND ADAPTATION

The preceding sections have examined the UAE education system through three analytical lenses: long-horizon strategic vision, centralized governance architecture, and culturally legitimized leadership. Considered together, these dimensions reveal that the system does not fit neatly into conventional governance binaries, centralization versus decentralization, tradition versus modernity, stability versus reform. Rather, the UAE reflects a hybrid governance model in which centralized oversight coexists with institutional diversity, cultural legitimacy reinforces administrative modernization, and rapid reform is embedded within a structurally stable political framework. This hybridity does not imply an equal distribution of authority; instead, it reflects the structured coexistence of centralized standard-setting with bounded institutional discretion and contextual adaptation. Understanding this hybridity is essential to explaining how sustainability governance in the UAE achieves system-wide alignment while preserving institutional responsiveness within defined parameters.

6.1 Centralized Oversight and Institutional Diversity

At first glance, the UAE education system appears strongly centralized. National visions set long-term priorities. Inspection systems standardize definitions of quality. Accountability cycles shape school behavior across hundreds of institutions. Yet the system simultaneously accommodates considerable institutional diversity, particularly in the private education sector, where schools operate under varied international curricula, ownership models, and pedagogical traditions serving a highly multicultural population.

Comparative education research suggests that governance coherence in diverse systems requires regulatory governance, where the state steers through standards, evaluation, and performance information rather than through direct micromanagement of institutional practice (Verger et al., 2019). The UAE model aligns closely with this logic. Central authorities define performance expectations and monitor compliance through inspection and KPI frameworks, while day-to-day instructional choices and curricular emphases remain partially differentiated across institutions. Evaluation criteria are centralized; delivery mechanisms retain bounded variation.

This hybrid arrangement enables what might be called *alignment without uniformity*. Institutional diversity persists; it is structured by recurring inspection cycles and shared accountability instruments. Rather than suppressing pluralism, centralized governance structures it. This configuration resembles what Karlsen (2000) describes as the coexistence of decentralization reforms with continued central steering, whereby institutional autonomy operates within centrally defined accountability frameworks. Scalability emerges from this design: sustainability-oriented reforms can be diffused across hundreds of institutions through standardized instruments, regardless of structural differences between them. This scalability can be understood as an intentional feature of the governance architecture.

6.2 Cultural Legitimacy and Administrative Modernity

A second dimension of hybridity emerges in the interaction between culturally grounded leadership traditions and modern bureaucratic governance instruments. Inspection systems, KPIs, and strategic planning mandates reflect performance-oriented administrative logic aligned with global public management reforms (Pollitt & Bouckaert, 2017). At the same time, leadership discourse in the UAE consistently draws on consultative authority, stewardship narratives, and nation-building commitments that provide normative grounding for centralized governance. This duality matters.

Sociological institutionalism emphasizes that institutions derive stability and legitimacy from alignment with culturally embedded norms and shared belief systems (Hall & Taylor, 1996). In the UAE context, modernization is framed as continuity rather than rupture with tradition, presented as the responsible evolution of a governance model committed to national advancement and intergenerational welfare. This framing reduces resistance to regulatory oversight and enhances the social acceptability of accountability mechanisms that might otherwise generate institutional friction.

The result is a governance model that is modern in instrumentation yet culturally grounded in justification. Administrative authority derives not only from statutory mandate but from alignment with expectations of responsible leadership and intergenerational stewardship. Hybridity here strengthens durability but does not eliminate tension. Administrative instruments ensure consistency, but they can also intensify compliance pressures at lower levels of the system. The hybrid model, therefore, operates within a dynamic equilibrium between authority and adaptability.

6.3 Rapid Reform and Structural Stability

A third and particularly significant hybrid dimension concerns the pursuit of accelerated reform within a stable political and institutional order. Governance theory emphasizes that institutional stability can become a liability under conditions of rapid and unpredictable change, requiring governance systems to balance flexibility with the capacity for coordinated collective action (Duit & Galaz, 2008). The UAE's governance architecture appears calibrated to manage this balance, enabling reform intensity while preserving structural continuity.

Long-horizon frameworks such as UAE Centennial 2071 and UAE Net Zero 2050 provide temporal continuity. By projecting a consistent strategic direction across decades, they insulate reform trajectories from short-term political and economic fluctuations. Inspection systems and accountability cycles create procedural regularity that persists across administrative transitions. Leadership narratives emphasizing stability and national continuity reinforce institutional confidence during periods of rapid change. Together, these elements allow the system to absorb curriculum modernization, digital transformation, innovation mandates, and sustainability initiatives without generating structural instability. Change is continuous, but direction remains consistent.

6.4 Comparative Reflection

This hybrid configuration becomes analytically clearer when contrasted with alternative governance models. Comparative OECD case studies highlight the need to balance centralized accountability mechanisms with flexibility and local responsiveness in governance models (Burns et al., 2016). Decentralized systems that emphasize school-level decision-making authority, community governance, and professional discretion may foster contextual responsiveness, but they also risk variation in quality and alignment with national goals. Highly centralized command models may ensure uniformity but suppress institutional initiative.

The UAE model diverges from strong decentralization paradigms and mitigates fragmentation through centralized oversight while still encouraging innovation within bounded parameters. Authority over standards, inspection, and performance definitions remains concentrated. However, it also differs from rigid command-and-control systems. Steering occurs through performance frameworks, evaluation cycles, and strategic planning mandates rather than direct instructional prescription. The result resembles what governance theorists describe as *centralized steering operationalized through decentralized enactment* (Verger et al., 2019), a configuration in which national direction is clear and enforceable, but institutional actors retain discretion in contextual implementation.

The comparative insight is not that one model is superior, but that governance design reflects contextual priorities. In a state pursuing post-oil economic transition, policy coherence and long-term alignment may be prioritized over extensive decentralization. Sustainability is framed as a collective national project requiring coordinated steering rather than localized experimentation. Hybridity, in this sense, is a deliberate governance architecture for managing the tension between alignment and adaptation that large-scale sustainability governance inevitably entails.

7. STRUCTURAL TENSIONS AND RISKS

Having articulated the hybrid governance model, this section examines the structural tensions that arise from its design. The UAE's sustainability-oriented governance model demonstrates coherence, scalability, and long-horizon coordination. Yet these strengths generate structural tensions. These tensions do not invalidate the model; rather, they reveal the inherent trade-offs embedded within sustainability governance in rapidly transforming, resource-dependent states. Four interrelated stress points merit critical examination: the hydrocarbon paradox, the balance between centralization and autonomy, global benchmarking pressures, and the gap between sustainability ambition and development realities.

7.1 The Hydrocarbon Paradox

A foundational structural tension is what political economy scholarship describes as the diversification paradox. Resource revenues finance modernization and reform, yet the same revenue model can entrench the dependency that sustainability strategy seeks to overcome (Hertog, 2011; Krane, 2018). Commitments such as UAE Net Zero 2050 signal environmental ambition, yet hydrocarbon exports continue to underpin fiscal capacity and state investment. The UAE's educational expansion, digital transformation, and long-horizon planning are materially sustained by revenues from the hydrocarbon sector, even as national sustainability strategies aim to reduce long-term dependence on that sector.

This is not a contradiction in policy intent but a temporal asymmetry. Short- to medium-term economic stability remains tied to revenues from fossil fuel production, while long-term resilience depends on gradual decoupling from that dependency. Education becomes central in this equation, positioned as the long-term exit pathway through human capital development. Yet educational reform itself unfolds within an economy still substantially structured by hydrocarbon wealth, creating pressure for visible and rapid progress in innovation, STEM, and knowledge economy sectors. Sustainability is framed as future-oriented resilience, but it is financed and stabilized through present resource structures. This asymmetry between funding source and strategic ambition is a structural condition that no governance architecture can fully resolve, but only manage.

7.2 Centralization and Bounded Autonomy

A second tension concerns the balance between centralized alignment and institutional autonomy. Centralized governance enhances coherence and scalability, yet it constrains the discretionary space for schools and leaders to innovate. OECD analyses highlight that school autonomy can enhance local responsiveness and flexibility in policy implementation, particularly when combined with clear accountability frameworks (OECD, 2015). However, excessive autonomy risks fragmenting quality standards and weakening equity.

In the UAE, inspection systems, KPI requirements, and strategic planning mandates limit variation in how quality is defined. School leaders retain operational agency, but their decision-making occurs within tightly structured accountability environments. Research on policy enactment suggests that high-stakes accountability can encourage compliance performances, including the strategic production of evidence that aligns with inspected domains, while more complex forms of pedagogic or organizational transformation are deferred or displaced (Ball et al., 2012). When performance indicators define what counts as success, institutional effort may flow toward measurable outputs rather than creative experimentation that sustainability innovation genuinely requires.

This tension is structural rather than incidental. Alignment requires defined standards and monitoring mechanisms. Yet sustainability discourse often emphasizes creativity, innovation, and adaptive capacity, which tightly bound compliance environments can suppress. If the balance tilts excessively toward procedural alignment, reform energy may risk becoming performance-oriented rather than substantively transformative. The governance challenge, therefore, lies not in choosing between alignment and innovation, but in calibrating their interaction so that regulatory coherence supports, rather than constrains, adaptive capacity.

7.3 Global Benchmarking Pressures

A third tension arises from the UAE's active engagement with international assessment frameworks and global education rankings. Participation in benchmarking frameworks signals competitiveness and global integration, and provides external reference points that can reinforce quality standards broadly aligned with SDG principles. However, comparative education research suggests that global metrics subtly reshape national priorities by intensifying focus on quantifiable, cross-nationally comparable outcomes (Rizvi & Lingard, 2010).

For a system pursuing long-horizon sustainability governance, this creates a temporal mismatch. International benchmarking cycles operate on relatively short intervals, emphasizing immediate measurable performance gains. Sustainability governance, by definition, requires patience, the slow accumulation of institutional capacity, cultural transformation, and systemic resilience. Pressure to demonstrate rapid improvement in global standings may conflict with the longer, less linear processes through which genuine sustainability is embedded. When ranking positions become highly visible in policy discourse and public narratives, movement up or down may be interpreted as a proxy for system success or underperformance.

This dynamic is not unique to the UAE. Many education systems navigate similar pressures. Yet in a rapidly reforming, globally oriented state, reputational considerations may amplify benchmarking sensitivity. In such environments, unmet targets can be experienced not merely as delayed progress but as symbolic setbacks, potentially compressing reform timelines and intensifying scrutiny across system levels. The governance challenge lies in ensuring that comparative visibility strengthens quality without allowing rank position itself to become the primary measure of legitimacy or reform success.

7.4 Sustainability Ambition and Development Realities

A final tension concerns the gap between ambitious sustainability narratives and uneven institutional capacity. Sustainability discourse frequently assumes institutional readiness to integrate environmental literacy, innovation, inclusion, and digital transformation simultaneously and at scale. In practice, schools may vary considerably in resources, leadership expertise, and staff preparedness. Policy ambition can outpace implementation infrastructure, producing symbolic compliance in under-resourced settings.

Governance theory emphasizes that reform in complex systems must be calibrated to institutional capacity in order to sustain resilience (Duit & Galaz, 2008), while organizational theory underscores the need for sufficient absorptive capacity, defined as the ability to recognize, assimilate, and exploit new knowledge (Cohen & Levinthal, 1990). In policy contexts, this implies that institutions require sufficient prior knowledge and internal learning structures to internalize and enact reform rather than merely comply with external mandates (Alkaabi, 2023). When reform velocity consistently exceeds institutional readiness, fatigue and performative alignment may emerge, resulting in the appearance of sustainability without substantive change. These risks are documented in the UAE context: empirical research on inclusive education implementation has documented this pattern, where inclusive education policy frameworks clarify expectations while simultaneously exposing capacity disparities across schools that regulatory instruments alone cannot bridge (Massouti et al., 2024).

These four tensions are not isolated from one another. The hydrocarbon paradox shapes the economic expectations placed on education. Centralization influences how benchmarking pressures are internally interpreted. Capacity disparities affect how sustainability ambitions are operationalized at the school level. Sustainability governance in transitional economies, therefore, requires ongoing calibration to manage these interacting stressors. The UAE model addresses some of these risks through long-horizon planning and structured coordination. The durability of the UAE model will depend not only on its capacity to coordinate reform, but also on its ability to recalibrate governance instruments in response to the tensions embedded within its design.

8. CONCLUSION: LONG-HORIZON LEADERSHIP AS SUSTAINABILITY INFRASTRUCTURE

This chapter has argued that sustainability within the UAE education system is institutionalized not primarily through environmental programming or isolated reform initiatives, but through governance architecture. Long-horizon strategic planning, centralized oversight mechanisms, culturally legitimized leadership traditions, and structured accountability frameworks collectively position education as a strategic instrument of national transformation. Sustainability, in this framing, extends beyond ecological reform to encompass economic diversification, institutional coherence, and intergenerational resilience.

National frameworks such as UAE Centennial 2071 and UAE Net Zero 2050 operate as temporal anchors, embedding sustainability within extended developmental trajectories rather than short-term reform cycles. Centralized inspection systems, performance indicators, and strategic planning mandates translate these trajectories into measurable institutional expectations. Leadership culture, shaped by nation-building narratives and stewardship ethics, provides legitimacy to this coordination. At the school level, leaders mediate between macro-level ambition and institutional practice, rendering sustainability actionable through planning cycles, innovation initiatives, curriculum alignment, and accountability management. Collectively, these elements form what may be described as sustainability infrastructure. Infrastructure, here, does not refer to physical assets, but to durable governance arrangements that coordinate action across time and scale, with long-horizon leadership functioning as the integrating principle.

The chapter has also situated this model within a hybrid governance logic. Centralized oversight coexists with institutional diversity; culturally grounded leadership traditions intersect with administratively modern performance systems; rapid reform unfolds within a stable political framework. Hybridity operates not as a compromise but as a structured governance design, enabling policy coherence and scalability while preserving bounded adaptation. Sustainability governance, as observed in the UAE, emerges from the deliberate management of this tension between alignment and discretion rather than from its elimination.

At the same time, the analysis in this chapter has emphasized that this governance model operates within structural constraints. The hydrocarbon paradox, the limits of bounded autonomy, global benchmarking pressures, and uneven implementation capacity remain embedded features of the system. Their management, rather than their resolution, will determine the long-term durability of the architecture. Sustainability governance in a resource-dependent, rapidly modernizing state requires continuous calibration between alignment and adaptability, ambition and absorptive capacity.

Education occupies a central role within this architecture. It is positioned as a strategic instrument of national resilience, expected to generate human capital for diversification, reinforce institutional coherence, and sustain long-term stability. The durability of the broader sustainability agenda is therefore linked to the capacity of the education system to balance coordination with innovation, accountability with professional agency, and strategic foresight with grounded implementation.

Ultimately, the UAE case contributes to wider debates on sustainability governance by illustrating how centralized coordination, cultural legitimacy, and long-horizon planning can function as mutually reinforcing components of systemic transformation. It demonstrates that governance architecture itself can operate as sustainability infrastructure. Whether this model sustains its coherence under evolving economic, demographic, and geopolitical pressures remains an empirical question that longitudinal research will need to address. What the analysis establishes, however, is that governance design—rather than any singular curriculum initiative or environmental program—can serve as the primary mechanism through which sustainability is institutionalized within a national education system. This insight extends beyond the UAE context, contributing to comparative debates on how states navigating structural economic transition may institutionalize sustainability through strategic governance design.

REFERENCES

Ahmed, G., Al Amiri, N., & Abudaqa, A. (2024). Strategic leadership and economic transformation: The United Arab Emirates (UAE) model. *Journal of Global Business Research and Practice, 1*(1), 60–77. DOI: 10.70273/RSFK3414

Al-Taneiji, S., & McLeod, L. (2008). Towards decentralized management in United Arab Emirate (UAE) schools. *School Effectiveness and School Improvement, 19*(3), 275–291. DOI: 10.1080/09243450802246384

Alkaabi, A. M. (2023). Revitalizing Supervisory Models in Education: Integrating Adult Learning Theories and Stage Theories for Enhanced Teaching and Learning Outcomes. In Abdallah, A., & Alkaabi, A. (Eds.), *Restructuring Leadership for School Improvement and Reform* (pp. 253–277). IGI Global., DOI: 10.4018/978-1-6684-7818-9.ch013

Areepattamannil, S. (2024, December). Building a knowledge economy: higher education as a catalyst for the United Arab Emirates' visionary growth. In *Frontiers in Education* (Vol. 9, p. 1510421). Frontiers Media SA. https://doi.org/DOI: 10.3389/feduc.2024.1510421

Badawy, H. R., Alkaabi, A. M., Qablan, A. M., Badawy, H. R., Almaazmi, K. M. M. H., & Ndlela, N. P. (2025). Nurturing future innovators integrating computational thinking, AI, and STEM in early education. In *Empowering early education with computational thinking, AI, and STEM* (pp. 97–134). IGI Global Scientific Publishing.

Ball, S. J., Maguire, M., & Braun, A. (2012). *How schools do policy: Policy enactments in secondary schools*. Routledge.

Ben Jaafar, S., Alzouebi, K., & Bodolica, V. (2022). Accountability and quality assurance for leadership and governance in Dubai-based educational marketplace. *International Journal of Educational Management, 36*(5), 641–660. DOI: 10.1108/IJEM-11-2021-0439

Biermann, F., Abbott, K., Andresen, S., Bäckstrand, K., Bernstein, S., Betsill, M., Bulkeley, H., Cashore, B., Clapp, J., Folke, C., Gupta, A., Gupta, J., Haas, P. M., Jordan, A., Kanie, N., Kluvankova-Oravska, T., Lebel, L., Liverman, D., Meadowcroft, J., & Zondervan, R. (2012). Navigating the anthropocene: Improving earth system governance. *Science, 335*(6074), 1306–1307. DOI: 10.1126/science.1217255 PMID: 22422966

Blaik Hourani, R., & Litz, D. (2016). Perceptions of the school self-evaluation process: The case of Abu Dhabi. *School Leadership & Management, 36*(3), 247–270. DOI: 10.1080/13632434.2016.1247046

Bray, M. (1996). *Decentralization of education: Community financing* (Vol. 36). World Bank Publication. DOI: 10.1596/0-8213-3724-6

Burns, T., Köster, F., & Fuster, M. (2016), Education Governance in Action: Lessons from Case Studies. Educational Research and Innovation, OECD Publishing, Paris. http://dx.doi.org/DOI: 10.1787/9789264262829-en

Cohen, W. M., & Levinthal, D. A. (1990). Absorptive capacity: A new perspective on learning and innovation. *Administrative Science Quarterly*, *35*(1), 128–152. DOI: 10.2307/2393553

David, S. A., & Abukari, A. (2020). Perspectives of teachers' on the selection and the development of the school leaders in the United Arab Emirates. *International Journal of Educational Management*, *34*(1), 56–69. DOI: 10.1108/IJEM-02-2019-0057

Davidson, C. M. (2009). The United Arab Emirates: Prospects for Political Reform. *The Brown Journal of World Affairs*, *15*(2), 117–127. https://www.jstor.org/stable/24590846

Duit, A., & Galaz, V. (2008). Governance and Complexity—Emerging Issues for Governance Theory. *Governance: An International Journal of Policy, Administration and Institutions*, *21*(3), 311–335. DOI: 10.1111/j.1468-0491.2008.00402.x

Ehren, M. C. M., & Visscher, A. J. (2006). TOWARDS A THEORY ON THE IMPACT OF SCHOOL INSPECTIONS. *British Journal of Educational Studies*, *54*(1), 51–72. DOI: 10.1111/j.1467-8527.2006.00333.x

Evans, P. B. (1995). *Embedded autonomy: States and industrial transformation*. Princeton University Press. DOI: 10.1515/9781400821723

Government, U. A. E. (2017). *UAE Centennial 2071*. https://u.ae/en/about-the-uae/strategies-initiatives-and-awards/strategies-plans-and-visions/uae-centennial-2071

Government, U. A. E. (2021). *UAE Net Zero by 2050 Strategy*. The Official Platform of the UAE Government. https://u.ae/en/about-the-uae/strategies-initiatives-and-awards/strategies-plans-and-visions/environment-and-energy/the-uae-net-zero-2050-strategy

Government, U. A. E. (2025, August 4). *Regulatory authorities of K-12 education*. Retrieved March 8, 2026, from https://u.ae/en/information-and-services/education/school-education-k-12/regulatory-authorities-of-k-12-education

Hall, P. A., & Taylor, R. C. R. (1996). Political science and the three new institutionalisms. *Political Studies*, *44*(5), 936–957. DOI: 10.1111/j.1467-9248.1996.tb00343.x

Herb, M. (1999). *All in the family: absolutism, revolution, and democracy in Middle Eastern monarchies*. State University of New York Press.

Hertog, S. (2011). *Princes, brokers, and bureaucrats: Oil and the state in Saudi Arabia*. Cornell University Press. DOI: 10.7591/9780801458774

Honig, M. I., & Hatch, T. C. (2004). Crafting Coherence: How Schools Strategically Manage Multiple, External Demands. *Educational Researcher, 33*(8), 16–30. DOI: 10.3102/0013189X033008016

Hvidt, M. (2013). Economic diversification in GCC countries: Past record and future trends. *The London School of Economics Kuwait Programme on Development, Governance and Globalisation in the Gulf States*, 27.

Ibrahim, H. R., Alghfeli, A. H., Alnuaimi, F. S., Alshamsi, N. N., & Alkaabi, A. M. (2023). STEM and Leadership in the Future: A Path to Innovation, Sustainability, and Entrepreneurship. In Abdallah, A., & Alkaabi, A. (Eds.), *Restructuring Leadership for School Improvement and Reform* (pp. 420–439). IGI Global., DOI: 10.4018/978-1-6684-7818-9.ch021

Jordan, A., & Lenschow, A. (2010). Environmental policy integration: A state of the art review. *Environmental Policy and Governance, 20*(3), 147–158. DOI: 10.1002/eet.539

Karlsen, G. E. (2000). Decentralized centralism: Framework for governance in education. *Journal of Education Policy, 15*(5), 525–538. DOI: 10.1080/026809300750001676

Krane, J. (2018). *Energy kingdoms: Oil and political survival in the Persian Gulf.* Columbia University Press. DOI: 10.7312/kran17930

Leithwood, K., Harris, A., & Hopkins, D. (2020). Seven strong claims about successful school leadership revisited. *School Leadership & Management, 40*(1), 5–22. DOI: 10.1080/13632434.2019.1596077

March, J. G., & Olsen, J. P. (1989). *Rediscovering institutions: The organizational basis of politics*. Free Press.

Massouti, A., Al-Rashaida, M., & Alhosani, M. (2024). A Qualitative Study on Dubai's Inclusive Education Policy from School Leaders' Perspectives. *Sustainability (Basel), 16*(3), 1252. DOI: 10.3390/su16031252

Meadowcroft, J. (2007). National sustainable development strategies: Features, challenges and reflexivity. *European Environment, 17*(3), 152–163. DOI: 10.1002/eet.450

Meuleman, L. (2018). *Metagovernance for sustainability: A framework for implementing the sustainable development goals*. Routledge. DOI: 10.4324/9781351250603

Meuleman, L. (2021). Public administration and governance for the SDGs: Navigating between change and stability. *Sustainability (Basel), 13*(11), 5914. DOI: 10.3390/su13115914

Mulholland, E., & Berger, G. (2017). Multi-level governance and vertical policy integration: Implementation of the 2030 agenda for sustainable development at all levels of government. *ESDN quarterly report, 43*.

OECD. (2015). *Education Policy Outlook 2015: Making Reforms Happen*. OECD Publishing., DOI: 10.1787/9789264225442-

OECD. (2016). *Policy coherence for sustainable development 2016: Towards sustainable and resilient societies*. OECD Publishing., DOI: 10.1787/9789264256996-

OECD. (2019). *Policy coherence for sustainable development 2019: Empowering people and ensuring inclusiveness and equality*. OECD Publishing., DOI: 10.1787/a90f851f-

OECD. (2023). *Driving policy coherence for sustainable development: Accelerating progress towards the SDGs*. OECD Publishing., DOI: 10.1787/a6cb4aa1-

Ozga, J. (2009). Governing education through data in England: From regulation to self-evaluation. *Journal of Education Policy, 24*(2), 149–162. DOI: 10.1080/02680930902733121

Pollitt, C., & Bouckaert, G. (2017). *Public management reform: A comparative analysis—into the age of austerity*. Oxford University Press.

Powell, W. W., & Snellman, K. (2004). The knowledge economy. *Annual Review of Sociology, 30*(1), 199–220. DOI: 10.1146/annurev.soc.29.010202.100037

Rai, J., & Beresford-Dey, M. (2023). School leadership in the United Arab Emirates: A scoping review. *Educational Management Administration & Leadership*. Advance online publication. DOI: 10.1177/17411432231218129

Rizvi, F., & Lingard, B. (2010). *Globalizing education policy*. Routledge.

Sachs, J. D. (2015). *The age of sustainable development*. Columbia University Press. DOI: 10.7312/sach17314

Spillane, J. P., Reiser, B. J., & Reimer, T. (2002). Policy Implementation and Cognition: Reframing and Refocusing Implementation Research. *Review of Educational Research, 72*(3), 387–431. DOI: 10.3102/00346543072003387

Steiner-Khamsi, G. (2016). Cross-national policy borrowing: Understanding reception and translation. *Asia Pacific Journal of Education*, *34*(2), 153–172. DOI: 10.1080/02188791.2013.875649

Suchman, M. C. (1995). Managing Legitimacy: Strategic and Institutional Approaches. *Academy of Management Review*, *20*(3), 571–610. DOI: 10.2307/258788

United Nations. (2015). *Transforming our world: The 2030 agenda for sustainable development*. https://sdgs.un.org/2030agenda

Verger, A., Fontdevila, C., & Parcerisa, L. (2019). Reforming governance through policy instruments: How and to what extent standards, tests and accountability in education spread worldwide. *Discourse (Abingdon)*, *40*(2), 248–270. DOI: 10.1080/01596306.2019.1569882

Weber, M. (1978). *Economy and society: An outline of interpretive sociology* (Roth, G., & Wittich, C., Eds.). Vol. 2). University of California Press.

Chapter 6
Social Justice, Self-Determination, and Sovereignty in Educational Reform

Zahid Ahmad Thoker
http://orcid.org/0009-0002-6897-5835
Lovely Professional University, India

Lovleen Marwaha
http://orcid.org/0000-0002-5175-2590
Lovely Professional University, India

ABSTRACT

This chapter, "Social Justice, Self-Determination, and Sovereignty in Educational Reform," examines how Indigenous knowledge systems and place-based pedagogies contribute to equitable and sustainable education. Through the lens of decolonial and critical pedagogical inquiry, this chapter explores the interconnections between social justice, self-determination, and educational sovereignty, contending that Indigenous communities' rights to design and govern their own education systems hold the key to sustainable and resilient learning. The authors provide examples from Māori schools in Aotearoa/New Zealand, Tribal Colleges in North America, and community-led models in India and Latin America to show that sovereignty enables cultural revitalization, ecological stewardship, and social empowerment. The chapter identifies links to SDGs 4, 13, and 15 but also recognizes constraints including policy, funding inequities, and epistemic tokenism.

DOI: 10.4018/979-8-3373-7650-9.ch006

INTRODUCTION

Indigenous knowledge systems are profoundly rooted place-based knowledge systems that maintain human and ecological health over time. These knowledge traditions are spiritual, ecological, cultural, and social aspects that have developed over time through ongoing communication between individuals and their surroundings. Nevertheless, there are still significant obstacles to the implementation of Indigenous epistemologies in general education, which are, in part, caused by the presence of the legacies of colonialism, the prevailing epistemic paradigms, and the systems of standardized learning that prefer Western knowledge paradigms (Battiste, 2019; Alam, 2025). The necessity to contextualize Indigenous knowledge in the context of sustainability thus requires an overhaul of the educational purpose and practice. To the Indigenous view, sustainability is not related to management of resources or environmental conservation, but also the moral, cultural, and spiritual sustenance of existence. Indigenous epistemologies focus on the principles of balance, reciprocity, and respect for the interconnectedness of all living systems, which are necessary in the long-term ecological and social resilience (Berkes, 2017; Whyte, 2018).

The relationship between education reform and justice, sovereignty, and ecology offers a powerful, transformative prism in which a new definition of sustainability can be revised. In this respect, educational justice suggests breaking epistemic hierarchies and establishing Indigenous pedagogies as valid and essential systems of thought (Berkes, 2017). Sovereignty is political as well as epistemic self-determination, the entitlement of the Indigenous communities to define, create, and pass their knowledge based on the cultural protocols and local ecologies. Relating sovereignty to the ecological consciousness makes education a place of opposition and regeneration in which learners will be able to develop ethical relationships with the environment and community (Fig.1). The ecological crisis that humanity faces today makes it clear why the incorporation of the Indigenous knowledge systems with the representation of sustainable coexistence and generational responsibility is as urgent as it is. This integrates the need to have an education system that goes beyond extractive learning models and fosters a sense of relational ethics, ecological stewardship, and group well-being (Cajete, 2016).

Although an increasing number of sustainability discourses across the world are recognizing Indigenous epistemologies, there are still research gaps. To begin with, the absence of the extensive frameworks that would operationalize the Indigenous knowledge in the formal education systems without diminishing it to tokenism and cultural symbolism is still present (Macfarlane et al., 2014). Second, there is a lack of empirical research on how Indigenous pedagogies can be used in curriculum development, teacher training, and community-based learning to achieve sovereignty and ecological justice (Kalidas, 2025). Third, there is a lack of exploration in the manner in which educational reform may be able to simultaneously intervene around historical injustices, strengthen local systems of governance, and promote ecological literacy (Wilson, 2008). The three objectives of this chapter are thus three-fold: (1) to critically analyze how indigenous knowledge can be contextualized in the sustainability education; (2) to address the interconnections between the educational reform, social justice, and ecological sovereignty, and (3) to recognize the approaches to reconsidering the education systems to become culturally grounded and ecologically responsive. This chapter provides contributions to the discourse of decolonizing education by addressing those dimensions to build Indigenous knowledge as the pillar of achieving social and ecological sustainability.

1. HISTORICAL CONTEXT OF INDIGENOUS KNOWLEDGE AND EDUCATIONAL REFORM

The history of Indigenous knowledge and education reforms is closely connected with colonization and cultural oppression, and the following decolonization and self-determination movements. Indigenous education systems were community-based and whole-bodied and had a direct relationship to the land before the introduction of colonial influence. The transmission of knowledge was carried out in orality, storytelling, rituals, apprenticeship, and experience in the natural environment (Cajete, 2016). These pedagogies also focused on relational responsibility, ecological ethics, and building individual and group responsibility to maintain living systems. The process of learning was contextual, ongoing, and oriented to the well-being of the community and the environment. But these local epistemologies were gradually marginalized or washed out with the introduction of colonial education systems in the name of civilization and modernization (Battiste, 2019).

Policies of colonial schooling, including the residential and missionary forms of education in North America, the British colonial education system in Africa and South Asia, and the assimilationist policies in Australia and New Zealand, aimed to ensure that the West had imposed its epistemologies and suppressed Indian languages and traditions, and governmental systems (Berkes, 2017; Alam, 2025). These systems served as instruments of epistemic control and cultural absorption and replaced the values of reciprocity, stewardship, and communal learning of the Indigenous people. The result was the break-up of cultural identity, the disappearance of linguistic diversity, and the alienation of Indigenous people from their ancestral land and ecological knowledge (Wilson, 2008). Education was used as a control mechanism and not a liberation mechanism, as it reinforced the hierarchy of knowledge that gave Western rationalism a privilege and pushed indigenous worldviews out of formal learning environments (Sund & Pashby, 2020).

The end of the 20th and the beginning of the 21st centuries were the time when the Indigenous movements revived their demands to be recognized in their rights to autonomous decision making, land, and cultural integrity. New discourse around the sustainability and education reform unfolded as a result of the emergence of global mechanisms like the United Nations Declaration on the Rights of Indigenous Peoples (UNDRIP, 2007) and the Decade of Education for Sustainable Development (UNESCO, 2014). Such trends made it clear that it is necessary to rebrand Indigenous knowledge as an essential element of environmental management as well as social justice (Macfarlane et al., 2014). Educational institutions started to consider their curricula, teaching methods, and institutional organization as Indigenous scholars and educators started to reclaim the epistemic sovereignty. Decolonization of education was thereby not only a process of content inclusion, but also an epistemic transformation, of redefining what is recognized as knowledge, its producers, and its use in the common good (Battiste, 2019; Kalidas, 2025).

When applied to the area of sustainability education, Indigenous knowledge provides an alternative, historically based approach to the reductionist paradigm that prevails in Western science. In contrast to industrial paradigms of education, where the environment is regarded as the resource to be exploited by humans, the Indigenous traditions consider it as the relational system that is not only alive but is also being created through human action (Berkes, 2017). This relational worldview represents what has been described as ecological sovereignty, in which environmental and cultural existence is regarded as inseparable (Whyte, 2018). To Indigenous people, sustainability is not a recent term; it is an old value that finds its way into the practice of seasonal migration, resource distribution, ceremonial exchange, and responsive management of ecosystems (Kimmerer, 2013). However, the disempowerment of these systems, caused by colonization throughout history, necessitated the revival of the Indigenous knowledge that is important to educational reform in the modern world.

Reforms in the education system that incorporate Indigenous ways of thinking have been slowly being seen as powerful means of reconciliation and ecological justice. Culturally responsive teaching, land-based education, or community-led curriculum design are some of the initiatives that may be viewed as attempts to incorporate Indigenous knowledge into the education system. As an example, the Truth and Reconciliation Commission (2015) suggested education as a tool of redressing the wrongs imposed by residential schools in Canada and encouraging intercultural understanding. Likewise, in Australia, there has been the introduction of policies that stressed the aspect of two-way learning to incorporate indigenous ecological knowledge in science and environmental studies. In Aotearoa New Zealand, Te Wharaiki early childhood curriculum provides a preview of the Māori whanaungatanga (relationships), manaakitanga (care), and kaitiakitanga (guardianship), as a result of a profound interrelation between learning and land stewardship (Hemara, 2000). These illustrations show that Indigenizing education can help to nurture not only cultural renewal but also sustainability literacy.

Irrespective of these developments, there are still hurdles in the institutionalization of Indigenous knowledge in the mainstream systems of education. The alternative epistemologies are still marginalized by structural obstacles like Eurocentric curriculum development design, standardized testing, and little representation of Indigenous peoples in educational leadership (Alam, 2025). Furthermore, the incorporation of Indigenous material is often merely surface-level, centered on the symbolic incorporation as opposed to the actual or epistemic revolution (Gunawan & Indrawan, 2025). To overcome these issues, it is necessary to go beyond multicultural tokenism and explore the idea of the radical transformation of the structural level at which the Indigenous ways of thought and learning transform the principles of education per se. It includes co-building curricula with Indigenous communities, making sure that there is a revitalization of language, and making Indigenous teachers the leaders in policy development and decision-making (Battiste, 2019; Kalidas, 2025). Fig.2 shows a good representation of the Historical Trajectory and Decolonization of Indigenous Education to Sustainability.

Traditionally, indigenous education has been a place of colonization and possibly a place of decolonization. The renewal of Indigenous education systems is not a failure to move backward to the past but a transition to epistemic justice, a schooling that serves as a means of cultural survival, environmental sustainability, and political strength. Reforming education in Indigenous paradigms is in the trend of global sustainability, specifically the movements focused on equity, biodiversity conservation, and climate adjustment. With Indigenous knowledge being incorporated into the modern educational frameworks, societies will be able to develop a type of learning that respects interdependence, facilitates resilience, and enhances justice towards people and the planet. It is based on this historical knowledge that the role of social justice, self-determination, and sovereignty can be redefined in furtherance of education as a mechanism of sustainable futures.

2. SOCIAL JUSTICE, SELF-DETERMINATION, AND SOVEREIGNTY: CONCEPTUAL FRAMEWORK

It is important to determine how social justice, self-determination, and sovereignty relate to each other and redefine education as a transformative instrument of cultural resurgence and ecological sustainability. Although these three ideas are different, they are closely interconnected in the Indigenous worldviews and decolonization struggles. Their combination gives the holistic structure of repositioning the educational reform along the lines of equity, cultural continuity, and ecological balance. With respect to Indigenous education, social justice is no longer understood as a form of equality but instead as having epistemic, cultural, and ecological aspects of justice. It acknowledges the entitlement of the Indigenous people to reclaim, renew, and even pass on their systems of knowledge as a key to personal empowerment and survival of the people collectively (Battiste, 2019). Equally, self-determination refers to the natural right of Indigenous people to control the educational, political, and cultural activities based on their traditions, values, and ecological environments (Corntassel, 2008). The sovereignty, in the meantime, is also not limited to the political level but includes epistemic and ecological sovereignty, the ability to maintain and manage knowledge, land, and community relationships according to the Indigenous cosmologies (Whyte, 2018; Wilson, 2008).

Indigenous Education and Social Justice

Social justice in Indigenous education entails the reallocation of power and acknowledgment in the systems that were historically meant to sideline the voices of the Indigenous people. This does not only involve fair access to education, but also the revision of the epistemological models that favor Western rationalism to the disadvantage of the Indigenous knowledge. The core of this process is epistemic justice: the acknowledgment of Indigenous ways of knowing as valid and rigorous, and crucial knowledge. In reality, it means that Indigenous values, languages, and environmental ethics have to be incorporated into the curricula, pedagogy, and assessment. In addition, it requires an inclusive educational governance through which Indigenous educators, elders, and knowledge holders should co-lead the decision-making processes.

In the context of sustainability, social justice likewise extends to ecological justice the right of Indigenous peoples to continue their association with their ancestral lands and ecosystems. The Indigenous people are disproportionately affected by environmental degradation, resources, and climate change, as the livelihood and identity of Indigenous people are strongly connected with the environmental well-being (Kimmerer, 2013; Whyte, 2018). Thus, Indigenous social justice in educational reform should also include land-based education in which these relationships are reasserted. These strategies foster in learners contemplation of ecological stewardship and relational ethics, helping to bridge the gap between cultural perpetuation and environmental conservation (Cajete, 2016).

Self-Determination and Educational Autonomy

Self-determination is also a political and educational requirement among the Indigenous people. It guarantees that they have a right to design, establish, and assess educational frameworks that express their cultural worldviews and priorities (Corntassel, 2008). In the past, this autonomy was not allowed by colonial education systems, which imposed Eurocentric education systems that disrupted the continuity of Indigenous epistemics (Berkes, 2017). Modern trends in Indigenous educational self-determination, including tribal colleges in North America, Maori immersion schools in Aotearoa New Zealand, and Aboriginal community-controlled education in Australia, are exemplary forms of reclaiming control over knowledge production and distribution (Kalidas, 2025; Alam, 2025).

Essentially, educational self-determination pertains to regaining the right to learn personal way. This will include language revitalization, curriculum development based on culture, and the role of elders and community leaders as educators. The reinstitution of these traditional roles and structures makes education a way of cultural renaissance instead of assimilation. Moreover, self-determination is applied to research methodologies, which is why it is recommended to use participatory, relational, and land-informed methods, which are based on Indigenous epistemic values (Wilson, 2008). With these structures, indigenous populations re-establish control over what will be recognized as legitimate knowledge and its role in the general well-being of the community.

In sustainability education, self-determination is in line with educational sovereignty, the capacity of communities to formulate their own learning agendas to achieve ecological and cultural objectives. Self-determination-led indigenous education not only resolves the past inequalities but also creates resilience and adaptability in response to environmental change. It is a necessary reassertion of autonomy to make sure that sustainability initiatives are not foisted upon people, but rather are organically derived within the boundaries of local knowledge systems and values (Macfarlane et al., 2014).

Sovereignty: From Political to Epistemic and Ecological Dimensions

Conventionally, sovereignty has been perceived as the political autonomy of a state or nation. But in Indigenous discourse, sovereignty is multidimensional, and it includes the ability to control land-based, material, and cultural resources, as well as the control of knowledge, culture, and ecological relationships (Simpson, 2017). Epistemic sovereignty is a term used to refer to the entitlement of the Indigenous peoples to regulate the representations and access to their knowledge and its application in educational and research institutions (Alam, 2025). This encompasses defense against cultural appropriation and the commodification of sacred or community knowledge.

It is closely connected with ecological sovereignty that acknowledges the interdependence of knowledge, land, and identity. The connections between indigenous peoples and their ecosystems are the basis of their epistemologies, so the decrease of ecological sovereignty through land dispossession also reduces the knowledge systems (Berkes, 2017; Whyte, 2018). The school reform accommodating this relationship must, consequently, reinstate the connection between the learning and the place. An example of epistemic and ecological sovereignty put into practice is land-based education programs, which place the environment as a teacher and classroom (Cajete, 2016).

Sovereignty also suggests decolonization of the academic and policy institutions. This includes breaking down hierarchical systems where Indigenous academics and views do not dominate through education. The models of collaborative governance where indigenous decisions are embedded in universities, ministries of education, and sustainability efforts are the major points of achieving actual sovereignty in education (Battiste, 2019). Besides, sovereignty implies protection and renewal of the Indigenous languages as epistemic continuity. Cultural identity and environmental ethics are maintained through language as a way of transmission of knowledge and as a way of perceiving the world. Therefore, linguistic sovereignty becomes part and parcel of the realization of holistic educational sovereignty (Gunawan & Indrawan, 2025). Table 1 below also reveals some of the key concepts that support Indigenous approaches to educational reforms.

Table 1. Interrelationships among Social Justice, Self-Determination, and Sovereignty in Indigenous Education and Sustainability

Concept	Core Meaning	Key Educational Implications	Role in Sustainability & Decolonization
Social Justice	Redistribution of power and recognition of Indigenous epistemologies and rights beyond equality.	Embedding Indigenous values, languages, and governance in curricula and decision-making promotes ecological justice.	Ensures equitable access and ecological stewardship; supports cultural and environmental rights.
Self-Determination	Indigenous communities' authority to govern educational, political, and cultural affairs independently.	Control over curriculum design, language revitalization, community-led education, and Indigenous research methodologies.	Empowers communities to set learning agendas that reflect cultural and ecological priorities.
Sovereignty	Multidimensional control over land, knowledge, culture, and environment aligned with Indigenous cosmologies.	Protection of knowledge systems from appropriation; institutional inclusion; language and land-based education as core pillars.	Reclaims epistemic and ecological sovereignty, re-establishing Indigenous relationships with land and knowledge.

Toward an Integrated Framework for Sustainable Education

When viewed together, social justice, self-determination, and sovereignty constitute an integrated framework for decolonizing and indigenizing education for sustainability. Social justice ensures recognition and redistribution; self-determination restores agency and autonomy; and sovereignty safeguards the continuity of land, culture, and knowledge. This triadic framework challenges conventional sustainability education, which often treats environmental issues as technical or apolitical. Instead, it positions sustainability as an inherently moral and relational project rooted in justice and sovereignty (Whyte, 2018).

This conceptual synthesis calls for a paradigm shift in how educational institutions engage with Indigenous knowledge. Rather than inserting Indigenous content into Western frameworks, institutions must restructure their epistemic foundations to embrace pluralism, reciprocity, and relational accountability. This transformation aligns with the *United Nations Sustainable Development Goals* (SDGs), particularly SDG 4 (Quality Education) and SDG 13 (Climate Action), by promoting inclusive and locally grounded learning that advances both equity and ecological resilience (UNESCO, 2020). The ultimate goal is to create education systems that not only impart knowledge but also cultivate wisdom—nurturing learners who act with respect, responsibility, and reciprocity toward the Earth and all its beings.

3. EMPIRICAL INSIGHTS AND CASE STUDIES

The Māori Education Reforms in Aotearoa/New Zealand

Mai te Ako Ako and Mahutga Māori Education, as the changes in the education system, have become one of the most effective attempts at cultural sovereignty and linguistic renewal in Aotearoa/New Zealand. These reforms are based on the Treaty of Waitangi (1840) and are a progressive decision to correct the colonial marginalization of Māori epistemologies and pedagogical systems (Berkes, 2017). Indigenous learning spaces have been redefined by the establishment of Kura Kaupapa Māori (Masifika-based schools) and Wharekura (secondary-based schools that are of Maori medium), where Maori language (te reo Māori), customs (tikanga), and values (kaupapa) are the keystones to education (Bishop, 2005). Empirically, it has been found that students who study in these schools have a higher sense of cultural identity, community interaction, and academic confidence than those who study in mainstream schools (Macfarlane et al., 2014). Moreover, place-based models of learning, which entail linking curriculum to local ecosystems, local history, and indigenous culture, have enhanced environmental stewardship and social unity. Nevertheless, there remain issues such as the fact that there are still fewer trained Māori teachers, structural funding disparities, and, as a result of standardized national testing, which does not always measure indigenous success (Greenwood and Te Aika, 2010).

North American Tribal Colleges and Universities

In the United States and Canada, Tribal Colleges and Universities (TCUs) have emerged as the focus issue in indigenous self-determination in education. TCUs are the products of the Native American self-determination movement of the 1960s and are community-based organizations that have incorporated Western academic standards with Native worldviews, languages, and ecological ethics (Stein, 2021). Not only do these colleges offer access to higher education in the rural regions that have been underserved, but they also preserve systems of traditional ecological knowledge (TEK) through which local environmental management practices are informed (Cajete, 2016). Land-based science, Native studies, and indigenous business development are some of the programs that support the dual purpose of cultural revival and economic viability. Empirical studies show that TCU alumni tend to go back to their communities as teachers, policy-makers, and environmental activists, to ensure there is sustainable local governance (Shotton et al., 2023). However, TCUs are persistently challenged with inadequate funding parity, infrastructure building, and accreditation practices, which indicate more systemic inequities between Aboriginal and mainstream institutions of education (Brayboy et al., 2014).

Community-Based Education in India and Latin America

In the Global South, the indigenous and rural education movements in India and Latin America offer an insight into how a community-based learning model can change the situation. Other models of alternative schooling, including the Lok Jumbish and Ekal Vidyalaya initiatives in India, focus on participatory learning based on the inclusion of traditional ecological knowledge and local governance with gender inclusivity. Such programs empower the rural communities by enhancing their livelihoods based on context-relevant pedagogies that establish literacy connections with agricultural sustainability, water conservation, and biodiversity management (Govinda, 2002). On the same note, in Latin America, schooling in places like Bolivia, Ecuador, and Peru has been re-imagined through an intercultural bilingual education (IBE) model, which has incorporated indigenous languages and cosmologies (sumak kawsay or "good living") into formal education (Arias-Gutierrez & Minoia, 2023). Empirical evidence indicates that the reforms can increase the retention rates of the schooling system and the sociocultural bonds, as well as question the epistemic privilege of Eurocentric paradigms. The implementation of these initiatives is, however, often constrained by inconsistent policy implementation, urban-rural inequalities, and political instability.

Table 2. Comparative Overview of Indigenous Education Models

Region	Education Model	Governance Structure	Key Pedagogical Features	Notable Outcomes
Aotearoa/New Zealand (Māori)	Kura Kaupapa Māori	Community-led boards with elders, parents	Māori language immersion, land-based learning, cultural continuity	Increased language retention, cultural identity reinforcement, and environmental stewardship
North America (Tribal Colleges & Universities)	TCUs	Tribal governance & charters	Place-based curricula, intergenerational knowledge, and STEM integration	Higher engagement, leadership development, and cultural preservation
India (Community-based education)	Village/community schools	Local committees & NGOs	Experiential learning, indigenous ecology, and multilingual instruction	Biodiversity awareness, social cohesion, and local sustainability practices
Latin America (Indigenous NGOs)	Community Learning Centers	Indigenous councils & participatory governance	Land-based education, traditional agriculture, and climate adaptation	Food security knowledge, climate resilience, empowerment of youth, and elders

Comparative Analysis of Outcomes and Challenges

Relatively, these world-case studies illustrate that indigenous reforms in education are united in a core pledging to cultural sovereignty, environmental reciprocity, and epistemic justice. In Māori Kura Kaupapa as well as North American TCUs and community-based schools in India and Latin America, one can find a common point: a focus on education as a form of ecological regeneration and self-determination (Battiste, 2019). But there are still a few obstacles: structural underinvestment, the under-appreciation of indigenous knowledge in policy-making, and the discord between local autonomy and state control. Institutionalization of bilingual and bi-cultural systems has been achieved in Aotearoa/New Zealand and is not the case in the Global South, which is typically faced with a lack of centralization and political backing (Santos, 2015). However, the fact that these varied experiments are bringing together emphasizes the universality of the indigenous pedagogies insofar as they are sustainable, justice-based educational paradigms. All these lessons prove that the reforms in the future should not supplement the indigenous knowledge with the curriculum but just reorganize the educational process as the epistemological plurality and the place of environmental ethics (Parsons et al., 2021).

4. INTEGRATING SOCIAL JUSTICE AND SUSTAINABILITY

Ecological Literacy and Place-Based Pedagogy

Ecological literacy as an integrated approach to indigenous learning models is one of the critical strategies in achieving sustainability and social equality. According to Capra (2007), ecological literacy entails having an understanding of how natural systems are interdependent on one another, the cultural, ethical, and social aspects that support them. This literacy is supported by indigenous pedagogies based on local ecologies and learning based on experience and involvement of the community. Place-based pedagogy is an increasingly popular concept in environmental education that conforms to indigenous worldviews that place more importance on relationality, that is, human beings as part of an ecosystem instead of observers. As an example, Māori ideas of kaitiakitanga (guardianship) and Native American ideas of seven generations sustainability are the integration of ecological awareness and moral agency (Kimmerer, 2013). The evidence is that students with place-based indigenous education have superior environmental ethics, problem-solving, and community stewardship, compared to students in conventional curriculum models. In addition, the methods will help to rebuild the cultural memory of the land and empower the local people in their endeavors to solve the immediate ecological crisis, like the loss of biodiversity, water scarcity, and climate disturbance. Therefore, ecological literacy, in place in indigenous pedagogies, can be not just a cognitive skill, but a moral and civic necessity in order to be sustainable.

Aligning Indigenous Education with the Sustainable Development Goals (SDGs 4, 10, 13, 15)

The transformative concepts of the United Nations Sustainable Development Goals (SDGs), in particular SDG 4 (Quality Education), SDG 10 (Reduced Inequalities), SDG 13 (Climate Action), and SDG 15 (Life on Land), are inherent in the indigenous education systems. SDG 4 promotes inclusive and equitable education, enabling all people to pursue lifelong learning; indigenous education adds to the requirements of inclusive education by focusing on epistemic inclusivity and cultural renewal (UNESCO, 2020). Indigenous learning models such as community-based schools, Tribal Colleges, and Māori immersion schools are an alternative gap in education and empower marginalized communities, which fulfil the SDG 10 goal by confronting systemic inequities (UNDP, 2021). Regarding climate action (SDG 13), indigenous knowledge brings subtle information on local climate response, agroecological sustainability, and biodiversity (Reid et al., 2021). Moreover, SDG 15, which emphasizes the importance of life on land, is close to the indigenous land stewardship ethics that emphasize balance, reciprocity, and intergenerational sustainability (Berkes, 2017). Nevertheless, to make indigenous education compatible with SDG implementation, it is necessary to adopt a critical decolonial prism, not one that will allow instrumentalizing the indigenous knowledge into the technocratic system, but one that will take into account its independent epistemic power (Santos, 2015). Therefore, it is necessary to regard SDGs in native education as the dialogue of knowledges (ecologies of knowledge) to guarantee global sustainability and cultural sovereignty.

Culturally Responsive Sustainability Education.

The notion of culturally responsive sustainability education (CRSE) comes into the limelight as one of the methods of pedagogy that incorporates cultural diversity, ecological integrity, and social justice in the form of supportive learning systems. CRSE is based on the principles of the culturally responsive pedagogy (**Higgins**, 2026) because it views sustainability education as an activity that needs to embody the sociocultural backgrounds of learners and the environmental history of the community in which the sustainability education takes place. This involves grounding teaching in indigenous stories, in environmental day-to-day practices, and intergenerational education in indigenous contexts (Bang et al., 2013). CRSE also disrupts the Western monopoly of sustainability discourses that usually revolve around technological innovation, disregarding relational ethics (Parsons et al., 2021). To illustrate, the incorporation of indigenous agricultural techniques, including the milpa in Mexico or the zabo water-harvesting system in Northeast India, is a physical channel of integrating biodiversity protection and livelihood security. Moreover, CRSE promotes identity affirmation and community well-being, which are necessary factors against ecological alienation and cultural erosion. The studies indicate that sustainability education based on culture can improve the motivation, cross-cultural empathy, and ecological citizenship in students. Thus, CRSE is a pedagogical approach as well as a kind of epistemic justice - allowing education to be a point of connection between cultural continuity and planetary responsibility.

5. CHALLENGES AND BARRIERS TO EDUCATIONAL SOVEREIGNTY

The right of Indigenous peoples to establish, manage, and maintain their own learning systems, now known as educational sovereignty, remains a pillar of decolonizing education and achieving sustainability. Nevertheless, despite the international recognition of Indigenous knowledge and self-determination in policy agendas such as the United Nations Declaration on the Rights of Indigenous Peoples (UNDRIP, 2007) and the Sustainable Development Goals (SDGs), their implementation remains hampered by deep structural inequities. Such issues are not only due to historical colonial pasts but also to present-day policy, budgets, and governance practices alienating Indigenous epistemologies. Educational sovereignty is therefore confronted with numerous obstacles across the institutional, epistemic, and leadership dimensions that limit the Indigenous-led reform and integration of cultural values into education systems.

Policy, Funding, and Institutional Barriers

Policy fragmentation and inadequate funding mechanisms are among the most enduring problems to educational sovereignty. Education systems in many countries worldwide often introduce standardized curricula and assessment schemes that do not align with Indigenous knowledge systems or languages (Battiste, 2019). Policy systems in most situations view Indigenous education as an ancillary or curative issue instead of a source of sustainable learning. For example, the so-called Closing the Gap strategy, intended to improve educational outcomes for Indigenous people in Australia, has been criticized for focusing on statistical equality while neglecting cultural integrity (Brown, 2019). Likewise, in the United States and Canada, governmental support for Tribal Colleges and Indigenous language immersion programs is underrepresented relative to mainstream institutions, which makes it difficult to build an autonomous governmental structure and a culturally based curriculum (Battiste, 2019).

Bureaucratic systems are institutional barriers that support Western epistemologies and pedagogical norms. The processes of accreditation, quality assurance, and teacher certification are often constructed around Eurocentric standards, which leave little room for the autonomy of Indigenous pedagogy (Pidgeon, 2016). Although Indigenous views are also incorporated, they are often relegated to the periphery of curricula, offered as optional courses or cultural awareness classes, rather than serving as the backbone of the curriculum. Such marginalization of the structure is a manifestation of an even greater policy paradox: even though governments advise on promoting the inclusion of Indigenous people, the systems of regulation still recreate colonial hierarchies of knowledge and power. In addition, the lack of sustainability due to the absence of long-term funding undermines sustainability. Because most community-based educational projects rely on short-term grants, projects are discontinued and staff turnover occurs (Carney, 2022). Policy reforms should go beyond inclusionary discourses and institutionalize resource equity, community control, and curricular pluralism to achieve substantive educational sovereignty.

Epistemic Tokenism and Cultural Misappropriation

The other significant obstacle to Indigenous educational sovereignty is the persistence of epistemic tokenism, that is, the superficial inclusion of Indigenous material without accompanying changes in epistemological logics. Tokenism can also be observed in individual classes on Indigenous culture or environmental care added to the mainstream curriculum, which is essentially Western in its organization and goals (Sund & Pashby, 2020). These practices jeopardize the status of Indigenous knowledge as cultural artifacts or moral supplements rather than as living, dynamic systems of knowledge. Such a reductionist practice undermines not only Indigenous epistemological practices but also supports extractive relationships between academia and Indigenous communities.

This is further worsened by cultural misappropriation. Whether intentionally or not, Indigenous symbols, stories, or ecological concepts are used in most educational settings to enhance environmental and sustainability education programs (Gunawan & Indrawan, 2025). As an example, non-Indigenous teachers tend to use Indigenous ecological metaphors like Mother Earth or Two-Eyed seeing in decontextualized forms and deprive them of their cultural and ethical contexts (Reid et al., 2021). This exploitation of knowledge continues the colonial processes of extracting knowledge, whereby native wisdom is converted into pedagogical objects without acknowledging the political and spiritual sovereignty within which it is produced (Santos, 2015).

Additionally, tokenistic inclusion is highly likely to ignore the heterogeneity of Indigenous knowledge systems and instead consider them as homogeneous at the expense of local specificity and inter-community diversity (Ahenakew, 2016). This homogenization not only distorts the diversity of the Indigenous epistemologies but it also limits their flexibility in other cultural and ecological settings. In order to overcome tokenism, the educational systems should not only adopt Indigenous knowledge as content but also as an epistemic framework. This involves co-curriculum development with the Indigenous people, ethically based protocols of knowledge sharing, and teacher education based on relational responsibility and cultural competence (Wilson, 2008). Authentic partnership also means that Indigenous knowledge integration should be grounded in mutual respect, reciprocity, and sovereignty, rather than cultural appropriation.

Governance and Leadership Gaps

The difference between the possibility of actualizing Indigenous sovereignty and its failure lies in governance and leadership structures within educational systems. Nevertheless, in most countries, the number of Indigenous people in administrative bodies, including ministries of education, curriculum boards, and university senates, is small (Brayboy et al., 2014). This lack of Indigenous voices in policy and leadership means that existing inequities within the system remain centralized, because decisions about what should be included in the curriculum, who should be recruited as teachers, and whether schools should be funded are frequently made without consultation. This exclusion undermines the principle of self-determination and perpetuates hierarchical, rather than relational, forms of governance.

The openings of leadership can also be observed in the Indigenous institutions themselves, where the community-based schools and colleges often fail in their capacity-building and leadership training as well as administrative support (Barnhardt and Kawagley, 2008). Most of the leaders among the Indigenous people have to strike a balance between adherence to the standards imposed by the state and cultural integrity. This working relationship puts a lot of strain on sustaining the institutions. Additionally, failure to provide professional development opportunities for Indigenous teachers in structured ways obstructs succession plans and undermines long-term leadership pipelines (Kirkness and Barnhardt, 1991).

On a more macro-level, the issue of governance also overlaps with colonial land dispossession and cultural repression. As educational autonomy is directly influenced by sovereignty over territory, the disruption of Indigenous education is directly caused by violations of territorial sovereignty (Simpson, 2017). Educational sovereignty is limited without authority over ancestral lands and local governmental bodies. Thus, to redefine leadership in Indigenous education, systemic decolonization, i.e., the transition from the top-down administrative systems to the community-based and relational governance systems based on the ideas of trust, collective responsibility, and land-based authority, is necessary. The social fabric of educational reform can be further enhanced through collaboration-based leadership in which the voices of Indigenous women, elders, and youth are valued, ensuring continuity across generations.

Altogether, the process of educational sovereignty is multifaceted, characterized by policy limitations, hierarchies of knowledge, and a lack of governance (Table 3). Solving these issues will require not only legislative change but also epistemological redefinition: education as a process of relational renewal rather than institutional compliance.

Table 3. Key Barriers to Educational Sovereignty

Barrier	Description	Implications	Example
Policy & Funding	Fragmented policies, low funding	Limits curriculum autonomy, infrastructure development	Australia's "Closing the Gap" strategy
Epistemic Tokenism	Superficial inclusion of Indigenous knowledge	Reduces knowledge to cultural artifacts	Westernized environmental education with minor Indigenous content
Cultural Misappropriation	Misuse of symbols, stories, and ecological concepts	Loss of cultural integrity undermines sovereignty	Adoption of Māori metaphors without consultation
Governance & Leadership Gaps	Lack of Indigenous representation in leadership	Weakens decision-making, succession planning, and accountability	Limited Indigenous voices in curriculum boards

6. PATHWAYS FOR TRANSFORMATIVE REFORM

The Community-Led Governance Models

Community-led governance models are key to transformative reform in Indigenous and place-based education. Top-down mandates will never bring about educational sovereignty; instead, they require governance arrangements that place Indigenous communities at the center as key decision-makers. Community-based governance aligns curricula, pedagogy, and administrative structures with local epistemology, cultural practices, and ecological knowledge (Barnhardt and Kawagley, 2008). The Kura Kaupapa Maori schools in Aotearoa/New Zealand exemplify such an approach, with school boards composed of elders, parents, and local knowledge holders responsible for language immersion, cultural programming, and land-based learning (Bishop, 2005). Such boards are custodians of knowledge, and their roles entail that their educational goals are based on community priorities, while at the same time they are subject to cultural expectations.

Equally, the case of Tribal Colleges and Universities (TCUs) in North America, which illustrate that community-based governance increases institutional resilience and relevance, shows that the same applies to other institutions. TCUs have frequently been chartered to give local tribes control over the curriculum design, faculty recruitment, and student services (Stein, 2021). These arrangements not only improve educational outcomes but also facilitate intergenerational knowledge transfer, helping to preserve language, traditions, and environmental stewardship practices. Comparative analyses demonstrate that community-based governance frameworks are always associated with increased prevalence of student interest, cultural validation, and leadership growth as compared to traditional, centralized systems of education (Shotton et al., 2023).

The main aspects of effective governance systems are participatory decision-making, open accountability, and sustained engagement with elders and knowledge holders (Pidgeon, 2016). The practices enable responsive, culturally based education that is adaptive to ecological and social issues. Moreover, community-based governance offers a framework for bargaining state assistance and federal funding while preserving Indigenous epistemic power, thereby eliminating the threat of tokenistic assimilation or co-optation.

Digital Sovereignty and Emerging Innovations

In the digital age, there are unprecedented opportunities to expand the reach of Indigenous education and enhance its effectiveness, but questions of digital sovereignty also arise. Digital sovereignty is the ability of communities to exercise control over information, digital platforms, and online pedagogical tools in ways that do not violate cultural practices and knowledge property. Interactive e-learning platforms, virtual reality, and mobile applications are emerging technologies that can be used to document and transmit Indigenous knowledge, support multilingual instruction, and establish connections among learners across geographic and political borders.

An example is the digital storytelling projects employed in the preservation and pedagogy of traditional ecological knowledge that connects local histories and sustainability practices (Wilson, 2008). Equally, mobile applications for language revitalization can enable students to interact with ancestral languages and, at the same time, help keep oral traditions alive in digital form (Hornberger, 2014). Notably, innovations should be developed in accordance with the principles of Indigenous data governance to ensure they cannot be misused, protect privacy, and respect the right to community decision-making (Barnhardt, & Kawagley, 2008).

Digital sovereignty also plays a key role in the resilience of education systems in the face of ecological crisis. Online learning environments and other collaborative web-based learning formats are capable of maintaining learning through natural disasters, pandemics, or limited resources, and maintain Indigenous pedagogical strategies. By integrating digital with place-based learning, hybrid approaches are possible in which experiential and land-based learning is integrated with interactive and adaptive digital learning materials (Parsons et al., 2021). This kind of integration is more ecologically literate, facilitates cross-communal cooperation, and develops learners' sustainability-focused action competence.

Table 4. Pathways for Transformative Reform

Pathway	Description	Potential Outcomes
Community-led Governance	Participatory school boards, elders' councils	Cultural relevance, accountability, and enhanced engagement
Digital Sovereignty	Control over digital platforms, e-learning, storytelling	Knowledge preservation, intergenerational learning, and wider access
Policy & Capacity Building	Equitable funding, professional development, and legal frameworks	Institutional sustainability, teacher preparedness, and curricular autonomy
Leadership Development	Mentorship, relational governance training	Resilient institutions, succession planning, collaborative problem-solving

Policy Recommendations and Capacity-Building Strategies

To enact transformative reform, it is necessary to develop holistic policy frameworks that promote Indigenous autonomy, ensure sustainable funding, and institutionalize culturally responsive pedagogy. The policy prescriptions should focus on the systematic reformation as well as the empowerment of the community. To begin with, the governments are encouraged to enact legislative acts that acknowledge Indigenous education as a separate constitutionally-affirming field, which offers legal safeguards of the language, pedagogy, and land-based curriculum (Carney, 2022). Second, policies must require a fair funding formula that takes into consideration geographical remoteness, lack of infrastructure, and culturally unique resource requirements, thus facilitating long-term institutional consistency.

Capacity-building is also necessary to perpetuate transformative reform. Educators should be equipped with Indigenous pedagogies, ecological literacy, and culturally responsive practices through teacher education programs that address complex socio-ecological realities (Higgins, 2026). Administrator leadership development programs need to focus on relational governance, participative decision-making, and strategic planning that is consistent with the community prioritization and sustainability (Kirkness and Barnhardt, 1991). Capacity-building can also be encouraged with the help of funding agencies and academic institutions, which can promote mentorship opportunities, exchange between institutions, or research projects based on ethical and communal practices.

Policy frameworks should also focus on epistemic equity by establishing mechanisms that will establish the validity of Indigenous knowledge, its authority, and actionability. This involves the incorporation of Indigenous research methodology in national standards of education assessment and advocacy of the importance of community-defined learning outcomes alongside conventional measures (Battiste, 2019). There will also be international cooperation, which will be based on the principles of mutual respect and cultural integrity, and help to exchange knowledge without extracting epistemology (Santos, 2015).

Lastly, monitoring and evaluation based on culturally sensitive and ecologically informed strategies should be combined with capacity building and policy reform. Participatory evaluation models, in which elders, educators, students, and community members co-assess outcomes, lead to sustained improvement and enhanced accountability and cultural relevance (Barnhardt and Kawagley, 2008). Indigenization of every step of governance and digital innovation, as well as policy implementation, would enable transformative reform at multiple scales, with changes that simultaneously focus on social justice, ecological sustainability, and educational sovereignty.

Finally, the avenues to transformative reform of Indigenous and place-based education demand a synergistic strategy that entails community-based governance, digital sovereignty, and capacity building informed by these policies. Models of community-led governance demonstrate that educational decision-making can be grounded in cultural values and ecological responsibility. When guided by the principles of digital sovereignty, digital innovations benefit access, continuity, and knowledge preservation. The structural support, resource equity, and professional development opportunities are provided through policy frameworks and capacity-building strategies that sustain these reforms over the long term. Together, these pathways help make education a means of achieving social justice, eco-stewardship, and Indigenous self-determination, thereby promoting global sustainability agendas and the realization of the United Nations Sustainable Development Goals (SDGs 4, 10, 13, 15).

7. FUTURE DIRECTIONS

The changing nature of Indigenous and place-based education indicates the need to develop proactive strategies to enhance sustainability, social justice, and epistemic sovereignty. The key elements of transdisciplinary integration and the adoption of Indigenous knowledge systems, which combine with ecological science, sustainability studies, and digital innovation, should assume a primary role in future research and practice (Bang et al., 2013). These strategies acknowledge that no single knowledge domain can address complex environmental and social issues, such as climate change, biodiversity loss, and social imbalances. The integration of Indigenous epistemologies and modern scientific approaches to education may produce students who are adaptive and context-specific in problem-solving, grounded in cultural awareness and ecological literacy (Kimmerer, 2013).

Emerging technologies create additional avenues for expansion. Indigenous knowledge can be accessed globally and across time through virtual reality, immersive digital platforms, mobile learning applications, and oral traditions for intergenerational learning (Wilson, 2008). Notably, these innovations should operate within the principles of digital sovereignty, in which communities retain control over their cultural and intellectual property and pedagogical content. Future projects must also prioritize the development of digital curricula in collaboration with Indigenous stakeholders and incorporate storytelling, place-based knowledge, and ecological ethics to strengthen learning outcomes and cultural integrity.

Another important direction is to align policies with global sustainability agendas. The Sustainable Development Goals (SDGs) can be achieved through indigenous education, especially SDG 4 (Quality Education), SDG 10 (Reduced Inequalities), SDG 13 (Climate Action), and SDG 15 (Life on Land) (UNESCO, 2020). Future policies must consolidate structures to entrench Indigenous epistemologies within formal curricula, teacher education, and school governance, and ensure equal funding, professional development, and community involvement. By integrating the values of cultural and ecological care into policy structures, education can be seen as a policy instrument for local empowerment and global sustainability.

Lastly, the most important aspect is the creation of resilient leadership models. Curriculum design, institutional management, and community engagement should be spearheaded by indigenous educators, elders, and youth. The theories of decolonialism, ecological literacy, and participatory governance must be incorporated into leadership programs, and the teaching profession must be prepared to address socio-ecological complexities in a culturally sensitive and strategic manner (Barnhardt and Kawagley, 2008; Brayboy et al., 2014). Regional and cross-national leadership networks can be used to share knowledge, mentor, and develop common solutions, thereby strengthening the global shift towards Indigenous educational sovereignty. Fig. 3 presents a conceptual map highlighting transdisciplinary knowledge, digital sovereignty, policy alignment with global sustainability goals, and resilient leadership development as paramount pathways for transforming Indigenous education.

8. CONCLUSION

The study on Indigenous and place-based education underscores the undeniable connection among cultural self-determination, ecological stewardship, and social justice. Indigenous pedagogies are more than a cultural supplement, as evidenced by the experience of Māori immersion schools and Tribal Colleges and Universities, as well as community-based programs in India and Latin America, which show that Indigenous pedagogies are entire epistemic systems capable of solving environmental and social issues and providing a sense of identity, resilience, and community integration (Berkes, 2017; Macfarlane et al., 2014). Challenges remain, including policy fragmentation, epistemic tokenism, inequitable funding, and gaps in governance. To overcome such barriers, systemic changes are needed: participatory governing models, an equal policy and funding system, culturally responsive pedagogy, and leadership development that places Indigenous knowledge and needs at its core (Santos, 2015; Pidgeon, 2016). Moreover, the use of digital technologies, grounded in digital sovereignty, introduces new opportunities for preserving knowledge, intergenerational learning, and cooperation with other countries (Wilson, 2008).

The avenues of transformative reform described in this chapter focus on the governance synergy, pedagogy, policy, and technology. Education can serve as a means of decolonization, social justice, and ecological responsibility by enhancing community-based decision-making, instilling ecological literacy and sustainability in the curriculum, and cultivating culturally grounded leadership. Notably, such reforms align with international sustainability objectives, indicating that Indigenous education is not solely locally transformative but also globally applicable to addressing complex socio-ecological problems.

As practitioners, this chapter can help them recognize the necessity of co-designing curricula with Indigenous communities, focusing on land-based and culture-grounded learning, and using digital tools to ensure that knowledge is transmitted and that sustainability actions are taken. Educators should be equipped with culturally responsive, ecologically literate, and participatory governance skills during teacher training. Leadership development efforts should prioritize Indigenous voices and employ strong, collaborative, and adaptive strategies.

For researchers, future research should focus on similar studies that explore the longitudinal outcomes of Indigenous-led education models, the effects of digital innovations on knowledge retention and ecological behavior, and comparative studies of Indigenous settings worldwide. Indigenous epistemologies and participation protocols must be used alongside participatory and decolonial methodologies to ensure that research processes are respectful of Indigenous epistemologies and empower local communities. After all, the convergence of Indigenous knowledge, sustainability education, and social justice provides a blueprint for transformative educational reform. Education can serve as an effective tool for transformation by prioritizing epistemic sovereignty, ecological responsibility, and community empowerment, ensuring that future generations possess the knowledge, values, and abilities to navigate an increasingly uncertain world.

REFERENCES

Ahenakew, C. (2016). Grafting Indigenous Ways of Knowing Onto Non-Indigenous Ways of Being: The (Underestimated) Challenges of a Decolonial Imagination. *International Journal of Qualitative Research, 9*(3), 323–340. https://www.jstor.org/stable/26372211. DOI: 10.1525/irqr.2016.9.3.323

Alam, M. (2025). Beyond Eurocentrism: Toward a philosophy of indigenous Fijian knowledge. *Cogent Arts & Humanities, 12*(1), 2588469. Advance online publication. DOI: 10.1080/23311983.2025.2588469

Arias-Gutierrez, R. I., & Minoia, P. (2023). Decoloniality and Critical Interculturality in Higher Education: Experiences and Challenges in Ecuadorian Amazonia. *Forum for Development Studies, 50*(1), 11–34. DOI: 10.1080/08039410.2023.2177562

Bang, M., Warren, B., Rosebery, A., & Medin, D. L. (2013). Desettling Expectations in Science Education. *Human Development, 55*(5-6), 302–318. DOI: 10.1159/000345322

Barnhardt, R., & Kawagley, A. O. (2008). Indigenous knowledge systems and education. *Teachers College Record, 110*(13), 223–241. DOI: 10.1177/016146810811001318

Battiste, M. (2019). *Decolonizing Education: Nourishing the Learning Spirit*. University of British Columbia Press., DOI: 10.59962/9781895830972

Berkes, F. (2017). *Sacred Ecology* (4th ed.). Routledge., DOI: 10.4324/9781315114644

Bishop, R. (2005). Changing Power Relations in Education: Kaupapa Mäori Messages for "Mainstream" Education in Aotearoa/New Zealand. *Counterpoints, 275*, 253–269. https://www.jstor.org/stable/42978789

Brayboy, B. M. J., Solyom, J. A., & Castagno, A. E. (2014). Looking into the hearts of Native peoples: Nation building as an institutional orientation for graduate education. *American Journal of Education, 120*(4), 575–596. DOI: 10.1086/676908

Brown, P. R. (2019). Understanding how deficit discourses work against implementing participatory approaches in Australian Indigenous policy. *The Australian Journal of Social Issues, 54*(4), 401–417. DOI: 10.1002/ajs4.78

Cajete, G. A. (2016). Indigenous education and the development of indigenous community leaders. *Leadership, 12*(3), 364–376. DOI: 10.1177/1742715015610412

Carney, S. (2022). Reimagining our futures together: A new social contract for education. *Comparative Education, 58*(4), 568–569. DOI: 10.1080/03050068.2022.2102326

Corntassel, J. (2008). Toward sustainable self-determination: Rethinking the contemporary Indigenous-rights discourse. *Alternatives*, *33*(1), 105–132. DOI: 10.1177/030437540803300106

Govinda, R. (Ed.). (2002). *India education report: A profile of basic education*. Oxford University Press.

Greenwood, J., & Te Aika, L. H. (2010). *Hei tauira: Teaching and learning for success for Māori in tertiary settings*. Ako Aotearoa – National Centre for Tertiary Teaching Excellence.

Gunawan, K. D. H., & Indrawan, I. P. O. (2025). Decolonising Science Education: A Bibliometric Analysis of Indigenous Knowledge Integration in Global STEM. *Journal of Posthumanism*, *5*(5), 2598–2622. DOI: 10.63332/joph.v5i5.1651

Hemara, W. (2000). *Māori pedagogies: A view from the literature*. New Zealand Council for Educational Research.

Higgins, B. J. (2026). Learning From Faculty Before Implementing a Culturally Responsive Teaching Program at a Community College. *New Directions for Community Colleges*, *2026*(213), 71–77. DOI: 10.1002/cc.70049

Hornberger, N. H. (2014). Bringing our languages home: Language revitalization for families. Ed. by Hinton Leanne. Berkeley, CA: Heyday Books, 2013. Pp. xx, 264. ISBN 9781597142007. $20. *Language*, *90*(2), 540–543. https://www.jstor.org/stable/24672189. DOI: 10.1353/lan.2014.0033

Kalidas, A. J. E. (2025). Challenges in implementing a culturally sustaining pedagogy in Indigenous formal education. *Journal of Indigenous Social Development*, *13*(3), 141–174. DOI: 10.55016/ojs/jisd.v13i3.81550

Kimmerer, R. W. (2013). *Braiding Sweetgrass: Indigenous Wisdom, Scientific Knowledge and the Teachings of Plants* (1st ed.). Milkweed Editions.

Kirkness, V. J., & Barnhardt, R. (1991). First Nations and Higher Education: The Four R's—Respect, Relevance, Reciprocity, Responsibility. *Journal of American Indian Education*, *30*, 1–15.

Macfarlane, A.H., Webber, M., Cookson-Cox, C., & McRae, H.S. (2014). Ka Awatea: An iwi case study of Māori students' success.

Parsons, M., Fisher, K., & Crease, R. P. (2021). Environmental Justice and Indigenous Environmental Justice. *Decolonising Blue Spaces in the Anthropocene: Freshwater management in Aotearoa New Zealand*, 39–73. https://doi.org/DOI: 10.1007/978-3-030-61071-5_2

Pidgeon, M. (2016). More than a checklist: Meaningful Indigenous inclusion in higher education. *Social Inclusion (Lisboa)*, *4*(1), 77–91. DOI: 10.17645/si.v4i1.436

Reid, A. J., Eckert, L. E., Lane, J. F., Young, N., Hinch, S. G., Darimont, C. T., Cooke, S. J., Ban, N. C., & Marshall, A. (2021). "Two-Eyed Seeing": An Indigenous framework to transform fisheries research and management. *Fish and Fisheries*, *22*(2), 243–261. DOI: 10.1111/faf.12516

Shotton, H. J., Lowe, S. C., & Waterman, S. J. (2023). *Beyond the asterisk: Understanding Native students in higher education*. Taylor & Francis.

Simpson, L. B. (2017). *As we have always done: Indigenous freedom through radical resistance*. U of Minnesota Press. DOI: 10.5749/j.ctt1pwt77c

Stein, S. (2021). Critical Internationalization Studies at an Impasse: Making Space for Complexity, Uncertainty, and Complicity in a Time of Global Challenges. *Studies in Higher Education*, *46*(9), 1771–1784. DOI: 10.1080/03075079.2019.1704722

Sund, L., & Pashby, K. (2020). Delinking global issues in northern Europe classrooms. *The Journal of Environmental Education*, *51*(2), 156–170. DOI: 10.1080/00958964.2020.1726264

Truth and Reconciliation Commission of Canada. (2015). *Honouring the truth, reconciling for the future: Summary of the final report of the Truth and Reconciliation Commission of Canada*.

UNESCO. (2014). *Shaping the future we want: UN decade of education for sustainable development (2005–2014) final report*. United Nations Educational, Scientific and Cultural Organization. https://unesdoc.unesco.org/ark:/48223/pf0000230171

UNESCO. (2020). Education for sustainable development: A roadmap. https://doi.org/.DOI: 10.54675/YFRE1448

United Nations. (2007). United Nations declaration on the rights of Indigenous peoples. United Nations.https://www.un.org/development/desa/indigenouspeoples/wpcontent/uploads/sites/19/2018/11/UNDRIP_E_web.pdf

United Nations General Assembly. (2007). United Nations Declaration on the Rights of Indigenous Peoples (A/RES/61/295). United Nations. https://undocs.org/A/RES/61/295

Whyte, K. P. (2018). Indigenous science (fiction) for the Anthropocene: Ancestral dystopias and fantasies of climate change crises. *Environment and Planning. E, Nature and Space*, *1*(1-2), 224–242. DOI: 10.1177/2514848618777621

Wilson, S. (2008). *Research is ceremony: Indigenous research methods.* Fernwood Publishing.

Chapter 7
Neurodiversity in Inclusive Education: Supporting Twice-Exceptional Learners

Ashraf Moustafa
http://orcid.org/0000-0001-6730-4680
United Arab Emirates University, UAE

Mohammad Al Rashaida
http://orcid.org/0000-0001-6354-1187
United Arab Emirates University, UAE

Al Reem Al Yaqoubi
United Arab Emirates University, UAE

ABSTRACT

Twice exceptional (2e) learners' students who combine high intellectual or creative potential with learning or developmental challenges represent an underserved group and a key example of sustainable, equitable inclusion. This chapter explores how schools can shift from theoretical equity to evidence-based practice. In alignment with UN SDG 4, it argues that sustainable inclusion must recognize full cognitive diversity, especially learners whose profiles defy traditional categories. Due to the masking effect, many 2e students remain unidentified, leading to academic, social, and emotional challenges. Drawing on research in gifted and special education, the chapter highlights differentiated instruction as an effective approach. It integrates major developmental and talent models to explain the cognitive, emotional, and ecological dimensions of 2e learners and concludes with recommendations showing that embracing cognitive diversity is essential for sustainable human development.

DOI: 10.4018/979-8-3373-7650-9.ch007

1. INTRODUCTION

The global commitment to Education for Sustainable Development (ESD) calls upon school systems to reimagine not only what is taught, but who is served and how. Sustainable education, in its fullest sense, must be equitable, inclusive, and responsive to the full spectrum of human diversity. It is against this backdrop that the case of twice-exceptional (2e) learners becomes both urgent and illuminating. In the contemporary landscape of inclusive education, the population of twice-exceptional (2e) learners represents one of the most significant challenges to traditional pedagogical silos. Defined by the co-occurrence of high cognitive potential or giftedness with one or more neurodevelopmental disabilities including, but not limited to, Specific Learning Disabilities (SLD), Attention-Deficit/Hyperactivity Disorder (ADHD), and Autism Spectrum Disorder (ASD) 2e learners embody a cognitive profile characterized by extreme internal variance (Foley-Nicpon & Assouline, 2020). This duality creates a paradoxical dynamic where extraordinary intellectual strengths frequently mask significant functional challenges, while areas of struggle can obscure underlying giftedness a phenomenon widely recognized as the "masking effect" (Baldwin et al., 2015; Besnoy, 2018). Within this framework, the student exists in a state of perpetual compensation, using high-level cognitive reserves to navigate environments that were not designed for their unique neurological architecture.

Central to understanding the 2e experience is the concept of developmental asynchrony. Unlike neurotypical gifted students whose cognitive and social-emotional development often track in parallel, 2e learners exhibit a "jagged" developmental profile. Asynchronous development refers to the lack of synchronicity between chronological age, mental age, and physical/emotional maturity (Ronksley-Pavia, 2015). A 2e student may possess the abstract reasoning capabilities of a college student while demonstrating the fine motor control or social-emotional regulation of a child half their age. This misalignment results in a systemic failure of standard inclusive classrooms, which are often designed for a "middle-of-the-road" learner, leaving 2e students in a diagnostic and instructional "no-man's land" where their potential is ignored because their struggles are prioritized, or their struggles are dismissed because their potential is so evident (Reis et al., 2014).

This failure of identification is frequently rooted in the rigid limitations of psychometric testing. In many educational jurisdictions, identification for gifted services relies heavily on high Full Scale Intelligence Quotient (FSIQ) scores. However, the FSIQ is often an invalid measure for 2e populations because it aggregates "peaks" in fluid reasoning and verbal comprehension with "valleys" in processing speed and working memory areas where 2e students typically struggle due to their specific disability (Assouline & Whiteman, 2011). Consequently, their overall scores frequently fall into the “average” range, a statistical artifact that hides their urgent need for both high-level enrichment and specialized support. This “invisible underperformance” creates a “double disadvantage”: they are deemed “not disabled enough” for special education services and “not high-achieving enough” for gifted programming (Foley-Nicpon & Assouline, 2020).

The consequences of this continued neglect are both far-reaching and profound. Without tailored educational interventions that address their dual needs, 2e learners face heightened risks of chronic academic underachievement, debilitating anxiety, and a fractured sense of self-concept (Foley-Nicpon, 2018). From a socio-economic perspective, failing to cultivate the talents of 2e individuals represents a significant loss of human capital. These learners often possess the exact traits required for 21st-century innovation: divergent thinking, high-level pattern recognition, and unconventional problem-solving abilities that allow them to see connections where others see clutter (Subotnik et al., 2011).

Addressing this paradox requires a fundamental shift in pedagogical philosophy. It necessitates moving beyond the fragmented, deficit-based paradigms of traditional special education toward an integrated, neurodiversity-affirming model. This chapter examines the critical imperative of supporting 2e learners by exploring the role of dually differentiated instruction which provides simultaneous challenge and support analyzing the institutional barriers that perpetuate their invisibility, and proposing evidence-based recommendations to unlock their full potential as both learners and innovators.

2. EDUCATIONAL IMPORTANCE

Supporting twice-exceptional (2e) learners is fundamentally a matter of educational justice and equity. The persistent failure to accurately identify and appropriately serve these students represents a direct contradiction to the foundational promise of inclusive education: that every learner, regardless of their unique profile of strengths and challenges, deserves the opportunity to thrive in an environment that recognizes and cultivates their potential (Hughes, 2021). For 2e students, however, the rhetoric of inclusion often fails to translate into meaningful practice. Their paradoxical profile places them in a systemic blind spot, where neither gifted education nor special education fully claims them as their own. This neglect is not benign; it actively compounds their disadvantages and forecloses opportunities for academic, social, and emotional flourishing.

The imperative to address this gap is both ethical and pragmatic. Ethically, educational systems are obligated under frameworks such as the Individuals with Disabilities Education Act (IDEA) and the United Nations Convention on the Rights of Persons with Disabilities (CRPD) to provide Free Appropriate Public Education (FAPE) in the least restrictive environment. When 2e students are denied access to gifted programming due to their disability, or denied special education services because their giftedness masks their struggles, they are effectively being denied this fundamental right (Foley-Nicpon & Assouline, 2020). Pragmatically, the failure to cultivate the talents of 2e individuals represents a significant loss of human capital and innovation potential for society at large. When investing in targeted, dual-focused support systems, educational institutions fulfill their legal and moral mandates while simultaneously advancing a more equitable and effective model of education for all learners.

2.1 The High Costs of Misidentification and Neglect

The consequences of failing to identify and appropriately support twice-exceptional learners are severe, multifaceted, and well-documented across decades of research. These costs manifest across academic, psychosocial, and long-term life outcome domains, creating a cascade of disadvantage that can persist well into adulthood.

Invisible Underperformance and the Masking Trap. One of the most insidious consequences of the 2e profile is the phenomenon of "invisible underperformance." Because 2e students often possess superior cognitive abilities particularly in areas such as verbal comprehension, abstract reasoning, and problem-solving they frequently develop sophisticated compensatory strategies to navigate tasks that are compromised by their disability (Reis et al., 2014). A student with dyslexia and superior verbal ability, for example, may rely heavily on context clues and memory to comprehend text, achieving reading scores that appear "average" or even "above average" despite significant underlying deficits in phonological processing. This compensation, however, is cognitively exhausting and unsustainable as academic demands intensify. Moreover, it actively masks the need for both remediation and enrichment. The student appears competent, often leading teachers to conclude that no intervention is necessary. Consequently, the student is simultaneously excluded from gifted programming because their performance does not reflect their true intellectual capacity and denied special education services because their struggles are not readily visible. This "double exclusion" leaves the student stranded in a curricular no-man's land, perpetually under-stimulated and under-supported (Foley-Nicpon & Assouline, 2020). Legally, this constitutes a denial of Free Appropriate Public Education (FAPE), as the student's individualized needs are not being met (Assouline & Whiteman, 2011).

Psychosocial and Emotional Risks: The Toll of Being "Smart but Struggling." The internal dissonance experienced by twice-exceptional learners is profound and psychologically damaging. These students live with a persistent contradiction: they are acutely aware of their intellectual capabilities, often scoring in the superior range on measures of reasoning and verbal ability, yet they struggle with tasks that peers seem to accomplish with ease. This disconnect between perceived ability and actual performance generates intense frustration, confusion, and shame. Meta-analytic research by Townend and Brown (2016) found that 2e children exhibit significantly lower global self-concept and higher rates of internalized distress compared to both their gifted-only and disability-only peers. They are not simply "gifted students with a problem"; their dual identity creates a unique psychological burden.

This burden frequently manifests as maladaptive perfectionism. Unlike healthy perfectionism, which drives high achievement, maladaptive perfectionism in 2e learners is characterized by an intense fear of failure, excessive self-criticism, and avoidance of challenging tasks (Foley-Nicpon, 2018). A 2e student may refuse to attempt an assignment unless they are certain they can complete it perfectly, leading to incomplete work and further academic decline. Conversely, some 2e students develop task avoidance behaviors, procrastinating or disengaging entirely to protect themselves from the humiliation of failure. Anxiety disorders, particularly generalized anxiety and social anxiety, are disproportionately prevalent in this population, as is depression (Foley-Nicpon & Assouline, 2020). The emotional toll is compounded by social isolation; many 2e students report feeling that they do not fully belong with either gifted peers (who do not share their struggles) or peers with similar disabilities (who do not share their intellectual interests). This "double alienation" can lead to profound loneliness and withdrawal (Baldwin et al., 2015).

School Disengagement and Dropout. The cumulative weight of academic frustration, emotional distress, and social isolation places twice-exceptional learners at a disproportionately high risk for school disengagement and dropout. Longitudinal research by McCoach and Siegle (2003) identified several key factors differentiating achieving from underachieving gifted students, among which a mismatch between student ability and curriculum demands was paramount. For 2e learners, this mismatch is twofold: the curriculum fails to challenge their intellect while simultaneously failing to accommodate their disability. Over time, motivation erodes. School becomes a site of repeated failure and humiliation rather than growth and discovery. Behavioral issues may emerge as students act out their frustration or seek to escape aversive academic demands. Truancy, suspension, and eventual dropout are far more common among unidentified or underserved 2e students than among their well-supported peers (Reis et al., 2014). The long-term consequences of dropout are well-established: reduced employment opportunities, lower lifetime earnings, increased reliance on social services, and diminished quality of life. The failure to support 2e learners is thus not only an educational failure but a social and economic one as well.

2.2 Twice-Exceptional Learners and Sustainable Inclusion in the UAE Context

In the United Arab Emirates, the national commitment to inclusive and equitable education has expanded significantly over the past decade, particularly through the implementation of the National Policy for Empowering People of Determination (UAE Ministry of Community Development, 2017) and the School for All framework issued by the Ministry of Education (MOE, 2020). Within this evolving landscape, twice-exceptional (2e) learners' students who demonstrate both high cognitive potential and co-occurring learning or developmental challenges represent a uniquely important group whose needs intersect directly with the UAE's vision for sustainable human development. The neurodiversity paradigm, which recognizes neurological differences as natural variations of the human brain, aligns closely with the UAE's shift toward strength-based, person-centered educational models. Supporting 2e learners is therefore not only a matter of individualized intervention but a broader commitment to cultivating diverse forms of talent that contribute to national innovation, social cohesion, and long-term sustainability.

Despite these policy advances, research and school-level reports indicate that 2e learners in the UAE often remain under identified or misidentified due to the masking effect, limited teacher preparation, and the continued reliance on standardized psychometric tools that do not capture asynchronous development (Gaad, 2015; Al-Naimi & Khamis, 2021). Many students of determination with high potential are placed exclusively in support programs without access to enrichment, while gifted students with subtle learning challenges are overlooked for accommodations. This gap highlights the need for educational models that recognize neurodiversity as a resource and position twice-exceptionality within a sustainable, future-oriented vision of inclusive schooling.

2.3 Place-Responsive Pedagogy as a Sustainable Model for Supporting 2e Learners in the UAE

A place-responsive approach rooted in the UAE's cultural, social, and environmental context offers a powerful framework for supporting twice-exceptional learners in sustainable ways. Unlike standardized international models that rely heavily on classroom-based differentiation, place-responsive pedagogy draws on the UAE's rich heritage of experiential learning, community engagement, and environmental stewardship. Learning experiences connected to local culture, desert ecology, maritime heritage, and community traditions can provide grounding, sensory regulation, and authentic engagement for 2e learners, who often thrive in hands-on, real-world environments. Such approaches align with neurodiversity-affirming practices by reducing cognitive overload, supporting executive functioning, and enabling students to activate their strengths such as creativity, problem-solving, and divergent thinking within meaningful contexts (Ridge et al., 2020).

Moreover, the UAE's emphasis on sustainability, as articulated in UAE Vision 2030 and Education for Sustainability initiatives, positions inclusive education as a cornerstone of national development. When schools adopt place-responsive, culturally grounded strategies, they create learning ecosystems that honor cognitive diversity and promote long-term well-being. For twice-exceptional learners, this means environments where their talents are nurtured, their challenges are supported, and their identities are affirmed as contributors to the UAE's knowledge-based future. Embedding 2e support within sustainable educational frameworks ensures that inclusion is not merely a policy requirement but a lived practice that strengthens community resilience and human capital

3. THEORETICAL FRAMEWORKS FOR UNDERSTANDING TWICE-EXCEPTIONALITY

A robust theoretical foundation is essential for comprehending the unique developmental trajectories and educational needs of twice-exceptional learners. While the practical applications of differentiated instruction are well-documented in the literature, the underlying theoretical mechanisms that explain why twice-exceptional students learn and develop differently require deeper examination. This section presents four complementary theoretical perspectives that illuminate the cognitive, affective, and ecological dimensions of twice-exceptionality, each offering unique insights for researchers and practitioners.

3.1 The Talent Development Megamodel

The Talent Development Megamodel, proposed by Subotnik, Olszewski-Kubilius, and Worrell (2011) and subsequently refined in their later work (Subotnik, Olszewski-Kubilius, & Worrell, 2019), represents a paradigm shift in understanding giftedness and its relationship to disability. Unlike traditional static conceptions of giftedness as a fixed trait, the Talent Development Model conceptualizes giftedness as a dynamic trajectory from potential to expertise to eminent achievement, mediated by psychosocial factors and environmental opportunities.

This model is particularly valuable for understanding twice-exceptional learners for several reasons. First, it explicitly acknowledges that the path to talent development is nonlinear and may involve obstacles that require resilience and adaptation (Olszewski-Kubilius & Subotnik, 2023). For twice-exceptional students, their disabilities represent precisely such obstacles, requiring them to develop compensatory strategies and psychosocial strengths that their neurotypical gifted peers may not need. Second, the model emphasizes the critical role of opportunity and access precisely the areas where twice-exceptional students are often disadvantaged due to underidentification and exclusion from advanced programming. Third, by focusing on the transformation of domain-specific abilities into competencies and then expertise, the model provides a framework for understanding how twice-exceptional students can achieve at high levels when provided with appropriate, sustained support (Subotnik et al., 2019).

Recent applications of this model to twice-exceptional populations have highlighted the importance of "selective enrollment" strategies allowing twice-exceptional students to engage deeply with advanced content in their strength areas while receiving targeted support in areas of challenge (Olszewski-Kubilius & Subotnik, 2023). This selective approach prevents the common practice of withholding advanced opportunities until disabilities are "remediated," a practice that often leaves twice-exceptional students perpetually waiting for access to challenging content.

3.2 The Differentiated Model of Giftedness and Talent

Françoys Gagné's Differentiated Model of Giftedness and Talent (DMGT) offers another powerful lens for understanding twice-exceptionality. Originally developed in 1985 and continually refined over four decades, the DMGT makes a fundamental distinction between giftedness the possession and use of untrained natural abilities in at least one ability domain and talent the superior mastery of systematically developed competencies in at least one field of human activity (Gagné, 2021).

This distinction is particularly illuminating for twice-exceptional learners. A student may possess remarkable natural abilities (giftedness) in, for example, logical reasoning or spatial visualization, yet struggle to transform these gifts into academic talents due to the interference of a learning disability. The DMGT explicitly recognizes that the talent development process involves intrapersonal catalysts including physical and mental health factors that can either facilitate or hinder the expression of gifts as talents (Gagné, 2018). Learning disabilities, attention-deficit/hyperactivity disorder, and autism spectrum disorder function precisely as such catalysts, interacting with environmental factors and developmental processes to shape whether a student's gifts become fully realized talents.

Gagné's model also incorporates the role of chance including genetic predispositions and unexpected life events in talent development. This acknowledges the reality that for twice-exceptional learners, the interplay between their giftedness and disability involves elements beyond anyone's control, while still emphasizing the critical role of systematic, intentional support (Gagné, 2021). For educators, the

DMGT provides a vocabulary for discussing why two students with similar cognitive profiles might have vastly different academic outcomes, depending on how their disabilities interact with environmental supports and intrapersonal characteristics.

3.3 The Theory of Positive Disintegration

Dabrowski's Theory of Positive Disintegration, while developed decades ago, has experienced renewed attention in gifted and twice-exceptional scholarship for its explanatory power regarding the intense emotional experiences of many twice-exceptional learners (Mendaglio, 2021). This theory proposes that personality development occurs through a process of disintegration and reintegration, driven by innate developmental potential and stimulated by internal conflicts and external challenges.

Central to the theory is the concept of "overexcitability" a heightened physiological sensitivity to stimuli that manifests in five domains: psychomotor, sensual, intellectual, imaginational, and emotional (Dabrowski, 2016). Research has consistently documented that gifted individuals, including twice-exceptional learners, tend to exhibit higher levels of overexcitability than their neurotypical peers (Mendaglio, 2021). For twice-exceptional students, this heightened sensitivity interacts with the challenges posed by their disabilities, potentially intensifying their emotional responses to frustration, failure, and social difficulty.

Importantly, Dabrowski's theory frames this intensity not as pathology but as potential a sign of rich inner experience and capacity for deep development. This reframing is crucial for educators working with twice-exceptional students, who often present with anxiety, perfectionism, and emotional volatility (Foley-Nicpon, 2018). Understanding these characteristics through the lens of overexcitability rather than mere dysfunction allows teachers to respond with empathy and support rather than frustration or pathologization. Recent scholarship has begun exploring how Dabrowski's framework can inform counseling approaches for twice-exceptional youth, emphasizing the need to honor their intense emotional experiences while helping them develop self-regulation strategies (Mendaglio, 2021).

3.4 Dynamic Systems Theory

A more recent contribution to the theoretical understanding of twice-exceptionality comes from Dynamic Systems Theory (DST), which has been applied to neurodevelopmental conditions and giftedness in innovative ways (Steenbeek, van Geert, & van Dijk, 2022). DST conceptualizes development as emerging from the continuous, real-time interactions of multiple components within a complex system including cognitive processes, emotional states, environmental contexts, and social relationships all of which are constantly evolving and mutually influencing one another.

For twice-exceptional learners, DST offers a framework for understanding why their functioning may vary dramatically across contexts and over time. A twice-exceptional student with attention-deficit/hyperactivity disorder and giftedness in mathematics might demonstrate brilliant problem-solving in a one-on-one tutoring session but struggle to complete problems independently in a noisy classroom. From a DST perspective, this variability is not inconsistency or lack of effort, but rather a predictable outcome of the dynamic interaction between the student's neurocognitive profile and specific environmental conditions (Steenbeek et al., 2022).

This perspective has profound implications for assessment and intervention. Rather than seeking a static, context-free measure of a twice-exceptional student's abilities or challenges, DST encourages educators to examine the conditions under which the student thrives and those under which they struggle. It shifts the focus from "fixing" the student to optimizing the fit between the student and their learning environment (van Geert, 2023). This aligns well with contemporary inclusive education principles and provides a theoretical justification for the flexible, responsive practices that effective twice-exceptional education requires.

3.5 Synthesizing Theoretical Perspectives

Each of these theoretical frameworks contributes a distinct dimension to our understanding of twice-exceptionality. The Talent Development Megamodel illuminates the developmental trajectory from potential to achievement, emphasizing the role of opportunity and psychosocial factors. The Differentiated Model of Giftedness and Talent provides precise language for distinguishing natural abilities from developed competencies and for understanding how disabilities function as catalysts in the talent development process. Dabrowski's Theory of Positive Disintegration offers insight into the intense emotional lives of twice-exceptional learners, reframing sensitivity as potential rather than pathology. Dynamic Systems Theory explains the contextual variability in twice-exceptional students' functioning and points toward environmentally responsive interventions.

Together, these theories move the field beyond simplistic, categorical thinking about twice-exceptionality. They remind us that twice-exceptional learners are not simply "gifted students plus a disability" but rather complex, developing individuals whose trajectories are shaped by the ongoing interaction of their unique neurocognitive profiles with environmental opportunities, relationships, and supports across time. For educators, these theoretical perspectives provide both conceptual understanding and practical guidance for creating the conditions in which twice-exceptional learners can flourish.

These theoretical perspectives collectively demonstrate that twice-exceptional learners cannot be adequately understood through single-dimension approaches. Their development is shaped by the dynamic interplay of individual characteristics, environmental contexts, and developmental processes over time. With this theoretical foundation established, the chapter now turns to the practical role of teachers and the instructional practices that can effectively support twice-exceptional learners in inclusive classrooms.

4. NEUROBIOLOGICAL AND PSYCHOMETRIC COMPLEXITY

The twice-exceptional profile cannot be accurately understood as a simple sum of "giftedness" plus "disability." Rather, it represents a unique and dynamic cognitive architecture in which strengths and challenges interact in complex, non-additive ways. Contemporary research in developmental cognitive neuroscience has begun to illuminate the distinct neurobiological underpinnings of various 2e profiles, revealing that these learners are not merely "gifted children with problems" but individuals whose brains are organized and function differently from both typically developing peers and those with isolated giftedness or disability (Foley-Nicpon & Assouline, 2020). Understanding this complexity is essential for accurate identification, appropriate educational planning, and the cultivation of a strengths-based, neurodiversity-affirming approach to support.

Neuroimaging studies have begun to reveal distinct patterns of brain structure and function associated with specific 2e profiles. Preliminary evidence suggests that 2e individuals often exhibit high global connectivity in prefrontal cortical regions responsible for executive functions, abstract reasoning, and problem-solving, while simultaneously showing localized deficits in white matter integrity in regions subserving processing speed and automaticity (Foley-Nicpon & Assouline, 2020). This pattern of "peaks and valleys" in neural organization helps explain the characteristic psychometric scatter and paradoxical performance profiles observed in this population. The same brain that excels at complex, novel problem-solving may struggle profoundly with rapid, rote, or sequentially automated tasks. This is not a contradiction to be resolved but a fundamental feature of the 2e cognitive phenotype.

4.1 The Gifted/Learning Disabled (G/LD) Profile

Among the most extensively studied 2e profiles is the coexistence of giftedness with a specific learning disability, most commonly in reading (dyslexia) or mathematics (dyscalculia). The G/LD profile presents a striking cognitive paradox: these learners often demonstrate superior vocabulary, exceptional verbal comprehension, and sophisticated abstract reasoning abilities, yet they struggle profoundly with foundational academic skills such as phonemic decoding, spelling, or arithmetic computation (Assouline & Whiteman, 2011). This paradox has been termed the “phonological paradox” in the case of gifted/dyslexic learners.

The Phonological Paradox and Neuroconstructivist Insights. Neuroconstructivist theory provides a compelling framework for understanding this apparent contradiction. According to this model, cognitive development is not a linear, domain-general process but rather a dynamic, experience-dependent process of specialization within interconnected neural systems. In the case of gifted/dyslexic learners, functional magnetic resonance imaging (fMRI) studies have revealed a distinctive pattern of neural activation during reading tasks: the left hemisphere posterior regions typically associated with fluent word recognition (e.g., the visual word form area) show significantly reduced activation, while prefrontal regions associated with effortful, top-down reasoning and contextual prediction are hyperactivated (Foley-Nicpon & Assouline, 2020). In essence, these learners are not reading in the typical, automatic manner; they are solving reading as a problem, recruiting their superior reasoning abilities to compensate for deficient bottom-up processing. This compensatory strategy is remarkably effective in the early grades, allowing the student to achieve seemingly adequate reading comprehension. However, it is cognitively effortful and slow, and as text becomes more complex and voluminous in later grades, this compensatory system begins to break down, revealing the underlying disability that was previously masked.

Psychometric Scatter: The Signature of 2e. The cognitive heterogeneity of the G/LD profile is readily apparent on standardized intelligence batteries such as the Wechsler Intelligence Scale for Children, Fifth Edition (WISC-V). Assouline and Whiteman (2011) documented a characteristic pattern of significant intra-individual scatter, with G/LD students frequently scoring in the 95th percentile or above on indices of Verbal Comprehension and Fluid Reasoning while simultaneously scoring in the 25th percentile or below on indices of Processing Speed and Working Memory. This discrepancy often exceeding 30 standard score points is a hallmark of the 2e profile and a critical clue to its identification. Yet paradoxically, this very scatter often leads to the under-identification of 2e students when Full Scale IQ (FSIQ) is used as the sole criterion for gifted program eligibility. The FSIQ, as a composite of all subtest scores, averages the peaks and valleys together, producing a middling score that accurately captures neither the student's profound strengths nor their significant weaknesses. A student with a Verbal Comprehension Index of 145 (gifted range) and a Processing Speed Index of 80 (low average range) may achieve an FSIQ of 115 (high average), a score that disqualifies them from gifted services while failing to capture their extraordinary intellectual potential. This is the “ceiling effect” in operation, and it represents one of the most significant and preventable barriers to appropriate identification and service delivery for 2e learners (Assouline & Whiteman, 2011).

4.2 The Gifted/Autistic (G/ASD) Profile

The coexistence of giftedness and Autism Spectrum Disorder (ASD) presents a distinct and equally complex cognitive profile. G/ASD learners are often characterized by what Baron-Cohen (2002) termed a "systemizing brain" a powerful drive and capacity to analyze, understand, and predict rule-based systems, whether mechanical, numerical, natural, or abstract. This systemizing propensity can manifest as exceptional talent in domains such as mathematics, computer programming, music theory, or the memorization and categorization of factual information. However, this remarkable strength is frequently accompanied by significant challenges in domains requiring cognitive flexibility, social inference, and the integration of information across contexts.

The Executive Function (EF) Bottleneck. One of the most pervasive challenges for G/ASD learners is the presence of significant executive function deficits that operate as a bottleneck on their intellectual output. Executive functions including working memory, inhibitory control, and cognitive flexibility are the self-regulatory skills that enable goal-directed behavior, organization, and the strategic deployment of attention. Many G/ASD learners exhibit pronounced weaknesses in these areas, particularly in cognitive flexibility, which manifests as perseveration, difficulty shifting between tasks or mental sets, and resistance to changes in routine or expectation (Atmaca & Baloglu, 2022). The result is a striking dissociation between the complexity of the student's thinking and their ability to organize, plan, and execute tasks commensurate with that thinking. A student may articulate a sophisticated, multi-step plan for a science fair project but be entirely unable to initiate the first step without intensive adult scaffolding. This EF bottleneck is a primary driver of the academic underachievement observed in this population.

Weak Central Coherence and the Local/Global Processing Imbalance. A second defining feature of the G/ASD cognitive profile is a bias toward local over global processing, a phenomenon described as weak central coherence (Happé & Frith, 2006). While typically developing individuals tend to automatically process information in terms of its overall gist or meaning, individuals with ASD often show a preferential focus on local details, components, or parts. This processing style confers distinct advantages in domains requiring careful attention to detail, pattern recognition, and the detection of deviations from expected regularities. However, it also creates significant challenges in tasks requiring the integration of details into a coherent whole, the extraction of main ideas, or the generalization of skills and knowledge across different contexts. A G/ASD learner may be able to memorize every fact about the Roman Empire but struggle to write an essay synthesizing those facts into an argument about the causes of its decline. They may master a mathematical procedure in one classroom setting but fail to recognize its applicability to a superficially different problem in another context. This difficulty with generalization is a persistent challenge in inclusive settings, where the ability to transfer learning across environments is a key indicator of mastery (Hughes, 2021).

Sensory Overload and the Hidden Tax on Cognition. Inclusive classrooms, by design, bring together diverse learners in shared physical spaces. For G/ASD learners, however, these environments often present a significant and underrecognized barrier to learning: sensory overload. Hypersensitivity to auditory, visual, or tactile stimuli the hum of fluorescent lights, the visual clutter of bulletin boards, the incidental physical contact of peers in crowded hallways can consume substantial cognitive and emotional resources (Hughes, 2021). For a neurotypical learner, these background stimuli are easily filtered out and ignored. For a G/ASD learner, they may be experienced as salient, distracting, and even painful. The cognitive load imposed by sensory processing leaves fewer resources available for higher-order academic tasks, further exacerbating the gap between intellectual capacity and academic performance. Addressing this hidden tax on cognition through environmental modifications and proactive sensory supports is an essential, yet often neglected, component of inclusive practice for G/ASD learners.

5. CHALLENGES IN INCLUSIVE SETTINGS: SYSTEMIC BARRIERS

Inclusive education, at its philosophical core, aspires to create learning environments where all students, regardless of their diverse profiles, can learn together in meaningful and equitable ways. For twice-exceptional learners, however, the promise of inclusion frequently remains unrealized. This is not primarily due to a lack of teacher effort or goodwill, but rather to deeply embedded systemic barriers rooted in how schools conceptualize, categorize, and resource student need. These barriers are structural, not personal, and they require systemic, not individual, solutions.

5.1 The "Silo Effect": Fragmented Systems, Fragmented Support

One of the most formidable obstacles to effective 2e support is the entrenched administrative and philosophical separation between gifted education and special education. In most school systems, these two domains operate as distinct, parallel bureaucracies with separate funding streams, separate eligibility criteria, separate professional development pathways, and separate personnel (Foley-Nicpon & Assouline, 2020). Gifted education is typically framed as an enrichment service for students who have already demonstrated high ability, while special education is framed as a remedial service for students who exhibit documented deficits. The 2e student, who requires both enrichment and remediation simultaneously, falls into the chasm between these two silos.

This "silo effect" has profound consequences for service delivery. Funding models are often rigidly categorical, requiring that dollars be spent exclusively on either gifted programming or special education interventions. A student identified with both giftedness and a learning disability may be eligible for services from both departments, yet without intentional coordination, these services are delivered in parallel rather than in integration. The student may receive reading remediation from a special education teacher during the same block that their gifted peers are engaged in a humanities seminar, forcing a choice between addressing their disability and nurturing their talent. Too often, the remediation takes precedence, and the student's giftedness remains uncultivated. This is not inclusion; it is a fragmented, reductionist approach that fails to honor the whole child (Reis et al., 2014).

Moreover, the silo effect is reinforced by divergent philosophical paradigms. Special education has historically been grounded in a deficit-based medical model, focused on diagnosis, remediation, and the amelioration of discrete skill deficits. Gifted education, by contrast, has been grounded in a talent development model, focused on the identification and cultivation of exceptional potential. These paradigms are not inherently incompatible, but in the absence of deliberate integration, they pull practitioners in opposite directions. The special educator asks, "What is broken and how do we fix it?" The gifted educator asks, "What is exceptional and how do we amplify it?" The 2e student, who is simultaneously broken and exceptional in different domains, requires both questions to be asked in concert. Few school systems have developed the collaborative structures, shared language, and integrated programming models necessary to make this possible (Bianco & Leech, 2010).

5.2 The Identification Gap and the "Ceiling Effect"

Accurate identification is the gateway to appropriate service delivery. Yet for twice-exceptional learners, this gateway is frequently locked. The very tools and criteria used to identify both giftedness and disability are often ill-suited to capturing the complex, heterogeneous profile of the 2e student, resulting in systematic under-identification and misidentification.

The "ceiling effect" is a particularly pervasive and pernicious barrier to identification. As previously discussed, many 2e students, particularly those with G/LD profiles, exhibit significant intra-individual scatter on standardized intelligence tests. Their exceptional verbal or fluid reasoning abilities are often offset by slower processing speed or working memory weaknesses. When a Full Scale IQ (FSIQ) score is calculated as a composite of all index scores, these peaks and valleys cancel each other out, producing a middling overall score that obscures the student's true intellectual potential (Assouline & Whiteman, 2011). A student whose Verbal Comprehension Index places them in the top 1% of the population may nonetheless achieve an FSIQ of 115, a score that falls below the cutoff for gifted program eligibility in most districts. The student is thus disqualified from the very services they need to be appropriately challenged, not because they lack giftedness, but because the instrument used to measure giftedness was not designed to capture their particular cognitive architecture.

Compounding this problem is the widespread reliance on timed, speed-based assessments for both identification and progress monitoring. Processing speed deficits are among the most common co-occurring challenges in 2e populations, particularly among those with dyslexia, dysgraphia, or ADHD. When assessments are timed, these students are effectively penalized twice: first, by the difficulty of completing tasks within the allotted time, and second, by the misinterpretation of their slow performance as evidence of limited ability rather than as a manifestation of their disability. A student who can solve complex algebraic problems with 100% accuracy given sufficient time may produce a score in the 30th percentile on a timed computation test. Without contextual knowledge of the student's processing profile, this score is read as evidence of mathematical weakness, and the student is routed toward remediation rather than the advanced coursework their conceptual understanding would otherwise merit (Foley-Nicpon & Assouline, 2020).

The identification gap is further widened by the lack of universal screening protocols that are specifically designed to detect twice-exceptionality. Most school districts screen for giftedness and for learning disabilities through separate, non-integrated processes. Gifted screening typically occurs at a single point in time, often in second or third grade, and relies heavily on teacher nominations and standardized test scores. This process disproportionately overlooks students who are underachieving, who have disabilities that mask their ability, or who come from culturally and linguistically diverse backgrounds. Special education screening, meanwhile, is typically initiated in response to a documented pattern of academic failure a reactive, rather than proactive, approach. By the time a 2e student's disability is severe enough to trigger this response, years of missed opportunity and accumulated frustration have already taken their toll (Reis et al., 2014).

6. ROLE OF TEACHERS AND INSTRUCTIONAL PRACTICES

While systemic barriers are formidable, the classroom teacher remains the single most influential factor in the daily educational experience of twice-exceptional learners. Teachers are the primary architects of the learning environment, the gatekeepers of curriculum access, and the frontline identifiers of student need. Their knowledge, skills, beliefs, and practices collectively determine whether a 2e student's school experience is one of empowerment or erosion. Effective support for 2e learners therefore depends critically on the development of teacher capacity to recognize, understand, and respond to the complex and often contradictory needs of this population.

6.1 Moving Beyond "One-Size-Fits-All": The Imperative of Complex Differentiation

Traditional models of differentiation, while valuable, are often insufficient for meeting the needs of twice-exceptional learners. These students require what Tomlinson (2014) terms "complex differentiation" instruction that is simultaneously accelerated and scaffolded, enriched and remediated, challenging and supportive. This is not a matter of occasionally offering choice boards or flexible grouping; it is a fundamental orientation to teaching that recognizes learner variability as the norm rather than the exception and designs instruction proactively to accommodate a wide spectrum of strengths and needs.

Complex differentiation for 2e learners operates across Tomlinson's four domains of instructional adaptation:

Content: What students learn. For 2e learners, content must be simultaneously enriched to engage their advanced cognitive abilities and accessible to accommodate their disability-related challenges. This may involve, for example, providing a student with gifted verbal ability and dysgraphia access to college-level texts on a topic of interest (enrichment) while also providing those texts in audio format or with embedded vocabulary supports (accessibility). It requires that teachers look beyond grade-level standards to identify opportunities for depth, complexity, and novelty, while simultaneously removing barriers to content engagement.

Process: How students learn. The activities and strategies through which students make sense of content must be flexible and responsive. A 2e student with ASD and superior mathematical reasoning may benefit from explicit instruction in metacognitive strategies to monitor their comprehension and recognize when they are stuck (Morrison & Rizza, 2007). A student with ADHD and creative giftedness may need frequent movement breaks and opportunities for hands-on, kinesthetic learning. Process differentiation requires that teachers offer multiple pathways to understanding and provide strategic, temporary scaffolding that is gradually removed as the student develops independent capacity.

Product: How students demonstrate learning. Traditional assessments, particularly timed essays and standardized tests, systematically disadvantage many 2e learners. Product differentiation involves decoupling composition from transcription, allowing students to demonstrate mastery through diverse modalities that leverage their strengths and bypass their disabilities (Winebrenner, 2003). A student with dysgraphia may create a documentary film, design a website, or deliver an oral presentation with accompanying slides. A student with ASD may construct a detailed 3D model or write a computer program. These alternative products are not "easier" options; they are different pathways to demonstrating the same rigorous learning objectives.

Learning Environment: The conditions and climate for learning. For 2e learners, the physical, social, and emotional environment is not a backdrop but an active ingredient in learning. This domain encompasses everything from the sensory characteristics of the classroom (lighting, noise, visual clutter) to the psychosocial climate (psychological safety, belonging, respect for neurodiversity). A G/ASD learner may require access to noise-canceling headphones and a quiet workspace; a G/LD learner who has experienced years of academic struggle may require explicit cultivation of a growth mindset and opportunities to experience competence and success.

6.2 Universal Design for Learning (UDL): A Framework for Proactive Inclusion

Universal Design for Learning (UDL) provides a robust, research-validated framework for designing instruction that is accessible to the widest possible range of learners from the outset, minimizing the need for retroactive accommodations and modifications (CAST, 2018). UDL is organized around three core principles, each of which has particular relevance for supporting twice-exceptional learners.

Multiple Means of Engagement: The "Why" of Learning. Engagement is the affective dimension of learning the factors that recruit and sustain student interest, motivation, and persistence. For 2e learners, engagement is often contingent on the presence of meaningful challenge and personal relevance. UDL emphasizes the importance of leveraging student interests and strengths as entry points to learning. For a G/ASD learner with an intense special interest in paleontology, the study of geological time becomes not an abstract curriculum requirement but an opportunity for deep, self-directed inquiry. For a G/LD learner with a passion for hip-hop, the analysis of poetic devices can be grounded in the lyrics of their favorite artists. This is not a dilution of academic rigor; it is a strategic deployment of intrinsic motivation in service of rigorous learning goals (Marino et al., 2014).

Multiple Means of Representation: The "What" of Learning. Learners differ in the ways they perceive and comprehend information. For 2e learners, particularly those with dyslexia or language-based learning disabilities, traditional text-based instruction presents a significant barrier. UDL calls for the provision of information in flexible, multimodal formats text, audio, video, graphics, manipulatives ensuring that all learners have access to the same essential content regardless of their modality-specific challenges. A student with dyslexia should not be excluded from advanced coursework in history because they read slowly; providing the textbook in audio format removes the barrier without reducing the rigor (Marino et al., 2014).

Multiple Means of Action and Expression: The "How" of Learning. As discussed in the context of product differentiation, learners differ in their capacity to navigate learning environments and demonstrate what they know. UDL emphasizes the importance of decoupling skill demonstration from skill construction. A student may have constructed sophisticated knowledge of a scientific concept but be unable to demonstrate that knowledge through a traditional written lab report due to dysgraphia or EF deficits. Providing options for expression an oral presentation, a multimedia slideshow, a physical model allows the student to demonstrate their learning without being penalized for their disability (Winebrenner, 2003).

6.3 The Importance of Teacher Mindset and Professional Learning

Effective practice for 2e learners is not solely a matter of technical skill; it is equally a matter of teacher mindset. Research by Bianco and Leech (2010) demonstrated that teacher attitudes and beliefs significantly influence referral and identification decisions for 2e students. Teachers who hold deficit-oriented views of disability are less likely to recognize giftedness in students who also struggle, attributing their occasional insights or sophisticated questions to luck or rote memorization rather than genuine intellectual ability. Conversely, teachers who hold strengths-based, neurodiversity-affirming views are more likely to see the whole child, recognizing that the same student who cannot organize their backpack may be capable of extraordinary creative or analytical thinking.

Cultivating this mindset requires sustained, high-quality professional learning. One-time workshops and informational handouts are insufficient to shift deeply embedded beliefs or to build the complex pedagogical repertoire required for effective 2e support. Effective professional development for 2e education is ongoing, job-embedded, and collaborative. It involves not only the transmission of knowledge about 2e characteristics and identification but also opportunities for teachers to observe model practices, analyze student work, engage in co-planning with colleagues, and receive feedback on their own implementation (Bianco & Leech, 2010). Particularly powerful is blended professional development that brings together general educators, special educators, and gifted specialists to learn together, developing shared language and collaborative problem-solving skills that can bridge the silos described earlier.

7. RECOMMENDATIONS FOR EVIDENCE-BASED PRACTICE

Addressing the complex and persistent challenges in twice-exceptional education requires a coordinated, multi-level approach that spans policy, assessment, instructional practice, and professional learning. The following recommendations are grounded in the available empirical evidence and represent actionable strategies for districts, schools, and individual practitioners committed to more equitable and effective support for 2e learners.

7.1 Policy and Systems-Level Recommendations

Implement Strength-Based, Dual-Focused Individualized Education Programs (IEPs). The traditional IEP is a deficit-focused document, organized around the identification and remediation of discrete skill deficits. For 2e learners, this narrow focus is fundamentally insufficient and may even be harmful, communicating to the student that their disability is the only aspect of their learning profile worthy of systematic attention. Districts should adopt policies requiring that IEPs for identified 2e students include strength-based goals that explicitly leverage the student's giftedness to address their challenges (Reis et al., 2014). For example, a goal for a G/LD learner with advanced spatial reasoning and dysgraphia might read: "Given access to graphic design software, [student] will create visual representations of complex historical events, using this strength-based product to demonstrate understanding of causal relationships." This approach reframes the student's giftedness not as a separate domain to be addressed in isolation but as an integral component of their overall support plan.

Prioritize the General Ability Index (GAI) in Gifted Identification. The widespread reliance on Full Scale IQ (FSIQ) as the primary or sole criterion for gifted program eligibility is a significant and preventable barrier to appropriate identification of 2e learners. State and district policies should be revised to mandate the consideration of the General Ability Index (GAI) a composite score derived solely from verbal and nonverbal reasoning subtests, excluding working memory and processing speed for all students referred for gifted evaluation (Assouline & Whiteman, 2011). Furthermore, policies should explicitly prohibit the exclusion of students from gifted programming based on low FSIQ scores when GAI scores meet eligibility criteria.

This single policy change would dramatically increase identification rates for G/LD and G/ADHD learners whose cognitive strengths are obscured by processing deficits.

Mandate and Fund Blended Professional Development. The siloed structure of gifted and special education is reinforced by siloed professional development. State education agencies and districts should mandate cross-training requirements for all special educators and gifted specialists, ensuring that each group develops foundational knowledge of the other's domain (Bianco & Leech, 2010). Moreover, districts should allocate dedicated funding and protected time for collaborative planning and problem-solving between gifted and special education staff at the building level. This is not an additional responsibility to be squeezed into already overloaded schedules; it is a structural prerequisite for integrated, coherent service delivery.

7.2 Assessment and Identification Recommendations

Adopt Universal, Dynamic Screening Protocols. The current reactive model of identification waiting for students to fail before evaluating them for disabilities, or waiting for teacher nominations before screening for giftedness systematically disadvantages 2e learners. Districts should implement universal screening for both giftedness and learning challenges at multiple grade levels, using non-speed-based, culturally responsive measures that are capable of capturing high potential even in the presence of disability (Reis et al., 2014). This screening should be dynamic, incorporating ongoing data collection and analysis rather than relying on a single assessment event.

Include Multiple Measures and Multiple Informants. No single test score can capture the complexity of the 2e profile. Identification decisions should be based on a comprehensive body of evidence that includes intelligence testing, achievement testing, creativity measures, behavioral observations, work samples, and parent and teacher reports (Assouline & Whiteman, 2011). Moreover, this evidence should be gathered from multiple informants across multiple contexts. A student who appears withdrawn and disengaged in math class may be actively building complex simulations in their robotics club; a student who struggles to produce written work may be an avid and sophisticated reader at home. Identification protocols must be structured to capture this contextual variability.

7.3 Instructional and Environmental Recommendations

Design for Sensory and Executive Function Access. For many 2e learners, particularly those with ASD or ADHD, the physical and organizational environment is a primary determinant of learning access. Schools should conduct environmental audits to identify and mitigate sensory barriers excessive noise, visual clutter, harsh lighting that impede learning for hypersensitive students (Hughes, 2021). Classrooms should offer flexible seating options and quiet workspaces accessible to all students without requiring a formal accommodation plan. Executive function supports visual schedules, checklists, organizational systems, planning time should be universally available rather than reserved for students with identified disabilities.

Foster Psychological Safety and a Strengths-Based Classroom Culture. Academic engagement for 2e learners is contingent on psychological safety the belief that one can take intellectual risks, make mistakes, and be seen as competent and valued. Teachers should explicitly cultivate a classroom culture that normalizes neurodiversity and frames learning challenges as opportunities for growth rather than evidence of deficiency (Morrison & Rizza, 2007). This includes modeling a growth mindset, celebrating effort and strategy use, and publicly valuing diverse ways of thinking and problem-solving. For students who have experienced years of failure and frustration, the development of a positive academic identity is a prerequisite for, not a consequence of, academic achievement.

8. CONCLUSION

The journey of the twice-exceptional learner through the inclusive classroom is one of profound potential shadowed by systemic risk. These are students who possess the cognitive architecture to make transformative contributions to science, art, technology, and society yet whose educational trajectories are too often characterized by underidentification, underachievement, and unaddressed emotional distress. They are the students who ask questions their teachers cannot answer, who see patterns their peers cannot perceive, who imagine possibilities others cannot envision. And they are also the students who cannot organize their backpacks, who freeze when asked to read aloud, who dissolve into tears when the fluorescent lights hum too loudly. They are not two separate children sharing one desk. They are one child, whole and indivisible, whose strengths and struggles are not contradictions to be resolved but facets of a singular, complex identity.

To serve these students well, we must abandon the fragmented, siloed paradigms that have historically defined both gifted education and special education. We must recognize that the medical model of disability and the talent development model of giftedness are not incompatible but complementary, each illuminating a different dimension of the 2e experience. We must develop identification systems capable of seeing beyond the masking effect to recognize both exceptional ability and disabling challenge. We must design instructional environments that are simultaneously rigorous and supportive, challenging and accessible, standardized and flexible. We must prepare teachers not merely to accommodate 2e learners but to recognize them as a source of insight into the nature of learning itself.

Most fundamentally, we must redefine the purpose of education. The goal is not merely to remediate deficits or to celebrate isolated achievements. The goal is to provide a scaffolded, responsive, psychologically safe environment in which neurodiversity is viewed not as a problem to be solved but as a vital catalyst for innovation, creativity, and deep cognitive inquiry. When we succeed in creating such environments, we do more than support twice-exceptional learners. We demonstrate what inclusive education, at its best, can be: not a compromise between competing priorities, but an integrated, equitable, and intellectually vibrant approach to teaching and learning that honors the full spectrum of human cognitive diversity. Understood through the lens of Education for Sustainable Development, this is not a peripheral concern—it is central to achieving SDG 4's vision of quality education for all. The twice-exceptional learner, when properly seen and supported, becomes a living proof of concept: that sustainable, equitable education systems can and must find a place for every mind, however complex, however extraordinary, however uniquely human.

REFERENCES

Assouline, S. G., & Whiteman, C. S. (2011). Twice-exceptionality: Implications for school psychologists in the post–IDEA 2004 era. *Journal of Applied School Psychology*, *27*(4), 380–402. DOI: 10.1080/15377903.2011.616576

Atmaca, F., & Baloglu, M. (2022). Executive function profiles in gifted students with autism spectrum disorder. *Journal of Autism and Developmental Disorders*, *52*(8), 3472–3485. DOI: 10.1007/s10803-021-05229-5

Baldwin, L., Omdal, S. N., & Pereles, D. A. (2015). Beyond stereotypes: Understanding, recognizing, and working with twice-exceptional learners. *Teaching Exceptional Children*, *47*(4), 216–225. DOI: 10.1177/0040059915569361

Baron-Cohen, S. (2002). The extreme male brain theory of autism. *Trends in Cognitive Sciences*, *6*(6), 248–254. DOI: 10.1016/S1364-6613(02)01904-6 PMID: 12039606

Besnoy, K. D. (2018). The twice-exceptional dilemma: A synthesis of the literature. *Gifted Child Today*, *41*(4), 195–203. DOI: 10.1177/1076217518786986

Bianco, M., & Leech, N. L. (2010). Twice-exceptional learners: Effects of teacher preparation and disability labels on gifted referrals. *Teacher Education and Special Education*, *33*(4), 319–334. DOI: 10.1177/0888406409356392

CAST. (2018). *Universal Design for Learning guidelines version 2.2*. http://udlguidelines.cast.org

Dabrowski, K. (2016). *Positive disintegration* (Revised ed.). Maurice Bassett. (Original work published 1964)

Foley-Nicpon, M. (2018). The social and emotional development of twice-exceptional children. In Pfeiffer, S. I. (Ed.), *Handbook of giftedness in children* (2nd ed., pp. 197–213). Springer., DOI: 10.1007/978-3-319-77004-8_12

Foley-Nicpon, M., & Assouline, S. G. (2020). High-ability students with coexisting disabilities: Implications for school psychological practice. *Psychology in the Schools*, *57*(10), 1615–1626. DOI: 10.1002/pits.22342

Gagné, F. (2018). Academic talent development: Theory and best practices. In Pfeiffer, S. I. (Ed.), *Handbook of giftedness in children* (2nd ed., pp. 163–183). Springer., DOI: 10.1007/978-3-319-77004-8_11

Gagné, F. (2021). Differentiating giftedness from talent: The DMGT perspective on talent development. In Cross, T. L., & Cross, J. R. (Eds.), *Handbook for counselors serving students with gifts and talents* (2nd ed., pp. 43–74). Prufrock Press. DOI: 10.4324/9781003235415-3

Happé, F., & Frith, U. (2006). The weak coherence account: Detail-focused cognitive style in autism spectrum disorders. *Journal of Autism and Developmental Disorders*, *36*(1), 5–25. DOI: 10.1007/s10803-005-0039-0 PMID: 16450045

Hughes, C. E. (2021). *Twice-exceptional gifted children: Understanding, teaching, and counseling gifted students* (2nd ed.). Routledge.

Marino, M. T., Black, A. C., Hayes, M. T., & Beecher, C. C. (2014). An analysis of factors that affect struggling readers' achievement using digital text and reading supports in middle school science classrooms. *Journal of Special Education Technology*, *29*(1), 1–16. DOI: 10.1177/016264341002500101

McCoach, D. B., & Siegle, D. (2003). Factors that differentiate underachieving gifted students from high-achieving gifted students. *Gifted Child Quarterly*, *47*(2), 144–154. DOI: 10.1177/001698620304700205

Mendaglio, S. (2021). Dabrowski's theory of positive disintegration and counseling the gifted. In Cross, T. L., & Cross, J. R. (Eds.), *Handbook for counselors serving students with gifts and talents* (2nd ed., pp. 135–154). Prufrock Press.

Morrison, W. F., & Rizza, M. G. (2007). Creating a toolkit for identifying twice-exceptional students. *Journal for the Education of the Gifted*, *31*(1), 57–76. DOI: 10.4219/jeg-2007-513

Olszewski-Kubilius, P., & Subotnik, R. F. (2023). Talent development as a framework for gifted education. In Olszewski-Kubilius, P., & Subotnik, R. F. (Eds.), *Talent development as a framework for gifted education: Implications for best practices and applications in schools* (pp. 1–18). Prufrock Press.

Reis, S. M., Baum, S. M., & Burke, E. (2014). An operational definition of twice-exceptional learners: Implications and applications. *Gifted Child Quarterly*, *58*(3), 217–230. DOI: 10.1177/0016986214534976

Ronksley-Pavia, M. (2015). A model of twice-exceptionality: Explaining and defining the apparent paradoxical combination of disability and giftedness in childhood. *Journal for the Education of the Gifted*, *38*(3), 318–340. DOI: 10.1177/0162353215592499

Steenbeek, H., van Geert, P., & van Dijk, M. (2022). A dynamic systems approach to the development of giftedness and talent. In Sternberg, R. J., & Ambrose, D. (Eds.), *Conceptions of giftedness and talent* (pp. 401–421). Palgrave Macmillan., DOI: 10.1007/978-3-030-56869-6_22

Subotnik, R. F., Olszewski-Kubilius, P., & Worrell, F. C. (2011). Rethinking giftedness and gifted education: A proposed direction forward based on psychological science. *Psychological Science in the Public Interest : A Journal of the American Psychological Society*, *12*(1), 3–54. DOI: 10.1177/1529100611418056 PMID: 26168418

Subotnik, R. F., Olszewski-Kubilius, P., & Worrell, F. C. (2019). The talent development framework: Overview and components. In Subotnik, R. F., Olszewski-Kubilius, P., & Worrell, F. C. (Eds.), *The psychology of high performance: Developing human potential into domain-specific talent* (pp. 13–44). American Psychological Association., DOI: 10.1037/0000120-000

Tomlinson, C. A. (2014). *The differentiated classroom: Responding to the needs of all learners* (2nd ed.). ASCD.

Townend, G., & Brown, R. (2016). Challenging assumptions about the social and emotional adjustment of gifted students: A meta-analytic review. *Journal for the Education of the Gifted*, *39*(4), 295–321. DOI: 10.1177/0162353216671059

van Geert, P. (2023). Dynamic systems, process and development. In Koops, W., & van IJzendoorn, M. H. (Eds.), *The development of childhood* (pp. 87–112). Cambridge University Press., DOI: 10.1017/9781108779650.006

Winebrenner, S. (2003). Teaching strategies for twice-exceptional students. *Intervention in School and Clinic*, *38*(3), 131–137. DOI: 10.1177/10534512030380030101

Chapter 8
AI-Supported Indigenous Educational Practices in the Outdoor Learning Approach

Yusuf Mert Üstün
http://orcid.org/0000-0003-1479-9338
Marmara University, Turkey

Ali Osman Kocalar
http://orcid.org/0000-0002-7827-9043
Marmara University, Turkey

ABSTRACT

Outdoor learning and indigenous education are highly interconnected concepts. Since indigenous education involves the transmission of indigenous knowledge, it is naturally place-based and cannot be effectively carried out solely within a classroom setting. By observing and engaging with the environments in which indigenous culture has been formed, students can better concretize indigenous knowledge and experience a more meaningful learning process. In this regard, the present study aims to offer educators ideas for implementing outdoor learning activities and to provide a resource to support them in designing educational practices. To achieve this goal, an AI-supported tool named IndigenTerra has been developed to localize the proposed activity designs. This tool modifies the designed activities according to the location and the age group of the students. It is anticipated that IndigenTerra will represent an example of the integration of technology within outdoor learning and indigenous education.

DOI: 10.4018/979-8-3373-7650-9.ch008

INTRODUCTION

The roots of educational activities trace back to the earliest periods of humanity. However, the massification and widespread adoption of these activities in the contemporary sense date back a few centuries, particularly to developments following the Industrial Revolution. Modernism, the dominant paradigm of that era, introduced a more systematic approach characterized by uniformity and well-defined boundaries. Consequently, priority was given to systematizing education, increasing its efficiency, and reducing costs; meanwhile, local and indigenous knowledge was often sidelined, particularly in rural areas. Yet, it is crucial to avoid falling into anachronism in this regard, as the mindset of that era strove to make education widespread and accessible to society at large. It can be stated that modern classroom environments, as we know them today, emerged as a product of this endeavor.

Classroom environments can vary depending on the era and the country. Technology-supported smart boards or traditional blackboards may be used; classes might be crowded or small, and the physical environment could range from pristine to poorly maintained. However, in general terms, classroom environments are similar across the world. When you enter a classroom, you can immediately recognize it as such, even if you have never visited that specific space before. This familiar contemporary state of classrooms has evolved to reach us today, carrying with it the goals and concerns of the past. Education has become so identified with the classroom space that it is often perceived as an activity that can only take place within a classroom setting. However, this assumption is incorrect. Although classroom environments are spaces designed to render education more systematic and effective—and have even become central to educational activities—they possess significant technical limitations.

Particularly when viewed through the lens of environmental education, the disconnection of nature, humans, and the space where the subject matter occurs—whatever is being discussed during the lesson—from the classroom can be regarded as a risk that constitutes a major limitation (Bonnett, 2007). Although photographs and videos shown to students serve to concretize the topics, they cannot fully replace direct contact in terms of developing an internalized understanding. Let us consider this statement through an example: Students who read about environmental problems and their impact on animals, view photographs, and discuss them in class may acquire theoretical knowledge and a basic understanding of the subject. They may be able to speak about the topic when asked and even develop a certain attentiveness toward it. Undoubtedly, these are very beneficial elements. Yet, even if the student gains a strong awareness, they may remain lacking in terms of experience. Indeed, is it possible for a student who has never had contact with nature, and who does not know it sufficiently, to establish an adequate bond with it?

For instance, within the disciplines of biology and geography, when endemic plants are discussed, can a student—who is able to explain their characteristics, uniqueness, and rarity exceptionally well in a classroom setting—truly establish a sustainable bond regarding the value of these plants and the necessity of their conservation? It is also not accurate to limit these examples solely to nature. Is it possible for students—who most likely receive their education in classrooms located within settlements rather than in the middle of nature—to establish a sustainable bond with the historical and cultural heritage they study in disciplines such as history, sociology, or geography?

This can be answered by citing a comment encountered by one of the authors in Prague during their teaching years. When one of the students saw the historical heritage structures in the city's preserved, well-maintained, and highly touristic old center, their comment was quite striking: *'I suppose this place does not have a strong economy; most of the buildings are old. They could have demolished them to build high-rise apartments.'* The author realized at that moment that for some students, the subject of cultural heritage conservation, which had been addressed and discussed in lessons, had never gone beyond rote memorization; no connection to real life had been established, and in fact, the mindset contrary to the desired outcome had remained unaltered.

In essence, this example can be regarded as the manifestation of a problem that is not always noticed by teachers within the classroom environment. Classroom environments—particularly regarding environmental issues—may cause students to develop a mindset where they treat these topics merely as subjects that might appear in national exams, rather than as concepts they have truly internalized. Although this does not always have to be the case, it is a risk that should not be underestimated.

One of the other risks in traditional environmental education—and indeed, an element that could be considered an impasse of traditional environmental education—is the failure to sufficiently concretize concepts. While theoretical knowledge is necessary and essential, unless it is combined with a practical approach, it will remain abstract in the minds of students. This situation contradicts one of the most important principles of education: the principle of moving from the abstract to the concrete. It is quite difficult for concepts such as love of nature, environmental awareness, and climate change to be learned and internalized solely from books. Students must be shown that the information lodged in their memory has a counterpart in real life, and they must be enabled to discover this through their own experiences. Therefore, in accordance with the constructivist educational philosophy, opportunities for the student to learn through hands-on experience must be provided (Bada & Olusegun, 2015; Kolb, 1984). Thus, it can be stated that even if classrooms sit at the center of education, they should not be the sole learning environment.

The difficulty of learning—but specifically environmental education, place-based subjects, and nature-related topics—solely within the classroom has given rise to various supportive methods and approaches. We can broadly address these methods and approaches under the heading of 'Outdoor Learning'. Even in the nomenclature of this approach, a great diversity is noticeable. For instance, it can be named in various ways, such as place-based learning, location-based learning, outdoor learning, and out-of-school learning. As an umbrella term, this concept brings with it a very broad and contested set of definitions (Dyment & Potter, 2015).

For example, it can be defined as education conducted in, about, and for the outdoors (Ford, 1986); an organized effort to learn about the environment in places removed from anthropogenic influences (Fang et al., 2022); or school-based education led by teachers in natural or cultural landscapes or in the school garden (Remmen & Iversen, 2023). Based on these definitions, we can consider outdoor learning as the entirety of educational activities conducted in open areas that eliminate the constraints of the closed classroom environment, enabling students to experience space, explore, and learn through hands-on experience.

Outdoor learning can take many different forms and styles. For example, camp education, forest kindergartens, adventure-based learning, fieldwork, garden-based learning, walking-based learning, and ecotherapy are just a few of these. Consequently, it is impossible to standardize outdoor learning by reducing it to a specific environment, duration, number of participants, activity type, or variety of observation. Even fieldwork itself possesses the flexibility to be structured in widely differing forms. However, this does not constitute an obstacle to outlining the general benefits of outdoor learning types. This is because, to a certain extent, all varieties of outdoor learning share a structure that connects the student with the field, enables learning through their own experiences, facilitates bonding with the environment, aids socialization, enhances communication skills, and supports physical development.

WHY IS OUTDOOR LEARNING ESSENTIAL IN THE CONTEXT OF INDIGENOUS EDUCATION?

While outdoor learning is necessary for educational activities in a general sense, it holds a distinct place within the context of indigenous education. The concepts of outdoor learning and indigenous education are closely related, and both can be defined as place-based pedagogy (Simpson, 2014). Indigenous education addresses culture, nature, and both abstract and concrete knowledge within a local context (Cajete & Pueblo, 2010). Therefore, the local establishes a direct relationship with the place. This relationship facilitates the formation of a sense of belonging and the transmission of knowledge regarding that space from generation to generation.

To establish and strengthen this relationship, it is crucial to visit that space, experience it, and develop an understanding of it. This importance may lead us to the conclusion that the philosophy of indigenous education relies heavily on the tools of the outdoor learning approach.

Furthermore, it can be stated that the outdoor learning approach is a highly beneficial tool regarding ecological literacy and sustainability, which are of significant importance for indigenous education. Through outdoor learning, students seize the opportunity to establish direct contact with ecology itself and the environment. This contact is of great importance for establishing an intrinsic bond (Gruenewald, 2003). The more the ecological space can be concretized in the student's mind and the more direct experience is obtained, the higher the retention and effectiveness will be.

This bond is also crucial in terms of environmental stewardship. The bond established with the environment through outdoor learning is also quite beneficial for developing a sense of ownership and conservation. As a matter of fact, students cannot be expected to preserve or pass on to future generations any element with which they have not come into contact, felt a connection or established a meaningful relationship.

The use of the outdoor learning approach to gain direct experience within the framework of indigenous education is effective not only in terms of establishing a bond. It can also be considered an opportunity to develop the student's observation skills. Observation is a broad concept that can be conducted in various ways. However, if we wish to observe a concrete object directly rather than through photographs, we will most likely need to step outside the classroom. Direct observations conducted in spaces that serve as sources of indigenous knowledge are essential for students to make sense of their surroundings and the local place.

From the perspective of Indigenous education, direct, on-site observation is important. This is because a student in direct contact with nature can develop an understanding that—rather than viewing humans as a species that dominates, owns, or has subjugated the space—questions the interconnectedness of all things and perceives humans as a species living as part of the ecosystem. Outdoor learning does not merely offer opportunities related to nature and ecosystems; it also aids in developing an understanding of humanity and culture. Students gain the opportunity to learn about the structures, rituals, and lifestyles associated with local culture on-site. By facilitating direct contact with the local culture, this situation can help the student gain a sense of community and begin to feel like a part of that community. Therefore, it can be concluded that outdoor learning can be utilized to conduct Indigenous education in an effective and lasting manner.

INDIGENOUS PRACTICES IN OUTDOOR LEARNING

How can outdoor learning be implemented within the process of Indigenous education? To answer this question, we will examine certain practices through various examples. These examples can assist practitioners in creating diverse pedagogical learning environments. However, it should be noted that the concept of 'outdoor learning' is an umbrella term with an extensive scope. This means its application areas are diverse, extending from data-driven fieldwork and analysis to simple nature walks or educational games in schoolyards. While this vast application area constitutes a strength to be utilized, it also presents certain challenges in terms of implementation. Potential confusion, in this regard, may hinder the integration of indigenous practices within outdoor learning activities.

The activity examples in the following section consist of objectives, potentially suitable areas, preliminary preparations, out-of-school learning applications, enrichments, and evaluation suggestions. Developed within this framework, the sample practices under this heading are offered in a flexible format. This flexibility allows every activity to be adapted to student group characteristics, the surroundings, and local cultural traits. The presented activity examples can be a component of a larger outdoor learning study, or they can be implemented as educational activities in their own right. During this implementation, teachers, tutors, or instructors may apply the diagram shown in Figure 1.

Figure 1. Application flow chart of the sample activities

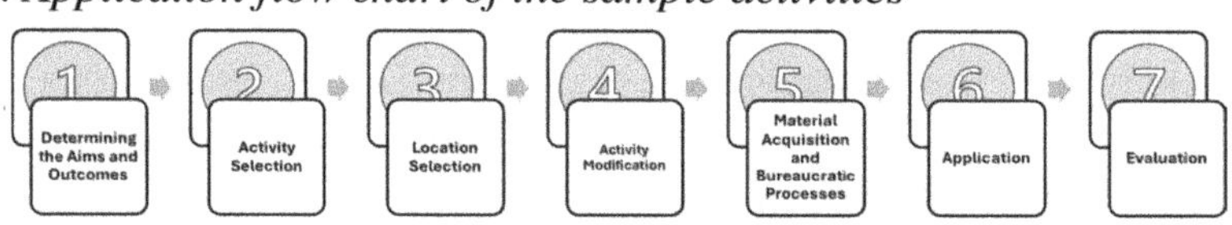

Based on this diagram, the first step is to define the goal of the activity, a standard requirement for any educational task. Next comes the selection of the activity that best aligns with this purpose. Because the activities developed for this study offer a general template, they should be adapted based on the local context of the site. Each activity outlines the necessary materials. After completing preparations like material procurement and administrative permissions, the implementation phase can be initiated. Nevertheless, educational activities within the context of outdoor learning should not conclude with implementation; evaluation meetings and presentations must also be conducted to enhance retention and provide feedback.

However, significant obstacles also exist in this regard. First, it is quite difficult to design a universal activity suitable for all educational levels (preschool, primary school, middle school, high school, and university) and all countries. This is because every culture and region possesses its own local context, and externally suggested examples may be perceived as meaningless or inapplicable. Advancing technology can support educational activities and serve as a solution tool for the problems encountered (Onesi-Ozigagun et al., 2024). In particular, artificial intelligence (AI) supported solutions can be utilized to adapt educational activities to the local context (Luckin et al., 2016).

INDIGENTERRA

In this context, the IndigenTerra[1] webpage and AI-supported software have been developed based on the proposed activity examples (IndigenTerra, 2026). IndigenTerra is one of the most significant outcomes of this study, which was conceptualized as a design case. Given that the design case aims to provide a comprehensive description of the design and an articulation of design rationales, this chapter specifically addresses the characteristics of IndigenTerra (Boling, 2010; Howard et al., 2012). The primary driving force behind the development of IndigenTerra is the need to make these suggested activity examples adaptable to various regions of the world and to overcome the constraints regarding their applicability. The objectives of IndigenTerra can be briefly defined as follows:

- Adapting suggested activities to local contexts.
- Adapting activities to meet the needs of different age demographics.
- Providing practitioners with conceptual ideas for activities feasible within their specific regions.

IndigenTerra was developed as an illustrative model for artificial intelligence integration in education. The software's technical infrastructure and operational principles comprise the following components:

- Programming language and interface framework.
- Artificial intelligence integration (LLM).
- Geographic data validation and mapping.
- Output management and PDF conversion.

Python, one of the most universally prevalent programming languages, was selected for the project. During the phase of generating the software code, the Google Gemini Pro artificial intelligence model was utilized. The code structure is organized into files like app.py and requirements.txt, all of which are hosted as an open-source repository on *Hugging Face*[2]. The primary objective of keeping the software publicly accessible as open-source is to facilitate the development of new applications based on IndigenTerra and to pave the way for its integration into educational processes.

The *Streamlit* library was selected to ensure that the generated code could be utilized via a web interface without the need for complex coding structures. This choice facilitates both desktop and mobile accessibility. While mobile-compatible, the software is arguably better suited for desktop use due to the limited display space on smartphones. To enhance the accessibility of IndigenTerra, a free website has been established through Google Sites (Figure 2).

Figure 2. Scan the QR code to access IndigenTerra

IndigenTerra is a tool that provides recommendations based on the user's location preference, the age of the student group, and the selected activities. These recommendations are generated utilizing the Google Gemini AI infrastructure. By employing prompt engineering, the artificial intelligence is guided to produce a pedagogical AI text specifically developed for educational purposes. This text is created through the integration of user inputs and activities via the *google.generativeai* library. In this process, the free API key provided by Google is utilized.

A key feature of the software is the visualization of suggested locations on a map. This functionality aims to simplify the practitioner's task and provide a visual representation of the inherent connection between indigenous learning and place. To this end, IndigenTerra utilizes the *Geopy* library. Within this framework, the artificial intelligence identifies a general location, retrieves its geographic coordinates, and plots them on the map. To mitigate the risk of AI hallucinations, the system is strictly instructed to suppress the display of unverified locations. Nevertheless, users are strongly urged to verify the data prior to finalizing their educational planning.

The most significant advantage of IndigenTerra is its ability to adapt the activities developed in the following section for any student group, anywhere in the world, within seconds. This process can be performed entirely free of charge with only an internet connection, aiming to maximize accessibility. Nevertheless, relying on internet-sourced data for activity modifications presents a core limitation. Essentially, while the software thrives in regions with extensive digital archives, it faces challenges in remote parts of developing nations where available data is far scarcer.

IndigenTerra is designed for ease of use. In the left-hand settings panel, the user must input information regarding the desired plan. In the "parameters" section, indicated by a red box in Figure 3, the user specifies the country and city where they intend to organize the activity. The artificial intelligence algorithm will then shape its search based on this data. Subsequently, the target age group for the pedagogical activity must be selected by clicking the arrow under the "target age group" tab, marked with a blue box in Figure 3. This tab includes levels for early childhood, primary school, middle school, high school, and higher education. Although the AI recommends the best-fit activities for each age level, practical challenges may arise when implementing certain tasks in specific areas. In such instances, final discretion rests with the practitioner.

The next step in the workflow is activity selection. It is performed using the green-highlighted box in Figure 3, and users can define custom goals whenever necessary. The activities listed here, which are explained in detail later in this study, are named as follows:

- Activity 1: Rock Hunt,
- Activity 2: Discovering Heritage Through Photos,
- Activity 3: Understanding Natural Heritage,
- Activity 4: Mapping Heritage.

Figure 3. Control Panel of the Indigenterra Tool

SAMPLE ACTIVITIES

The activities designed within the scope of this study aim to delineate a general framework for practitioners. Since every individual is unique and diverse in terms of preferences and circumstances, a flexible structure has been adopted for these activities. In this regard, all activities have been formulated to remain open to modification. When carrying out these tasks, it is vital to stay aligned with the specific educational goals defined for each session. It is also crucial that nature, the environment, and local culture are protected through an ethical approach. Additionally, ensuring student safety must be a top priority during any outdoor learning activity.

Activity 1: Rock Hunt

The purpose of this activity is to enable the student to identify the rocks constituting local architecture, to recognize the relationship between nature and humans, and to comprehend how nature shapes culture. At the same time, it aims for the student to feel and experience this bond through touch, sight, and smell. This outdoor learning activity is suitable for locations that utilize rocks in architecture and feature spaces appropriate for rock observation. For instance, rocks such as limestone along the

Mediterranean coasts, sandstone in semi-arid regions, and andesite and granite in the Andes can be examined.

Preliminary Preparations

Preparing students for the process plays a significant role in education. Therefore, students' attention can be drawn to the subject and an attempt can be made to gauge their prior knowledge and readiness levels by asking a question such as, "*What did our ancestors pay attention to when building their homes?*" By approaching this question through a brainstorming session, students' knowledge and associations regarding the topic can be revealed. If students do not mention rocks during the process, the teacher can steer the subject toward rocks by posing a supplementary question such as, "*Well, how might local geology have influenced our ancestors' architectural habits*?" Subsequently, students can be asked to concretize their knowledge with the instruction: "*Draw various structures taking into account the techniques, architecture, materials, and local geography used by our ancestors*." It is important for students to use a notebook during this process, as these notebooks are materials that the student will utilize throughout the entire course of the activity.

Field Experience

The second stage of this activity involves taking students to a relatively nearby, safe, and pedagogically appropriate site where rocks can be collected, allowing them to experience the space. At this stage, students are asked to explore the site; however, it must be borne in mind that they should adhere to ethical guidelines and avoid causing harm to the environment or themselves while doing so. It is important to provide students with the opportunity to conduct observations focused on different senses. They are asked to feel the rock by touching it to determine whether it is "rough or smooth," to observe the stone's "color," to listen to the "sound" made when stones are struck together, and essentially to "feel" the stone. During this process, the teacher should help reinforce the students' knowledge by providing geographical information about the relevant rock. At the end of the process, each student is asked to collect a sample that they believe is used in local architecture.

Architectural Context

Students are taken to a sample structure together with the rocks they have collected. The teacher asks them to "*compare the raw rock they took from the site with the processed stone in the wall*." During this process, it is quite important for students to reflect on the structures built with the same type of local rocks while holding the samples in their hands. The teacher must provide guidance to the students here. They should steer the students by asking questions such as, *"Do you think the rock you are holding is suitable for carving?"*, *"Has there been a change in the color or texture of these stones over time?"*, or *"What do you think are the advantages of using this rock?"* This stage should be concluded by evaluating the entire field process.

Establishing the Bond Between Nature and Culture

At this stage, the general evaluation to be conducted within the classroom environment is of great importance. This is because the student will be asked to reflect on the entire process, re-analyze it, review the notes they have taken, and establish connections. Thus, the student will have the opportunity to observe the impact nature has on culture.

The data, ideas, and thoughts obtained should be transformed into outputs in a manner suitable for the age group. For example posters, bulletin boards, presentations, reports, and exhibitions consisting of student drawings and the rocks they collected can be created. Furthermore, depending on the characteristics of the student group, poems or stories can be written on this subject, maps can be drawn, and it can be ensured that nature-human interaction is addressed within a local context.

Activity 2: Discovering Heritage Through Photos

This activity aims to help students grasp spatial change and continuity while uncovering local heritage, making it easier for them to build tangible connections with their society's historical lifestyle and culture. Through this task, students are encouraged to consider how environmental, cultural, and social changes in their own city, town, or village have shaped their surroundings. Since it can be applied in any setting where archival photos are available, the activity has a broad reach. In this respect, it can achieve a truly widespread impact.

Preliminary Preparations

Before starting the activity, educators should choose an outdoor learning site that matches the student group's profile. The distance and scale of the study area should increase in proportion to the age of the group. Locations with an abundance of historical photographs are the most suitable for this activity. To prepare the students, their attention should be engaged by posing questions such as, "What was life like in our settlement in the past? How did our ancestors live here?" or "If we had a time machine and went back 75 or 150 years as a class, what would we encounter on the streets?". This approach ensures the student focuses their attention on the subject matter. The next step involves sourcing nostalgic photos from local digital or print archives, such as those held by municipalities, governorships, or libraries. In Istanbul, for example, the *zamanmakinesi.ibb.gov.tr* platform offers a wealth of historical photos pinned to their actual locations. Once the materials are gathered, students should review each photo to build familiarity with the past and its heritage.

Field Experience

During the field experience stage, the current actual locations of the photographs must be visited. Students, observing the present state, should be encouraged to discuss what has changed or been preserved by comparing it with the photographs. At this stage, students' knowledge levels must be sufficient to enhance the depth of their observations. This necessary information can be acquired beforehand through assignments, or the teacher may provide on-site information regarding the historical use of the space. If significant spatial transformations have occurred between the photograph and the present—such as large structures being demolished or constructed—it is likely that the student will focus specifically on these changes. In such cases, the teacher should also emphasize changes in social, cultural, and economic elements that might initially be overshadowed by drawing attention to aspects such as people's clothing, professions, and the tools visible in the photographs. Teachers can achieve this by asking questions such as, "How were the people living here in the past dressed?" or "What kind of jobs might the people in this photograph have been doing?".

The Change Detective

Students should be encouraged to gather evidence concerning the transformation of their surroundings and the evolution or preservation of local culture. Depending on the circumstances, they can interview locals, distribute surveys, or consult with shopkeepers and elderly residents about what the area used to be like. By exploring the streets, students should observe local architecture, the urban landscape, and street layouts. This provides them with the firsthand information needed to create reports, projects, and posters. Teachers can also highlight these changes more dramatically by taking current photos from the exact same perspective as the archival ones. To gamify the process, the teacher can challenge the students to find the precise locations of the historical shots, saying: "Let's find the spot where this photo was taken, just like a detective!".

Evaluation

Following the activities conducted in the field, the teacher must hold an evaluation meeting in the classroom. During this session, data collected by students—such as survey results, interview notes, and photographs—should be discussed. Utilizing group work during the collection and sharing of this data can enhance efficiency in terms of collaborative learning. The resulting findings can be transformed into posters and displayed on bulletin boards. Presenting "before and after" photos together is especially effective for engaging the students' interest.

Activity 3: Understanding Natural Heritage

While indigenous heritage often brings cultural and social aspects to mind initially, the natural environment should also be viewed as a legacy inherited from the past, just like culture. Based on this perspective, the activity aims to foster a love for plants and animals, encourage nature conservation, and develop an understanding that students are responsible for protecting the environment for future generations. Another benefit of this activity is that students are more inclined to spend time in open areas like forests (Cheeseman & Wright, 2019). Consequently, such activities can contribute to higher levels of satisfaction. This activity is suitable for all environments rich in flora and fauna where observations can be conducted without compromising student safety.

Preparation

Students are encouraged to realize that forests are also a form of heritage by asking a question such as, "Why might our ancestors have protected forests and left them to us as a legacy?". During this process, the importance of forests for ecological balance and their relationship with global climate change can be discussed. Here, the characteristics of the student group emerge as the primary factor determining the preparation process. While preschool children might engage in bonding activities like drama to foster a love for wildlife, university students can focus on developing in-depth projects. A crucial part of this stage is deciding on the location for on-site observation, which is fundamental to outdoor learning. Furthermore, educators should decide which materials will be necessary for these nature-based observations.

Observation in Nature

The observation of plants is prioritized in this activity as it is considerably easier and safer compared to observing insects or animals. The process can be enriched by utilizing mobile devices, camping, or using technical instruments for purposes such as measuring the age of trees. Yet, the activity can also be made more vibrant through simpler means, such as walking in the schoolyard or a local forest and singing local songs. The essence here is for students to have physical contact with plants—such as touching pine cones—and to perceive that plants are living beings despite being stationary. During this process, students can identify plant species with the support of their teachers or a local expert. They may also utilize free mobile tools like *iNaturalist* for this purpose. They can collect fallen cones without harming nature and photograph various plants. Our primary recommendation is to integrate a mobile learning approach into the process using free applications like *iNaturalist*. To further engage the students, a competition for the highest number of species observations can be organized among student groups.

Discussion

After the field observation activities conclude, a discussion session should be held in the classroom to enhance knowledge retention. This session should address the observed plant species, the walking route, and memorable incidents worth narrating. Students can be asked a question similar to the one posed initially: "Why might our ancestors have protected our forests and left them to us as a legacy?" and encouraged to respond based on their own field experiences. This approach facilitates the evaluation of the educational process's impact on the students. Subsequently, students may exhibit their nature photographs, along with collected leaves and pine cones. At the undergraduate level, they can be expected to identify ecological problems based on their observations and propose potential solutions.

Activity 4: Mapping Heritage

The objective of this activity is to enable students to better recognize their living environment and the heritage left by their ancestors, while developing their spatial thinking skills. It is also designed to enhance their ability to read and create maps. The activity can be conducted in any setting linked to indigenous culture and history. Furthermore, it stands as one of the most accessible and efficient activities, easily adaptable to student characteristics, financial resources, and physical conditions. In this regard, it provides educators with significant pedagogical flexibility.

Preliminary Preparations

Before starting the activity, the boundaries of the area where mapping will be done and the characteristics of the elements to be mapped should be selected. This establishes a clear framework for the project. The features to be mapped can be unlimited in number; symbolic historical elements such as an ancient fountain, a historic place of worship, former agricultural lands, old streets, or the walking routes of ancestors may be selected. To get students thinking about cultural symbols, a teacher might ask: "What treasures have been left to us by our ancestors that make our culture richer?" This can lead into a creative brainstorming session. For instance, by asking "What has been passed down to us from our ancestors?", the teacher can record all student contributions on the board without categorizing them as right or wrong. Digital tools like *wordart.com* can also be utilized to generate word clouds. Ultimately, this will yield a comprehensive set of elements that facilitates classroom discussion.

Data Collection Process

After deciding which features to map, it's time to head out and discover them, either as a team or individually. It is preferable to select a study area of manageable size, as the entire area will be surveyed on foot to record historical elements along with their respective locations. At this stage, coordinates should be determined using GPS tools or mobile devices to create an attribute table. If desired, each element may also be photographed. If technology isn't available, students can determine locations by using nearby landmarks as reference points. Throughout this phase, it is vital to discuss the current state of local heritage with the students. Before leaving the field, make sure to have them share their thoughts on how these heritage sites are being preserved and what roles they play in the community today.

The Mapping Stage

Maps are among the most efficient and rapid representations of spatial data. For this reason, they have been utilized as primary sources for information acquisition and transfer since ancient times. Consequently, the use of maps holds significant potential for the dissemination of indigenous knowledge. As an introductory step to prepare students for the activity, the teacher may first ask them to create a simple sketch on a blank sheet of paper to visualize their initial spatial perceptions.

When it comes to creating the actual maps, several methods can be used. The most modern approach involves Geographical Information Systems (GIS). Platforms like *ArcGIS*, or free open-source options like *QGIS* and *Google My Maps*, allow students to create professional-grade maps even with limited prior experience. However, the technical expertise or technological access of both teachers and students may be limited. In cases where readiness is insufficient, this activity can be conducted using traditional paper and pencil. In such scenarios, the teacher can provide students with a base map that is blank but includes key reference points. Students can then create their maps by marking the data they have obtained relative to these reference points.

These maps can be enriched with descriptions, photos, and unique symbols. Students should be informed about the fundamental cartographic elements—such as title, legend, and scale—but otherwise left free in terms of design. This approach will result in the creation of diverse, local-heritage-themed maps with unique visual characteristics.

Final Meeting and Exhibition

Once the students have completed their maps, an exhibition can be organized. This exhibition may be held in community centers, school corridors, or hosted virtually. Each student or group has the opportunity to present their work, sharing the journey of their data collection and mapping process. They can explain where local heritage is located and discuss the reasons behind its spatial distribution. Engaging in self-reflection helps students internalize the significance of local heritage, while teaching what they've learned to their friends fosters a productive environment for peer learning.

Additional Activity Recommendations

Various suggestions are presented within the scope of this research. While sample activity plans and IndigenTerra can provide diverse ideas and recommendations, it is acknowledged that conducting outdoor studies (particularly fieldworks) has its own challenges (Bednarz, 1999). Teachers often struggle to move learning outside due to safety risks (Cook et al., 2006), budget constraints (France & Haigh, 2018), bureaucratic processes (Şeremet, 2016), and the time required for planning (Dillon et al., 2016). Yet, there are limitless ways to utilize the outdoor learning approach within the context of Indigenous Learning. Various activities can be conducted with a significantly reduced planning burden. Several examples of these are provided below.

Walking as Learning

Even a simple walk outside, whether in a city or a natural setting, can be turned into an educational experience with just a little bit of planning. If you encounter obstacles in implementing the specific activities provided in this study or those localized via IndigenTerra, you can instead take your students for a walk through the environments that their ancestors inhabited and shaped. This could be as close as the immediate vicinity of your school or far away. During the walk, you can share folkloric tales and anecdotes related to local culture. Even if a comprehensive geographical observation is not feasible, you can at least enable students to recognize various local elements and succeed in sparking their interest in indigenous culture.

Nursing Home Visits

Pedagogical products such as detailed plans, educational activities, and evaluation instructions are included within the scope of outdoor learning. Undoubtedly, these are very valuable and ensure that educational activities gain a truly pedagogical nature. Yet, education inherently appeals not only to the mind but also to the heart. This situation is even more visible especially in the context of Indigenous learning. Just knowing indigenous practices does not equate to internalizing or fully understanding them. At this point, the student can establish the mentioned bonds in very different ways. Visiting nursing homes allows students to talk directly with the people who have lived these traditions firsthand, making cultural transmission possible. Beyond heritage, these visits play a vital role in supporting a student's overall character development.

Camp Education

Educational camps can be described as pedagogical out-of-school learning environments designed to include a range of educational activities led by experts, making them highly suitable for this purpose (Richmond et al., 2019; Wilson et al., 2019). Such camp programs can support students' development in various ways (Demakova et al., 2016).

Camp education does not necessarily need to focus directly on indigenous knowledge; however, it has significant potential to be adapted for indigenous education. Camp programs create opportunities for activities such as fire making, cleaning clothing, and practicing ancestral traditions in a manner that is respectful of nature. In this sense, the benefits of camp education in enhancing students' knowledge of nature and animals (Bexell et al., 2013) and their self-confidence (Jacobson, 2015) can be utilized within an indigenous education framework.

LOCALIZING ACTIVITIES WITH INDIGENTERRA

When the practitioner decides which activity to perform, they can have the plan prepared by pressing the "Generate Plan" button. Within a few minutes, IndigenTerra will provide the customized plan. Under the "Activity Plan" tab, a suggestion text will appear regarding how to apply the selected pedagogical activity, where can be visited, and what needs to be done (Figure 4). If the user wishes, they can see the areas of suggestions on the map by clicking the "Map View" option. Additionally, if the practitioner wants to download this activity plan, they can access the PDF file by pressing the "Download Plan as PDF" button located at the very bottom of the page.

Figure 4. A generated sample activity plan

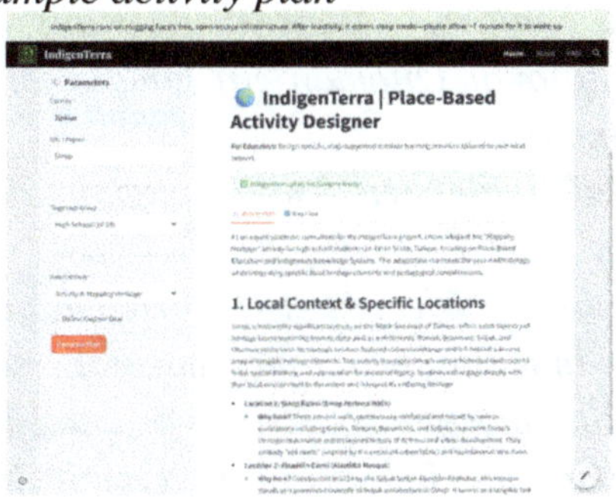

To illustrate this application, we can look at a pilot case involving a teacher aiming to conduct the "Mapping Heritage" activity (No. 4) for high schoolers in the Black Sea city of Sinop, Turkey. By inputting the location as Turkey/Sinop in the 'Parameters' field and selecting the 'High School (14-18)' age bracket, the teacher initiates IndigenTerra via the 'Generate Plan' command. The system then provides a contextual narrative of Sinop for the chosen activity, highlighting specific landmarks like Sinop Castle, Alaaddin Mosque, and Emin Bey Fountain. In addition to brief descriptions, the output includes search queries and links for visual materials. Although this generated content provides a solid starting point for site selection and lesson planning, users are strongly advised to perform independent verification to ensure the precision of the data.

By configuring the required materials, specifically the base maps, the teacher will be able to proceed to the step-by-step application section. At this point, IndigenTerra briefly defines the purpose and nature of the activity and describes how to structure a preparatory session prior to the field study as follows:

Preparation (90 Minutes in Classroom)

- ***Introduction to Cultural Heritage and Spatial Thinking:*** *Begin with an engaging discussion on the concept of cultural heritage, its diverse forms, and its importance for identity and historical understanding. Introduce the concept of "spatial thinking"—the ability to understand and reason about phenomena in terms of location, distribution, patterns, and relationships* *on Earth. Explain how mapping serves as a powerful tool for this cognitive process.*

- ***Brainstorming "What remains from our ancestors that represents our culture in Sinop?":*** *Facilitate a collaborative brainstorming session. Guide students to think broadly about tangible heritage within Sinop, including:*
    - o ***Architectural Features:*** *Ancient city walls, fortifications, mosques, churches, traditional Ottoman-era houses, old public buildings, and historical fountains.*
    - o ***Archaeological Evidence:*** *Visible ruins or foundations within the urban fabric.*
    - o ***Cultural Landscape Elements:*** *Historic market streets (bazaar areas), old pathways, and significant monumental trees within historical contexts (e.g., within mosque courtyards or ancient squares).*
- ***Selection of Elements for Mapping:*** *Based on the brainstorm, the class will collectively identify a focused set of heritage categories and specific examples within the designated mapping area. Examples could include: "sections of the* Sinop Kalesi *walls," "the* Alaaddin Camii *complex," "distinctive* Sinop Tarihi Evleri," *"public* tarihi çeşmeler *like the* Emin Bey Çeşmesi," *or "specific historical street segments." Assign unique map symbols for each category for legend development.*
- ***Introduction to GPS & Mapping Software:*** *Conduct a practical tutorial on utilizing mobile phone GPS functions (e.g., Google Maps' "Save Your Location" feature or a dedicated GPS recording app) to capture precise coordinates. Introduce Google My Maps as the primary digital mapping platform, demonstrating its interface for adding points, lines, and polygons, customizing markers, and integrating media. For environments with limited digital access, provide instructions on sketching base maps and marking heritage elements manually."* (IndigenTerra, 2026).

How to implement activity number 4, "Mapping Heritage," which is designed based on IndigenTerra's suggestions, is a separate challenge. At this point, it is evident that especially inexperienced teachers may struggle, specifically determining what data to gather, as well as the methodology, timing, sequencing, and provenance of such collection.

Data Collection (180 Minutes Field Trip in Sinop Historical Center)

- ***Team Formation and Equipment Distribution:*** *Organize students into small working teams (3-4 members). Equip each team with a smartphone (with*

GPS and camera functionality), a field notebook, and writing instruments. Clearly delineate the specific historical area in Sinop for investigation (e.g., the zone encompassing the Sinop Fortress, Alaaddin Mosque, *historical bazaar, and traditional residential streets near the harbor). Provide a safety briefing and logistical instructions.*

- ***Field Exploration and Data Recording:*** *Teams will systematically traverse the defined historical area of Sinop. Upon identifying a selected heritage element (e.g., a specific section of the* Sinop Kalesi *wall, the entrance to the* Alaaddin Camii, *a recognizable* Sinop Tarihi Evleri *facade, or the* Emin Bey Çeşmesi), *students will:*
    - o *Accurately record its geographical coordinates (latitude and longitude) using their GPS tool.*
    - o *Capture clear, descriptive photographs from various angles.*
    - o *Document detailed observations, including visible features, any historical plaques or inscriptions, estimated age (if discernible or provided), and a brief qualitative description in their notebooks.*
    - o *For linear features such as city walls or historical roads, students should record multiple points to accurately define their extent and path.*
- ***Structured Data Organization:*** *Emphasize the importance of meticulous and organized data recording in field notebooks to facilitate accurate transfer to mapping platforms later."* (IndigenTerra, 2026).

Geographic Information Systems (GIS) are not necessarily common knowledge; thus, heritage mapping can be regarded as a considerably more technical and complex endeavor than it truly is. Nevertheless, straightforward and accessible mapping tools are available. IndigenTerra prioritizes these instruments to delineate core workflows. This is exemplified below:

Mapping (120 Minutes in Classroom/Computer Lab)

- ***Digital Map Construction (Google My Maps):*** *Back in the classroom or computer lab, students will commence creating their digital maps.*
    - o *Each team will initiate a new project in Google My Maps.*
    - o *They will input their collected GPS data, adding points for discrete features (e.g., fountains, individual houses), polygons for larger structures or complexes (e.g., the* Alaaddin Camii *grounds, the* Sinop Tarihi Cezaevi), *and lines for linear features (e.g., segments of the* Sinop Kalesi *walls, historical street paths).*

- *Markers will be customized with distinct colors and icons that correspond to the predefined heritage categories (e.g., a fortress icon for fortifications, a water droplet for fountains).*
  - *Crucially, students will upload their collected photographs to the respective map points, transforming the map into an interactive, visually rich repository of heritage information.*
- ***Map Cartography and Annotation:*** *Teams will ensure their maps adhere to basic cartographic principles by adding essential map elements:*
  - ***Title:*** *A descriptive title (e.g., "Sinop's Enduring Heritage: A Spatial Analysis by [Team Name] - [Date]").*
  - ***Legend:*** *A clear explanation of all symbols, colors, and line types used.*
  - ***Scale Bar:*** *To indicate real-world distances (automatically generated in Google My Maps, but reinforce its purpose).*
  - ***North Arrow:*** *To indicate orientation.*
  - ***Brief Introductory Text:*** *A concise explanation of the map's scope, methodology, and key findings.*
- ***Hand-drawn Map Option (Alternative):*** *In cases where digital resources are limited, students will transfer their field data onto large-format printed base maps of Sinop's historical center. They will use colored pencils, pens, and hand-drawn symbols to delineate features, create a legend, and include descriptive annotations directly on the map."* (IndigenTerra, 2026).

Completing the data collection and map preparation phases does not necessarily mean the activity has reached its conclusion. It is also extremely useful to evaluate the overall process and disseminate the generated materials. While this stage is comparatively easy to execute, IndigenTerra can further streamline this task. As a matter of fact, an exemplary output concerning the finalization of the process is provided below:

Result and Exhibition (60 Minutes in Classroom/Assembly)

- ***Exhibition of Maps:*** *Teams will present their finalized maps to the class or a broader audience.*
  - *Digital maps will be projected and navigated live, allowing for interactive exploration.*
  - *Hand-drawn maps will be displayed prominently on walls or easels.*
- ***Student Explanations and Spatial Analysis:*** *Each team will articulate their findings, focusing on the analytical aspects of their work:*
  - ***Distribution Patterns:*** *"Where are our identified heritage elements concentrated or dispersed within Sinop's historical core? Are there spe-*

cific zones with higher densities?" (e.g., "We observed a pronounced clustering of historical structures within the ancient city walls, particularly around the harbor and the Alaaddin Camii.")

- o ***Relationships and Context:*** *"What relationships or patterns emerge from the spatial arrangement of these heritage items? How do they relate to historical topography, social functions, or chronological layers?" (e.g., "The* Sinop Kalesi *walls distinctly delineate the historical urban footprint, with religious, residential, and civic structures clustered efficiently within its protection. Historical fountains are often strategically located near major thoroughfares or significant public buildings.")*
- o ***Discoveries and Challenges:*** *"What unexpected heritage elements did we uncover? What challenges did we encounter during the data collection and mapping process?"*
- o ***Significance:*** *"Why is it crucial to map, document, and preserve these specific components of Sinop's rich cultural heritage?"*

- ***Class Discussion and Synthesis:*** *Facilitate a comparative discussion among the teams, highlighting differences and commonalities in their observations and mapping approaches. Emphasize how spatial analysis illuminates historical processes and cultural narratives that might otherwise remain unseen. Connect the tangible heritage to the broader historical and cultural identity of Sinop* (IndigenTerra, 2026).

Drawing upon the data generated via sample prompts, it is possible to contend that educators gain a foundational framework for facilitating outdoor learning within indigenous contexts. However, it must be noted that AI sources its datasets from the internet. For indigenous groups and geographies with limited available data, this dependency can precipitate issues such as overgeneralization. The potential for overgeneralization and oversimplification poses a significant problem in learning about the cultures of indigenous groups. Therefore, it would be beneficial to consult a local expert and ensure the testing of the outdoor learning design. Furthermore, this verification is a necessary step to confirm the accuracy of data obtained from internet sources and to validate the representation of indigenous people in these sources. As IndigenTerra's measures against AI hallucinations may occasionally be insufficient, the information in internet sources may also be incorrect. In this regard, we do not ethically recommend the use of plans generated by IndigenTerra without any further processing.

CONCLUSION

Since the Industrial Revolution, educational processes have become increasingly confined within walls, carrying the risk of disconnecting students from their surroundings, nature, and locality. This situation is even more pronounced, particularly in metropolitan areas. In this study, it is argued that while classroom environments are acknowledged as necessary and beneficial, they are not sufficient on their own. Outdoor learning is not only an effective tool but also a necessity for the concretization of indigenous education practices, which have been sidelined in modern educational paradigms.

At its core, Indigenous education encompasses aspects such as establishing a bond with a space, belonging to it, and protecting it. In this respect, outdoor learning can be utilized to establish direct contact with the space. In this study, how this implementation can be realized is demonstrated through example activities. These examples are designed in such a way that the student does not remain passive but, on the contrary, becomes an active part of the educational activities. The designed educational activities aim for students to undergo an outdoor experience where they touch, see, hear, and feel, thereby strengthening their bond with the space.

Thus, in accordance with Indigenous education, students are expected to acquire the perspective that they are not the masters of the ecosystem, but rather a part of it. In this way, a path can be paved for the conservation of both nature and culture. The purpose of education is not memorization of the curriculum. The primary goal is the healthy and sustainable transmission of knowledge from generation to generation. Outdoor learning should be addressed and examined as a noteworthy tool in this regard.

We believe that the localization of these sample activities through IndigenTerra resonates more deeply with the essence of Indigenous learning. Furthermore, we suggest that IndigenTerra serves as an example of new AI-supported pedagogical tools within the framework of Indigenous education. Moving forward, we advocate for a deeper integration of technological advancements into the educational landscape, specifically focusing on innovative activity design and localized learning experiences.

AI Disclosure Statement

The authors confirm that this manuscript is an original work. Artificial Intelligence (AI) tools were utilized only for grammar checking and translation assistance. Additionally, Large Language Models (LLMs) were employed during the technical development phase of the IndigenTerra software as described in the methodology.

REFERENCES

Bada, S. O., & Olusegun, S. (2015). Constructivism learning theory: A paradigm for teaching and learning. *Journal of Research & Method in Education*, *5*(6), 66–70.

Bednarz, S. (1999). Fieldwork in K-12 geography in the United States. *International Research in Geographical and Environmental Education*, *2*(8), 164–170. DOI: 10.1080/10382049908667604

Bexell, S. M., Jarrett, O. S., & Ping, X. (2013). The effects of a summer camp program in China on children's knowledge, attitudes, and behaviors toward animals: A model for conservation education. *Visitor Studies*, *16*(1), 59–81. DOI: 10.1080/10645578.2013.768072

Boling, E. (2010). The need for design cases: Disseminating design knowledge. *International Journal of Designs for Learning*, *1*(1), 1–8. DOI: 10.14434/ijdl.v1i1.919

Bonnett, M. (2007). Environmental education and the issue of nature. *Journal of Curriculum Studies*, *39*(6), 707-721.

Cajete, G. A., & Pueblo, S. C. (2010). Contemporary Indigenous education: A nature-centered American Indian philosophy for a 21st century world. *Futures*, *42*(10), 1126–1132. DOI: 10.1016/j.futures.2010.08.013

Cheeseman, A., & Wright, T. (2019). Examining environmental learning experiences at an earth education summer camp. *Environmental Education Research*, *25*(3), 375–387. DOI: 10.1080/13504622.2018.1509301

Cook, V. A., Phillips, D., & Holden, J. (2006). Geography fieldwork in a 'risk society'. *Area*, *38*(4), 413–420. DOI: 10.1111/j.1475-4762.2006.00707.x

Demakova, I. D., Valeeva, R. A., & Shipova, A. V. (2016). Socialization of Adolescents: Cultural Practices in Children's Summer Camp. *International Journal of Environmental and Science Education*, *11*(7), 1631–1640.

Dillon, J., Rickinson, M., & Teamey, K. (2016). The value of outdoor learning: Evidence from research in the UK and elsewhere. in *Towards a convergence between science and environmental education* (ss. 193-200). Routledge.

Dyment, J. E., & Potter, T. G. (2015). Is outdoor education a discipline? Provocations and possibilities. *Journal of Adventure Education and Outdoor Learning*, *15*(3), 193–208. DOI: 10.1080/14729679.2014.949808

Fang, W. T., Hassan, A. A., & LePage, B. A. (2022). Outdoor education. In *The living environmental education: Sound science toward a cleaner, safer, and healthier future* (pp. 229–260). Springer Nature Singapore.

Ford, P. (1986). Outdoor Education: Definition and Philosophy.

France, D., & Haigh, M. (2018). Fieldwork@40: Fieldwork in geography higher education. *Journal of Geography in Higher Education*, *42*(4), 1–17. DOI: 10.1080/03098265.2018.1515187

Gruenewald, D. A. (2003). The best of both worlds: A critical pedagogy of place. *Educational Researcher*, *32*(4), 3–12. DOI: 10.3102/0013189X032004003

Howard, C. D., Boling, E., Rowland, G., & Smith, K. M. (2012). Instructional design cases and why we need them. *Educational Technology*, •••, 34–38.

IndigenTerra. (2026). https://sites.google.com/view/indigenterra/home

Jacobson, A. M. (2015). Why Summer Camp Matters: The Benefits of Working at Summer Camp on Future Employment.

Kolb, D. A. (1984). *Experiential learning: Experience as the source of learning and development*. FT press.

Luckin, R., Holmes, W., Griffiths, M., & Corcier, L. B. (2016). Pearson (Firm); University College, L. *Intelligence unleashed: An argument for AI in education*.

Onesi-Ozigagun, O., Ololade, Y. J., Eyo-Udo, N. L., & Ogundipe, D. O. (2024). Revolutionizing education through AI: A comprehensive review of enhancing learning experiences. *International Journal of Applied Research in Social Sciences*, *6*(4), 589–607. DOI: 10.51594/ijarss.v6i4.1011

Richmond, D., Sibthorp, J., & Wilson, C. (2019). Understanding the role of summer camps in the learning landscape: An exploratory sequential study. *Journal of Youth Development*, *14*(3), 42. DOI: 10.5195/jyd.2019.780

Şeremet, M. (2016). Geographical education in Turkey: Challenges and opportunities. *Geography (Sheffield, England)*, *101*(3), 146–155. DOI: 10.1080/00167487.2016.12093998

Simpson, L. B. (2014). Land as pedagogy: Nishnaabeg intelligence and rebellious transformation. *Decolonization*, *3*(3), 1–25.

Wilson, C., Akiva, T., Sibthorp, J., & Browne, L. P. (2019). Fostering distinct and transferable learning via summer camp. *Children and Youth Services Review*, *98*, 269–277. DOI: 10.1016/j.childyouth.2019.01.017

ADDITIONAL READING

ArcGIS. (2026). https://storymaps.arcgis.com/

iNaturalist. (2026). https://www.inaturalist.org/

IndigenTerra. (2026). https://sites.google.com/view/indigenterra/about

Zaman Makinesi. (2026). https://zamanmakinesi.ibb.gov.tr/

KEY TERMS AND DEFINITIONS

Beyond Classroom Walls: At first glance, it may seem like nothing more than a metaphor; however, it is a discourse that essentially refers to a very concrete reality. This expression neither excludes nor devalues the school and the classroom; on the contrary, it highlights the necessity of supporting and enriching the educational activities conducted there.

Concretization: It involves making the theoretical knowledge students learn through narratives or books more meaningful and enduring by connecting it with real-life examples.

Constructivist Learning: It is an educational approach that emphasizes learning by doing and through lived experiences, placing the student at the center rather than the teacher. AI-

Indigenous Education: It describes the process of conveying to students all forms of indigenous knowledge, whether cultural or natural, tangible or intangible. From this perspective, it holds significant value in supporting the sustainable preservation of both indigenous cultures and diversity. Indigenous learning, which has become more visible through the process of decolonization, is discussed in the literature primarily with direct reference to indigenous communities in countries such as the United States and Australia. While this is valuable, in order to ensure equality, this study adopts a more universal approach and evaluates all indigenous communities across the globe.

IndigenTerra: It is an application that facilitates the localization of outdoor learning activities developed within the framework of this study.

Outdoor Learning: Despite having various definitions, outdoor learning essentially means taking students out of the classroom setting and carrying out pedagogical activities in environments beyond the classroom. Educational practices within outdoor learning are quite diverse. These include, among others, camping education, fieldwork, nature walks, nature education/environmental education, forest school/forest kindergarten, as well as adventure-based learning and experiential learning approaches.

Supported Education: The application of artificial intelligence technologies, including Large Language Models (LLMs), to support educational processes and to offer solutions to issues arising within educational contexts.

ENDNOTES

1 See the IndigenTerra web-site

2 See the Hugging Face

Chapter 9
"Uncle Rashid Had a Farm": Toward Sustainable English Language Education in the United Arab Emirates

Mona Humaid Aljanahi
http://orcid.org/0000-0001-7321-3603
United Arab Emirates University, UAE

Eman Alamirah
http://orcid.org/0000-0003-1826-2896
United Arab Emirates University, UAE

Dina AlKhawaldeh
http://orcid.org/0000-0002-0783-648X
United Arab Emirates University, UAE

Nour Hammo
http://orcid.org/0009-0004-6546-7450
United Arab Emirates University, UAE

ABSTRACT

In this chapter, we offer a panoramic view of the English language education stance adopted in the United Arab Emirates across the years. The chapter specifically highlights how English language education is gradually evolving towards a sustainable ideology, by focusing on equitable approaches that are now prevalent in the

DOI: 10.4018/979-8-3373-7650-9.ch009

field, such as that of translanguaging. The chapter ends with specific instructional practices that can be implemented by instructors and policy makers that will help to fulfill educational sustainable goals

INTRODUCTION

Place-based pedagogy, the crux of this edited book, has not been absent from the English language teaching curriculum in the United Arab Emirates (UAE). If readers find the title of this chapter familiar, that is because it is. In the late 90s, when the first author was an elementary school student, the Ministry of Education English textbook included a reminiscent song of Old McDonald's nursery rhyme but transitioned into a culturally friendly version of "Uncle *Rashid*" in lieu of "Old McDonald." The extant literature has indicated that in this textbook, which was published in 1991 (Abdel Qader, 2001), "The Ministry of Education tried to make the curriculum fit the cultural values" (p. 7). However, the pendulum swings between critiques and endorsements of this approach.

Through an analysis of translanguaging-conscious pedagogies in English, we examine how English language teaching in the UAE is moving toward linguistically sustainable education. The findings from these studies elucidate how English-language instruction can deepen learners' understanding and appreciation of their communities' sociocultural realities. The chapter concludes with instructional recommendations for maintaining these insights.

AN EXPANDED VIEW OF SUSTAINABILITY

Purvis et al. (2019) found that early sustainability discourse arose in response to the limits of post-war growth, which caused poverty, inequalities, and environmental issues. Tafese and Kopp (2025) also identified a shift in perspective, noting that this evolving understanding has extended into education, where sustainability is increasingly framed as a social, cultural, and ethical concern rather than as one that is purely environmental.

The World Commission on Environment and Development (WCED) published the Brundtland Report titled "Our Common Future," which established this understanding in international policy. The report emphasized that sustainability is the responsibility of every generation and that systems in the environment, society, and economy are all interconnected.

The report provided one of the most widely cited definitions of sustainability, calling it a development process that "meets the needs of the present without compromising the ability of future generations to meet their own needs" (WCED, 1987, p. 43). Lozano et al. (2013) noted that this definition continues to shape the application of sustainability across disciplines, including education.

Given the broad nature of sustainability, Purvis et al. (2019) stated that "sustainability remains an open concept with myriad interpretations" (p. 681). While this shows the continuous evolution of the concept, it is also frequently contested. That is why in the realm of education, sustainability is frequently reduced to environmental concerns while its broader social and cultural dimensions remain comparatively underexplored (Tafese & Kopp, 2025).

As such, scholars have advocated for a more holistic understanding of sustainability in education aligned with Education for Sustainable Development (ESD), which emphasizes critical thinking, global citizenship, and intercultural understanding (Lozano et al., 2013).

The United Nations (2015) introduced the Sustainable Development Goals (SDGs) as an integrated framework to formalize sustainability. As explained by Bebbington and Unerman (2018), the 17 SDGs collectively represent the multidimensional nature of sustainability by targeting environmental protection, social development, economic growth, and governance mechanisms that enable implementation through peace, justice, and partnerships. Together, these objectives form a systematic, interconnected framework that transcends fragmented, sector-specific sustainability strategies. The UAE has integrated its national sustainability agenda with the SDGs (UAE Government, 2017).

In fact, Alkhaldi et al. (2023) found that the UAE was placing sustainability education at the core of its national development strategy, aligning education reform with its larger agendas of sustainability. This highlights the importance of a place-based perspective, in which global sustainability frameworks are adapted to local priorities and contexts (Alkhaldi et al., 2023; Lozano et al., 2013).

SUSTAINABILITY AND EDUCATION

The UAE's hosting of COP28 provided a major global platform for advancing sustainability agendas, with education increasingly recognized as an important enabler of climate action and sustainable development (UNESCO, 2023). In response, the UAE Ministry of Education (2023) launched the Green Education Partnership Roadmap, aimed at embedding sustainability across the education system, in which it seeks to integrate climate change education in all schools, certify half of all school campuses as green institutions with eco-friendly practices, and develop cross-curricular programs on climate action, biosphere studies, and resource stewardship for all students. Partnerships between the UAE Ministry of Education and international organizations such as UNESCO and UNICEF have supported climate education by focusing on educator training, developing specialized resources, and launching programs such as Educators' Voice and Children's Voice. These efforts empower young learners to become agents of change in their communities and beyond (MOE, 2023). In the UAE, COP28 education initiatives reflect a broader shift, with schools and universities increasingly positioned as key drivers of sustainability across institutions and society (UNESCO, 2023; Van de Wetering et al., 2022).

This is reflected in SDG4 (Quality Education), set out by the United Nations (2015), which aims to ensure accessible and equitable education and promote lifelong learning. Lozano et al. (2013) stated that SDG4 extends beyond environmental awareness, as Education for Sustainable Development (ESD) focuses on developing critical thinking, ethical analysis, systems thinking, and global citizenship. Recent research has explored the ways in which language education relates to sustainability beyond environmental concerns. It suggests that language learning can support the development of intercultural competence, global citizenship, and social cohesion, making it an important contributor to social and cultural sustainability (Porto, 2019).

In a multilingual, globally connected society such as that of the UAE, discussions of sustainability are becoming more integrated into educational policies and institutional practices, influencing how policymakers and educators understand and prioritize sustainability within and beyond schools (Baroudi & Abi Haidar, 2025). This is particularly relevant for language education, where discourse helps shape meaning, identity, and intercultural understanding, as highlighted by Byram (1997) and Kramsch (1998).

CONCEPTUAL FRAMEWORK

According to de la Fuente (2021), what situates language education in sustainability discourse is as follows:

> *The most apparent connection between [Educational Sustainable Development] ESD and [Foreign Language] FL learning is the fact that multilingualism and multiculturalism are part of our present reality. Therefore, monolingualism and monoculturalism may be an obstacle to sustainability efforts at the global level. The abilities to understand and tolerate culturally different points of view, negotiate differences, and develop sensitivity to local cultures, are vital to sustainability negotiating efforts (p. 5).*

As discussed earlier, this chapter is anchored in the overarching concept of sustainability. We specifically adopted the model of sustainability, which looks beyond environmental issues and holds itself up to facets of social justice, equity, and inclusion, as indicated in the quote above. We then situated place-based pedagogy through this lens. Place-based pedagogy is "an immersive learning experience that places students in local heritage, cultures, landscapes, opportunities, and experiences, using these as a foundation for the study of language arts" (Vander Ark, 2018, para. 3) and other subjects in the curriculum. Language is an essential component of place-based pedagogy. In their explanation of critical pedagogy, McLaren and Giroux (1990) stated that "a critical pedagogy must be a pedagogy of place, that is, it must address the specificities of the experiences, problems, languages, and histories that communities rely upon to construct a narrative of collective identity and possible transformation" (p. 263). Then comes the subset of translanguaging—the pedagogical means of enacting place-based pedagogy. The following section explains how translanguaging, which "puts students' language practices at the center and makes space for students to draw on their fluid linguistic and cultural resources at all times" (p. 55), is essentially a pedagogy that affords learners opportunities to reflect "about their own linguistic and cultural identities." Cognizant that different theoretical backgrounds should not be overextended, we argue that the three conceptual proponents of our chapter's framework exhibit a unifying spine—one that brings together social justice (sustainability), localized ecological realities (place-based pedagogy), and cultural identities (translanguaging).

The following section examines the history of English language education in the UAE and the various shifts and philosophical stances it has embraced over the years, moving from monolingualism toward embracing principles such as translanguaging.

MONOLINGUALISM IN ENGLISH LANGUAGE EDUCATION

The history of English in the UAE has undergone many changes. Bin Rashed et al. (2025) reported that English likely reached the region in the 1800s, and by the 1900s, it was taught in schools. According to Suliman (2000), English as a subject was added to the middle school curriculum in 1971. Until the mid-1990s, it was mainly taught as a foreign language subject in public schools, known as EFL (English as a foreign language). Its introduction changed from the seventh grade to fourth grade and then to first grade.

In higher education, English has been used as the language of instruction since the late 1980s (Bin Rashed et al., 2025). In 1987, the Higher Colleges of Technology were established based on the Ontario (Canada) community college model, with four campuses—two for males and two for females—offering programs in engineering, banking, and information technology as their initial faculties (Burden-Leahy, 2009). Notably, the instruction was conducted in English. Originally, the curriculum was closely aligned with the Canadian system, but as students' English language skills became evident, it was gradually adapted to better suit their second-language learning, a process often referred to as "Emiratization" (Burden-Leahy, 2009).

In 2010, Abu Dhabi introduced bilingual education, with subjects such as science and math taught in English (Baker, 2016). Gallagher (2011) highlighted several reasons why introducing bilingual Arabic-English education in UAE state schools was considered "monumental" (p. 63). First, all subjects had previously been taught only in Arabic, with English as a foreign language, so Arabic now had to share dominance with English. Second, language-in-education issues had not been decisively addressed before, and this explicit policy was unusually direct. Third, the consideration of an equal standing of English was controversial because of threats that English would undermine Arabic sensitivities in the region, and thus, such policies were charged. Fourth, bilingual education was compulsory and general; every child in the Abu Dhabi school system was obliged to learn both languages in school, regardless of their parents' wishes.

English-language education in the UAE at the time had developed in a sociolinguistic environment characterized by multilingualism (Siemund et al., 2021). The UAE is a heterogeneous nation whose official language is Arabic, but English is the lingua franca in education, technology, business, and interaction. Siemund et al. (2021) argued that English in the UAE occupies a complex position, as it is not purely a foreign or second language given the country's demographics and global economic orientation.

As the significance of the English language increased, the UAE instituted policies of English-mediated instruction (EMI) and internationalization in schools and federal higher education at the beginning of the 21st century. This monolingualism has been greatly advanced by the expansion of EMI, particularly in tertiary education. Al-Bataineh (2020) noted that higher education language policy in the UAE has elevated English proficiency as a significant requirement for academic success and positioned English as a universal path to accessing international knowledge and improving the nation's competitiveness. English is commonly employed as the primary instructional language in this policy setting, as well as the primary language of evaluation and institutional communication and measure of academic credibility.

These positions indirectly support monolingual policies, although students continue to engage in multilingualism. In a more general conceptual sense, this norming of English goes hand-in-hand with what Troudi and Jendli (2011) referred to as a "choiceless choice" (p. 41), an unchallenged fact of Arab life whereby quality education is synonymous with English-medium instruction.

Notably, monolingualism in this context cannot be considered an absence of other languages but as an attitude that influences institutional expectations and norms. García and Torres-Guevara (2010) assumed that a monoglossic approach "values only monolingualism and ignores multilingualism," viewing language as "an autonomous skill that functions independently of the context in which it is used" (p. 182). In the UAE, the prioritization of English will cause a mismatch between language policies and everyday communication practices and raise questions regarding how a monolingual system of education in a multilingual environment can be followed without conflicting with equity and sustainability.

Standardized Testing and Assessment Regimes

With the rise in the use of English-medium instruction in UAE higher education, institutions have placed greater emphasis on setting readiness criteria and regulating access to study in English-medium instruction. This means that standardized assessment functions as an instrument of placement, and admission.

Here, the issue of standardized English language testing is directly related to the idea of gatekeeping. Jenkins and Leung (2019) even referred to the role of large-scale standardized tests in university acceptance as an "overriding gatekeeping function" (p. 86) that affects the education and subsequent career choice of applicants. They also identified internationally administered exams, such as the International English Language Testing System (IELTS) and the Test of English as a Foreign Language (TOEFL), as the most widely used assessments for university admissions around the world (Jenkins & Leung, 2019). Such observations help explain why standardized testing is mandatory in institutions where university programs are taught in English.

One major issue in the language assessment literature is that several traditional language assessments presuppose a monolingual, native-speaker norm. Jenkins and Leung (2019) remarked that English tests taken by nonnative speakers have traditionally been designed around an idealized native English model and focus on correctness and abstract academic standards of English. This practice is likely to establish proficiency in compliance with such norms rather than the skill of communicating in a variety of multilingual academic environments. They found it ironic that these tests, such as the IELTS, expect candidates to produce English that is akin to that of native speakers rather than assessing their preparedness to communicate in the lingua franca in international universities. They criticized this as "unfair and unjust" (Jenkins & Leung, 2019, p. 97).

Standardized testing is a common aspect of education across the globe, as well as a national policy and provision in the UAE. Zoghbor (2023) found that students in the UAE are frequently required to prove that they are competent in English by taking international examinations, including the IELTS. The Emirates Standardized Test (EmSAT) is a computer-based entry-level test used by UAE universities for admissions (as an alternative to the IELTS or the TOEFL; Zoghbor, 2023). This test was first introduced in 2018 (and retracted in 2024). This shows that English proficiency standards are incorporated into the higher education framework, as there are bridges between foreign and local testing systems. Al-Bataineh (2021) also noted the use of the Common Educational Proficiency Assessment (CEPA) English score and IELTS results as data sources to examine EMI-related linguistic frictions at a state university in the UAE.

Standardized testing practices are historically connected to wider monolingual language ideologies that view standardized varieties as the legitimate standard. Flores and Schissel (2014) explained how nationalist language ideologies present monolingualism in a standardized form as the norm, linking standards-based reforms to the history of assessments, where early standardized tests were monoglossic. They also mention the ongoing presence of "monoglossic testing regimes" (p. 476), supporting the idea that assessment methods can reinforce monolingual standards even within multilingual educational contexts (Flores & Schissel, 2014). Similarly, Al-Bataineh (2021) suggested that a monolingual perspective underpinned institutional practices at UAE universities, helping to explain how assessment and EMI practices may align with monolingual expectations.

In 2024, and in an encouraging move towards equitable opportunities, the Ministry of Education along with the Ministry of Higher Education have cancelled EMSAT exams in an effort to "give universities more responsibility and flexibility in setting admission criteria [which] will enable them to attract students who might not meet the English language proficiency requirements" (WAM, 2024, para. 9). Universities were given the opportunity to set up their own criteria for admission with what suits the specializations' requirements. This step is in line with educational sustainable goals, as it alleviates the consequences of high-stake exams on students' opportunities and provides equitable opportunities for students to enter universities.

KEY DISADVANTAGES OF MONOLINGUALISM

Research on English-medium instruction and language policy in the UAE shows that English holds a dominant structural role in higher education. In an investigation at a UAE state university, Al-Bataineh (2021) found that the conceptualization of institutional practices was based on monolingualism, although informal recognition of multilingual competencies occurred.

This institutional arrangement is related to broader sociocultural concerns. Bridi and Al Hosani (2022) reviewed the growing literature on the effect of Englishization on Emirati culture, especially in light of students' low Arabic literacy rates. They also observed that some perceived the use of English as the medium of instruction in higher education to be a threat to local culture and identity, and that it is occasionally associated with the depiction of Arabic culture and language as the "other" in their own country. They found that Emirati students had varied attitudes as well: although some perceived English as progressive and a necessity, others feared potential cultural loss (Bridi & Al Hosani, 2022). The authors also cited other measures, such as the Kalima Project (Word Project), in which texts in other languages are translated into Arabic, demonstrating institutional action to overcome perceived linguistic differences.

In their article on multilingual, bilingual, and monolingual Arabic teachers in the Emirates, Tang and Calafato (2021) noted that the idea of a "monolingual approach" to language teaching lingers in pedagogical practice. They provided information indicating that this approach can limit the ongoing development of students' metalinguistic and cross-linguistic awareness. The authors also discussed the survival of English in schools and the influence of English on the use of Arabic (Tang & Calafato, 2021). These findings suggest that the use of teaching methods that address one language may impact the way students use their general linguistic skills.

In a multilingual society such as the UAE, questions arise regarding the definition of linguistic legitimacy as well as how multilingual resources are incorporated into formal education systems. The next section discusses the emerging pedagogical practice of translanguaging in the UAE and how it seeks to redefine the role of multilingual repertoires in English language teaching.

TRANSLANGUAGING AS A RELATIVELY NEW APPROACH

The echo of translanguaging has been heard in the extant literature in recent years with regard to language education in the UAE, and the term is one of the most prevalent in contemporary language research. Studies have been instrumental in describing the nature of the English language vis-a-vis Arabic in the UAE (Al-Bataineh & Gallagher, 2021; Coelho et al., 2024; Hopkyns, 2022; Palfreyman & Al-Bataineh, 2018). Thomas (2023) wrote that the reason for "current research that portrays multilingualism [and translanguaging]" is that these phenomena are "not only prevalent in the larger community but also within Emirati families" (p. 360). As such, translanguaging provides a platform for understanding the nuanced use of language in the UAE and depicting its "complex linguistic diversity" (p. 346).

The other reason we focus on translanguaging in particular is that its premise intersects closely with that of the sustainable development goals. Putra et al. (2025) held similar sentiments, arguing that "The relevance of pedagogical translanguaging extends beyond linguistic competence. It is deeply intertwined with educational equity and social justice, as it recognizes and validates students' linguistic identities" (p. 657). These are linked to the issues ESDGs call for. Finally, despite the prevalence of translanguaging in the literature, especially in the UAE and other Gulf Cooperation Council (GCC) countries, the existing literature indicates that it is not entirely embraced by various stakeholders (Bin Rashed et al., 2025; Hopkyns et al., 2021; Palfreyman & Al-Bataineh, 2018). Although the context of the UAE lends itself naturally to translanguaging practices, this is not enacted in pedagogical practice (Hopkyns et al., 2021). In the hopes of continuing discussion about it, we have made it the highlight of our chapter.

In contrast to the view of monolingualism, which treats languages as detached compartments, translanguaging values the learner's full linguistic repertoire. When we consider the meaning-making process of a bilingual person, it is important to acknowledge that the process does not occur through two separate systems but rather operates within a single toolkit that integrates the complete range of linguistic and social experiences (Cenoz & Gorter, 2020). Thus, this practice serves communication and understanding, which are the fundamental purposes of language.

The term *trawsieithu* was first used in the 1980s in the educational context of Wales to describe a specific practice of teaching and learning in Welsh secondary schools (Vogel & García, 2017). This practice refers to presenting information to students in one language, such as reading a text in Welsh, and asking them to write their responses in another language, such as writing an essay in English. Accordingly, bilingualism is reinforced and conceptual understanding is increased through timed, planned support that responds to students' progress (Vogel & García, 2017). Later, the use of the term broadened in scope from "the school to the street" (Cenoz & Gorter, 2017, p. 903), which led to a distinction between the two terms described by Cenoz and Gorter (2017); pedagogical translanguaging (the intentional use of instructional strategies) and spontaneous translanguaging (the natural discourse practices of bilinguals in everyday life, where linguistic boundaries are constantly shifting).

This strategy was then translated into English and has since developed as a classroom method to become a practical representation of sociolinguistic theory. Vogel and García (2017) stated that this contemporary shift recognizes that multilinguals do not switch between two autonomous systems but rather function within a single system of meaning that includes all their linguistic and social experiences. According to Canagarajah (2011), translanguaging refers to "the ability of multilingual speakers to shuttle between languages, treating the diverse languages that form their repertoire as an integrated system" (p. 401). This implies that an Emirati student does not turn off Arabic to study science in English but uses all the content of their mental library to make sense of new information. To illustrate, an English-medium science teacher can discuss a concept such as evaporation in English, have the students discuss the concepts in Arabic, and then ask them to explain what they have learned in English using scientific terms related to the subject.

In this manner, Arabic does not become an interference but a cognitive bridge that aids students in processing the content, testing their understanding, and engaging more confidently. For a large number of learners in linguistically mixed classrooms in the UAE, this process can also reflect their lived experiences of flowing naturally between Arabic and English and/or English and their native languages.

PEDAGOGICAL TRANSLANGUAGING IN PRACTICE

When teachers shift their attitude toward a strength-based model rather than a deficit model, they begin to perceive students not as speakers of limited English but as emergent bilinguals. According to Cenoz and Gorter (2020), this transformation is at the core of pedagogical translanguaging. It is not just a random combination of words but a strategic plan through which teachers guide students in employing their home languages to enhance their understanding of the curriculum.

In this approach, multilingual resources are used to scaffold learning, increase access to academic content, and strengthen participation, especially for students who might otherwise be silenced by English-only rules (Duarte, 2020). At the same time, Leimgruber et al. (2024) point out that translanguaging can build metalinguistic awareness, which can help students notice connections between their languages rather than treating those languages as barriers. In classrooms where the official language of instruction is English but students' strongest meaning-making resources may include Arabic, pedagogical translanguaging becomes a way to make learning more inclusive, more realistic, and more aligned with how bilingual minds actually work (Leimgruber et al., 2024). For instance, in an English lesson where students read an English text, students are encouraged to connect the text to their prior knowledge via discussions in Arabic, and then finally write a paragraph about the text in English.

ALIGNMENT WITH SUSTAINABILITY AND EQUITY

The UAE Linguistic Landscape

With changing perspectives in education, the theoretical base of translanguaging stands on the grounds of responding to diversity. Siemund et al. (2021) described the UAE as a unique linguistic context shaped by multiple variables, including the diglossic nature of Arabic, where two systematic varieties of the language coexist and serve different social functions and values. This is evident in the different uses of standard Arabic (*Fusha*) and the Emirati dialect (Al Suwaiyan, 2018).

As stated by Leimgruber et al. (2024), the large percentage of expatriates residing in the UAE also contributes to the dominance of English in many fields, including education, as a practical tool for communication among speakers of different native languages This has led to the implementation of EMI in both government and private schools in the UAE. In this context, translanguaging is especially important since it provides a framework for achieving equity in EMI classrooms without disregarding the fact that Arabic is the linguistic core of identity, community, and meaning-making (Rogier, 2012).

Translanguaging as Social Justice

Socially, translanguaging can be adopted to address certain negative impacts of accredited educational systems. For example, graduates of elite international schools usually meet English fluency requirements, whereas students who attend government or Arabic-curriculum schools may lose opportunities to acquire knowledge and demonstrate their skills (Siemund et al., 2021).

The question of social justice is more evident when we consider the way in which English proficiency is connected to socioeconomic status. As explained by Tupas (2019), those from higher social classes often have privileged access to the standard variety valued by elite institutions, while others are marginalized due to their distance from dominant forms. Piller (2016) argued that English could function as a tool that reinforced social gaps and favored those with higher social standing. Translanguaging, in contrast, serves as a tool for social justice by validating the cultural and linguistic "wealth" that students bring from home. It ensures that those who are often silenced by English-only rules are given a voice and are not judged against a native-speaker standard that may be irrelevant to their lived reality. In practical school terms, this means that a student who understands a historical concept, a scientific term, or a literary theme should be given adequate opportunities to demonstrate their understanding even when their level in English is still developing. Students in the UAE enter classrooms with unequal exposure to English outside school, that is why translanguaging can reduce the unjust consideration of English fluency as an indicator of intelligence or academic ability. A student who completed their primary schooling in an Arabic-curriculum school, for instance, and transferred to an English-medium secondary school may be fully capable of analyzing a historical event or solving a complex mathematics problem, but may struggle to articulate that understanding in English within a timed setting. Without translanguaging practices, this particular student's silence is easily misread as a lack of comprehension. Instructional practices in the proceeding section will highlight how teachers can avoid these mishaps.

Sustainability and Equity in Practice

Translanguaging contributes to efforts to foster equity within education. According to Cenoz and Gorter (2020), permitting students to use their own language to study content areas helps break down historic hierarchical notions that certain languages are more advantageous than others for studying science. This establishes what scholars such as Yilmaz (2021) refer to as a "third space" (p. 6), which is a classroom where the identities of students are validated and where they are perceived as competent thinkers, instead of passive consumers of a foreign culture.

To implement this sustainably in the UAE, we must be conscious of the imbalanced relationship between English and Arabic. Sustainable translanguaging means creating what Cenoz and Gorter (2020) termed "breathing spaces" (p. 307) for Arabic. When we allow English to dominate all interactions, the local language may be drowned out. Real equity requires safeguarding the practical use of Arabic as well as enjoying the global benefits of English.

Translanguaging needs to shift from being perceived as a short-term scaffolding tool to being a form of social justice that responds to educational inequalities. As García and Leiva (2013) pointed out, this involves abandoning the deficit perspective, which constructs students based on apparent limitations but appreciates their linguistic resources and possibilities. The only way to attain linguistic sustainability is to recognize that language use is economically determined in most cases. In celebrating the entire repertoire of students, we would disrupt hierarchies in classes that determine who has a voice. In addition to merely teaching English, this method also advocates the idea of linguistic justice and safeguards linguistic human rights by making diversity not an obstacle to success but a human right (García and Leiva, 2013).

Nevertheless, this principle can only be relevant to educators once it is converted into everyday classroom practice. Practices can comprise scheduling Arabic-supported understanding, bilingual word banks, enabling translanguaging throughout peer work, permitting students to draw on their funds of knowledge and lived experiences, acknowledging them as sources of knowledge, and designing assessments that distinguish between content understanding and linguistic precision. It is important to note that when students have to return to English to perform, teachers should make this clear. This balance will assist in keeping expectations high, as well as achieving more equity in participation.

Four commitments are then needed to ensure a sustainable model of classroom implementation. First, try to see Arabic and/or other languages as learning tools, not as classroom problems. Second, plan for the integration of translanguaging instead of letting it happen by chance. Third, make sure that the multilingual interaction between classmates excludes no one. Fourth, connect every practice of translanguaging to a definite academic objective. These practices can be used as a practical guide by teachers who would like to implement equity without compromising the curriculum.

THE FLIP SIDE: THE CONSTRAINTS OF TRANSLANGUAGING AND WAYS TO ADDRESS THEM

Like any other rose, translanguaging is not without its thorns. Translanguaging presents pedagogical and practical challenges, especially when transitioning from theory to classroom implementation. Research highlights a persistent gap between the growing theoretical support for translanguaging and the development of effective, teachable classroom strategies, indicating that its success depends on careful planning and pedagogical alignment rather than on it occurring automatically (Liu & Fang, 2020; Canagarajah, 2011). In fact, the pedagogical dimension of translanguaging remains underdeveloped, and guidance on systematically integrating it into instructional practice is limited (Canagarajah, 2011). Empirical studies further demonstrate that implementing translanguaging requires ongoing negotiation among teachers, students, and institutional expectations, rather than serving as a straightforward instructional approach (Ticheloven et al., 2019). At the classroom level, issues can arise when using multiple languages and can ensue discomfort or exclusion for students who lack shared linguistic resources. Additionally, teachers might feel unsure about how to monitor and support all interactions effectively (Ticheloven et al., 2019). The absence of clearly defined pedagogical goals can also lead to the superficial application of translanguaging, without effectively supporting learning outcomes (Ticheloven et al., 2019). Tensions also emerge regarding institutional norms, particularly in high-stakes academic contexts such as writing, where standardized language expectations may limit acceptance of multilingual practices (Canagarajah, 2011). Furthermore, Canagarajah (2011) cautions against "romanticizing" translanguaging if its complexities and limitations are not critically examined, emphasizing the need for a balanced, critical orientation. These considerations do not diminish the value of translanguaging; instead, they emphasize the need for context-aware, well-organized, and pedagogically solid approaches to ensure it works effectively and sustainably across various educational environments. In the UAE context, where English-medium instruction coexists with ongoing concerns about the role of Arabic in education and society, these challenges highlight the need for carefully designed approaches that balance multilingual practices with efforts to sustain Arabic's functional and cultural presence (Al-Bataineh, 2021; Bridi & Al Hosani, 2022).

To work around these challenges, we recommend the following solutions. Starting with teacher preparation programs. Many of the stances promoted in English

Teaching syllabi and textbooks call for immersive and monolingual views or even nonequal grounding of English when juxtaposed against other languages, with English being the language of power and prestige. To eliminate this challenge, it is necessary for preservice English teachers to be exposed to various philosophies and ideologies of English instruction and to be aware of their own positionality and reflective practices. Until 2021, English teachers held negative views toward the use of Arabic in the classroom (Al-Bataineh & Gallagher, 2021). According to Portolés and Martí (2020), this was because "widely spread language teaching approaches are still grounded on a monolingual bias and languages are conceived as separate entities and, consequently, taught separately" (p. 25).

Second, because it is implausible that teachers would embrace translingual practices if their school or institution did not support such embeddedness (Coelho et al., 2024), it is imperative that educational institutions establish policies that promote such inclusion. Without clear policies, very little can be expected of teachers. In addition, it is not fair to let teachers bear the brunt of these principles by themselves. One approach, as postulated by Hopkyns et al. (2021), is for policymakers and educational institutions to "provid[e] translanguaging spaces where students are encouraged to communicate naturally through any language or mix of languages they choose, rather than adhere to strict 'English only' classroom rules" (p. 188).

Third, professional development in the UAE should move beyond general endorsement of multilingualism and instead equip teachers with usable classroom tools. These may include sample lesson structures, examples of bilingual questioning techniques, models of translanguaging-friendly assessment, and guidance on when Arabic use supports learning and when it may need redirection. Peer observations that focus specifically on translanguaging moments rather than general classroom management can further support teachers in refining their practices. Without this level of pedagogical specificity, translanguaging will be at risk of remaining an attractive theory that teachers support in principle but struggle to enact in practice.

An important aspect to consider with regard to instructional practice is that "Some of the activities have been used in language classes for many years. The difference is that pedagogical translanguaging activities use the resources from the students' repertoire in a systematic way in the language classes" (Cenoz & Gorter, 2022, p. 350). This means that many of the activities included as implications or recommendations in the literature are not novel—they have been included in English teachers' expertise repertoire—but what has been provided unknowingly as translanguaging practice needs to be made explicit. These practices vary, but they could include activities such as students interviewing family members in Arabic and then transcribing it in English and reflecting on the meaning-making process of using their first language as a vehicle to complete the task in the second (Bin Rashed et al., 2025). García et al. (2017) described another translingual activity pertinent to writing in which students wrote public service announcements in the language of their choice, as well as using various modes through posters or performative tasks. In another study, an instructional activity was given that included peer feedback on essay writing using translanguaging in a shared document (Sun & Zhang, 2022). These are just a few examples of an exhaustive list of instructional activities that are beyond the scope of these pages. It would take an entire book (see Cenoz & Gorter, 2021) to list all the different ways teachers could incorporate translanguaging, and as argued by Lee (2021), many translingual moments can be found in teachers' mundane practices.

Policy Moves for Recentering Arabic

The shift from monolingualism toward a more plurilingual orientation in the UAE has not emerged as an explicit policy statement but as a gradual resetting shaped by research documenting the consequences of English-medium expansion.

Following the agendas of modernization and global competitiveness, Rogier (2012) contextualized English institutionalization as part of the reform of higher education that placed English as the main language of pedagogical practice. According to Badry and Willoughby (2016), this transformation caused some structural tensions in relation to cultural continuity and the role of Arabic in national systems of higher education. In this context of reform, Al-Bataineh (2021) unveiled how English-medium policies led to enacting proficiency demands in ways that led to stratification and limited student choice, and contributed to uncertainty regarding the academic future of Arabic. She also demonstrated that daily linguistic activities at the university went against the official monolingual orientation. This showed that there was a gap between policy and practice, where multilingual negotiation continued even when institutional English was dominant. These works follow a distinct pattern: the standardization of English-medium instruction, the emergence of its sociolinguistic and equity consequences, and the challenge to monolingual policy presumptions. Notably, this interrogation has not been confined to academic critique but has increasingly shaped the policy-making process, especially in revising the assessment procedures that permit access to higher education.

One of the most important points of this phase of implementation occurred in the substitution of the CEPA with the EmSAT in the 2016/2017 academic year. EmSAT was presented by the UAE Ministry of Education (2017) as a national computer-based testing system to measure competency in Arabic, English, mathematics, and science at various grade levels. EmSAT framed language assessment in a nationalized system in line with UAE Vision 2021 and global benchmarking practices in comparison with CEPA, which was a more or less English-based university admission requirement. This change reorganized the logic of assessment in both Arabic and

English into a single national assessment system without abandoning English as a major academic language.

Arabic is reinforced as an academic and cognitive tool and not an outsourced cultural tool in newer reforms. With the major change that the UAE is experiencing, the focus on learners' native language is elevated, while undermining the belief that English should function as the exclusive language of academic development. This reflects the growing recognition that fostering intellectual development in the native language is achievable, which justifies the shift toward policies that safeguard and encourage the use of Arabic alongside English (WAM, 2025).

Beginning with the 2025/2026 academic year, all private kindergartens are obligated by the Ministry of Education to teach Arabic on a daily basis, with about one hour of classroom time per day by 2027 (The National, 2025). Another indication of a policy shift is the *Loughat Al Daad* program being implemented in Dubai. The program strengthens the idea of early engagement by ensuring that Arabic is taught to children aged 0–6 in early childhood centers and private schools (KDHA, 2025). Support for these measures is evident in the tools of accountability, such as the National Identity Mark by ADEK, which evaluates schools based on their incorporation of local values and language (ADEK, n.d.). In informal schooling initiatives, "Bil Arabi" takes Arabic to digital and social domains to encourage children and community members to use Arabic in various settings. Teacher training is also enhanced in a progressive manner to increase the quality and access to pedagogical services (MBRF, n.d.). Consequently, the UAE is developing an environment in which proficiency in the national language is established as a pillar of identity formation and academic success. This is achieved by preserving Arabic as a mandatory subject through Grade 9 for all students, along with the introduction of skills-based competitions.

The benefits of translanguaging are acknowledged and understood, but misinterpretation may occur in practice. Vogel and García (2017) pointed out that teachers might deliberately set standards lower than those of the learner or simply fail to be consistent in implementing various linguistic elements. In addition, they may undermine translanguaging out of classroom discipline concerns or because it is easier to use well-described methods of evaluating student performance. Yilmaz (2021) also added institutional policies and regulations to the discussion of possible constraints. This includes standardized tests, such as the Program for International Student Assessment (PISA) and the EmSAT, which may hinder the natural flow of translanguaging. The question of social justice becomes even more urgent when a local Emirati student, whose native language is Arabic, seeks to enter an Arabic-based program yet is required to demonstrate English mastery by passing the IELTS or TOEFL exams.

Teaching mathematics and science through EMI in the lower classes of the primary level further limits language use to English. Students' linguistic diversity is disregarded when all ideas and classroom messages are conveyed in the same language. To be truly sustainable and equitable, the educational future of the UAE must overcome these policy gaps and enable translanguaging to be not just an idea on paper but a means to promote student success.

BRIDGING THE THREE PILLARS

The previous sections discussed ways in which educational stakeholders can implement translanguaging in real settings. In this section, we aim to conclude our discussion with a vignette that exhibits not only the implementation of translanguaging practices, but also one that bridges it with facets of sustainability and place-based pedagogy, the three pillars of this paper. Given that the chapter started with "Uncle *Rashid* Had a Farm," we thought it fitting to conclude with the same song, bringing the discussion full circle.

Vignette

Miss Awatef starts the class by playing the song "Uncle *Rashid* Had a Farm." Students listen and sing along. They then have discussions about whether they have ever been to a farm and which animals they have owned. Next, students work in groups to determine which animals are most likely to be found on a farm in the UAE and which are not. Learners then listen to the original "Old McDonald Had a Farm" song and sing along, making specific comparisons between the animals found in Eurocentric regions and those in the GCC, particularly the UAE. During these interactions, *Miss Awatef* goes around the groups and ensures that students are interacting organically in both English and Arabic. She helps students with the names of animals in English by using Arabic terms. They compare the animals mentioned in each version of the farm and the sounds made by each animal in each of the songs. Next, students create their own farms using presentation slideshows and digital storybook applications. Attention is given to the name of the farm, its location, and the animals and produce it cultivates. Finally, a discussion of animal welfare takes place through the game Animal Crossing as a vehicle for discussion. *Miss Awatef* plays with students and then have students play independently and discuss the needs of the animals. They discuss what happens when care is not given to the place itself.

CONCLUSION

The contribution of this chapter lies in providing a comprehensive analysis of language education in the UAE in relation to components of sustainable education. Although many publications have traced the history of English in education in this region, to our knowledge, no study has specifically scrutinized this rich history through this lens. As such, the significance of this study lies in addressing this gap.

In our final analysis, we reiterate what has been argued between the lines of this chapter: that the language scene in the UAE is very rich and complex. Such complexity calls for the implementation of a rather comprehensive framework that embraces the multi- and plurilingual repertoire that learners in this region of the world hold. It is only through a serious and systematic implementation of translanguaging in language classes that we can truly move toward the fulfillment of the educational SDGs and the implementation of sustainable education.

As indicated in the previous sections, the implication is not exclusively practical and pedagogical in nature but also essentially philosophical in stance. We implore policymakers to be cognizant of the philosophies and frameworks of thought to which teachers and trainees are exposed across a range of perspectives, including sustainability and translanguaging. We also call for researchers in the region to move away from investigating stakeholders' attitudes toward translanguaging and other sustainable language education approaches and to move toward empirical studies in which the pedagogical implications of these approaches are tested to better assess their impact in regional classrooms and on the use of both English and Arabic.

REFERENCES

Abdel Qader, M. M. (2001). *Developing the listening skills of Arab learners of English as a foreign language in the UAE* (Master of Philosophy thesis, University of Glasgow).

ADEK. (n.d.). School Performance Plaques. https://www.adek.gov.ae/en/Programmes-and-Initiatives/School-Performance-Plaques#:~:text=About%20National%20Identity%20Mark,belonging%2C%20volunteering%2C%20and%20conservation

Al-Bataineh, A. (2021). Language policy in higher education in the United Arab Emirates: Proficiency, choices and the future of Arabic. *Language Policy*, *20*(2), 215–236. DOI: 10.1007/s10993-020-09548-y

Al-Bataineh, A., & Gallagher, K. (2021). Attitudes towards translanguaging: How future teachers perceive the meshing of Arabic and English in children's storybooks. *International Journal of Bilingual Education and Bilingualism*, *24*(3), 386–400. DOI: 10.1080/13670050.2018.1471039

Al Suwaiyan, L. A. (2018). Diglossia in the Arabic language. *International Journal of Language and Linguistics*, *5*(3), 228–238.

Alkhaldi, M., Moonesar, I. A., Issa, S. T., Ghach, W., Okasha, A., Albada, M., Chelli, S., & Takshe, A. A. (2023). Analysis of the United Arab Emirates' contribution to the sustainable development goals with a focus on global health and climate change. *International Journal of Health Governance*, *28*(4), 357–367. DOI: 10.1108/IJHG-04-2023-0040

Badry, F., & Willoughby, J. (2016). Higher education revolutions in the gulf. In *Higher education revolutions: Short-term success Versus long-term viability?* (pp. 204–236). Routledge.

Baker, F. S. (2016). National pride and the new school model: English language education in Abu Dhabi, UAE. In *English language education policy in the Middle East and North Africa* (pp. 279–300). Springer International Publishing.

Baroudi, S., & Abi Haidar, H. (2025). Sustainability in the United Arab Emirates secondary schools: A policy practice analysis. *Sustainability (Basel)*, *17*(7), 3129. DOI: 10.3390/su17073129

Bebbington, J., & Unerman, J. (2018). Achieving the United Nations Sustainable Development Goals: An enabling role for accounting research. *Accounting, Auditing & Accountability Journal*, *31*(1), 2–24. DOI: 10.1108/AAAJ-05-2017-2929

Bin Rashed, F. M., Aljanahi, M. H., & Al Qenai, R. M. (2025). Promoting Translanguaging in the ESL Classroom: A Closer Look into Kuwait and the United Arab Emirates. In *Intercultural Communication Education and Research in the Middle East and North Africa* (pp. 104-121). Routledge.

Bridi, R. M., & Al Hosani, N. (2022). An examination of the status of higher education in the United Arab Emirates: From humble beginnings to future challenges. *Eurasian Journal of Educational Research, 2022*(99), 26–44.

Burden-Leahy, S. M. (2009). Globalisation and education in the postcolonial world: The conundrum of the higher education system of the United Arab Emirates. *Comparative Education, 45*(4), 525–544. DOI: 10.1080/03050060903391578

Byram, M. (1997). *Teaching and assessing intercultural communicative competence*. Multilingual Matters.

Canagarajah, S. (2011). Codemeshing in Academic Writing: Identifying Teachable Strategies of Translanguaging. *Modern Language Journal, 95*(3), 401–417. https://doi-org.uaeu.idm.oclc.org/10.1111/j.1540-4781.2011.01207.x. DOI: 10.1111/j.1540-4781.2011.01207.x

Canagarajah, S. (2011). Translanguaging in the classroom: Emerging issues for research and pedagogy. *Applied Linguistics Review, 2*(2011), 1–28. DOI: 10.1515/9783110239331.1

Cenoz, J., & Gorter, D. (2017). Minority languages and sustainable translanguaging: Threat or opportunity? *Journal of Multilingual and Multicultural Development, 38*(10), 901–912. DOI: 10.1080/01434632.2017.1284855

Cenoz, J., & Gorter, D. (2020). Teaching English through pedagogical translanguaging. *World Englishes, 39*(2), 300–311. https://doi-org.uaeu.idm.oclc.org/10.1111/weng.12462. DOI: 10.1111/weng.12462

Cenoz, J., & Gorter, D. (2021). *Pedagogical translanguaging*. Cambridge University Press.

Cenoz, J., & Gorter, D. (2022). Pedagogical translanguaging and its application to language classes. *RELC Journal, 53*(2), 342–354. DOI: 10.1177/00336882221082751

Coelho, D., Khalil, N., & Shankar, D. D. (2024). Teachers' perceptions and 'invisible'uses of plurilingual pedagogy in UAE K-12 schools. *International Journal of Multilingualism, 21*(2), 670–688. DOI: 10.1080/14790718.2022.2099400

De la Fuente, M. J. (2021). Introduction: toward education for sustainable development (ESD) in foreign languages. In *Education for sustainable development in foreign language learning* (pp. 1-13). Routledge.

Duarte, J. (2020). Translanguaging in the context of mainstream multilingual education. *International Journal of Multilingualism*, *17*(2), 232–247. https://doi-org.uaeu.idm.oclc.org/10.1080/14790718.2018.1512607. DOI: 10.1080/14790718.2018.1512607

Flores, N., & Schissel, J. L. (2014). Dynamic Bilingualism as the Norm: Envisioning a Heteroglossic Approach to Standards-Based Reform. *TESOL Quarterly*, *48*(3), 454–479. DOI: 10.1002/tesq.182

Gallagher, K. (2011). Bilingual education in the UAE: Factors, variables and critical questions. *Education, Business and Society*, *4*(1), 62–79. DOI: 10.1108/17537981111111274

García, O., Ibarra Johnson, S., & Seltzer, K. (2017). *The Translanguaging Classroom. Philadelphia*. Caslon.

García, O., & Torres-Guevara, R. (2010). Monoglossic ideologies and language policies in the education of Latinas/os. In Murillo, E., Villenas, S., Galván, R. T., Muñoz, J. S., Martínez, C., & Machado-Casas, M. (Eds.), *Handbook of Latinos and education: Research, theory and practice* (pp. 182–193). Routledge.

Hopkyns, S. (2022). Translanguaging in Gulf English-medium instruction higher education. In Al Rashdi, F., & Mehta, S. R. (Eds.), *Language and identity in the Arab world* (pp. 76–92). Routledge., DOI: 10.4324/9781003174981-6

Hopkyns, S., Zoghbor, W., & Hassall, P. J. (2021). The use of English and linguistic hybridity among Emirati millennials. *World Englishes*, *40*(2), 176–190. DOI: 10.1111/weng.12506

Jenkins, J., & Leung, C. (2019). From mythical 'standard' to standard reality: The need for alternatives to standardized English language tests. *Language Teaching*, *52*(1), 86–110. DOI: 10.1017/S0261444818000307

KDHA. February 21, 2025. Under the directive of HH Sheikh Hamdan bin Mohammed bin Rashid Al Maktoum, Dubai launches new policy for Arabic language education in early childhood education. https://web.khda.gov.ae/en/About-Us/News/2025/Under-the-directive-of-HH-Sheikh-Hamdan-bin-Mohamm#:~:text=The%20new%20policy%20encourages%20a,13%20in%20Dubai%20private%20schools

Kramsch, C. J. (1998). *Language and culture*. Oxford University Press.

Lee, E. (2021). Translingualism in the teaching of English: Theoretical considerations and pedagogical implications. In Vinogradova, P., & Shin, J. K. (Eds.), *Contemporary foundations for teaching English as an additional language* (pp. 121–130). Routledge.

Leimgruber, J. R., Al-Issa, A., Lorenz, E., & Siemund, P. (2024). Managing and investing in hybrid identities in the globalized United Arab Emirates. *Journal of Language, Identity, and Education, 23*(6), 955–972. DOI: 10.1080/15348458.2022.2070849

Liu, Y., & Fang, F. (2020). Translanguaging Theory and Practice: How Stakeholders Perceive Translanguaging as a Practical Theory of Language. *RELC Journal*. Advance online publication. DOI: 10.1177/0033688220939222

Lozano, R., Lukman, R., Lozano, F. J., Huisingh, D., & Lambrechts, W. (2013). Declarations for sustainability in higher education: Becoming better leaders, through addressing the university system. *Journal of Cleaner Production, 48*, 10–19. DOI: 10.1016/j.jclepro.2011.10.006

MBRF. (n.d.). Bil Arabi. https://mbrf.ae/en/reading-arabic/bil-arabi

McLaren, P., & Giroux, H. (1990). Critical pedagogy and rural education: A challenge from Poland. *Peabody Journal of Education, 67*(4), 154–165. DOI: 10.1080/01619569009538705

Ministry of Education. (2017). The Emirates Standardized Test (EmSAT) to replace CEPA at universities. https://www.moe.gov.ae/En/MediaCenter/News/pages/emsat.aspx#:~:text=News,The%20Emirates%20Standardized%20Test%20(EmSAT)%20to%20replace%20CEPA%20at%20universities,the%20MOE%20English%20language%20curriculum

Ministry of Education. (2023). Green Education Partnership Roadmap. United A rab Emirates Ministry of Education. https://www.moe.gov.ae/En/MediaCenter/News/Pages/Green-Education.aspx

Palfreyman, D. M., & Al-Bataineh, A. (2018). 'This is my life style, Arabic and English': Students' attitudes to (trans) languaging in a bilingual university context. *Language Awareness, 27*(1-2), 79–95. DOI: 10.1080/09658416.2018.1431244

Piller, I. (2016). *Linguistic diversity and social justice: An introduction to applied sociolinguistics*. Oxford University Press. DOI: 10.1093/acprof:oso/9780199937240.001.0001

Porto, M. (2019). Affordances, complexities, and challenges of intercultural citizenship for foreign language teachers. *Foreign Language Annals, 52*(1), 141–164. DOI: 10.1111/flan.12375

Portolés, L., & Martí, O. (2020). Teachers' beliefs about multilingual pedagogies and the role of initial training. *International Journal of Multilingualism, 17*(2), 248–264. DOI: 10.1080/14790718.2018.1515206

Purvis, B., Mao, Y., & Robinson, D. (2019). Three pillars of sustainability: In search of conceptual origins. *Sustainability Science, 14*(3), 681–695. DOI: 10.1007/s11625-018-0627-5

Putra, H. S., Sugiharto, S., Da Silva, A. M., & Syahrizal, T. (2025). Pedagogical Translanguaging in Language Education: Constructing Translanguaging Space to Reduce Cognitive Load in Support of Sustainable Development Goals (SDGs). *International Journal of Language Education, 9*(3), 656–675. DOI: 10.26858/ijole.v1i1.77403

Rogier, D. (2012). *The effects of English-medium instruction on language proficiency of students enrolled in higher education in the UAE*. University of Exeter.

Siemund, P., Al-Issa, A., & Leimgruber, J. R. E. (2021). Multilingualism and the role of English in the United Arab Emirates. *World Englishes, 40*(2), 191–204. DOI: 10.1111/weng.12507

Suliman, O. (2000). *A descriptive study of the educational system in the United Arab Emirates*. (Doctoral dissertation, University of Southern California, 2000). Retrieved from ProQuest Dissertations and Theses database.

Tafese, M. B., & Kopp, E. (2025). Embedding social sustainability in education: A thematic review of practices and trends across educational pathways from a global perspective. *Sustainability (Basel), 17*(10), 4342. DOI: 10.3390/su17104342

Tang, F., & Calafato, R. (2021). Multilingual, bilingual, and monolingual Arabic teachers' development of learner self-regulation and language awareness in the Emirates. *Foreign Language Annals, 54*(1), 233–254. DOI: 10.1111/flan.12515

The National. June 22, 2025. Arabic lessons daily in all UAE private schools for kindergarten pupils. https://www.thenationalnews.com/news/uae/2025/06/22/arabic-lessons-daily-in-all-uae-private-schools-for-kindergarten-pupils/

Thomas, S. (2023). English in the United Arab Emirates. *World Englishes, 42*(3), 344–370. DOI: 10.1111/weng.12560

Ticheloven, A., Blom, E., Leseman, P., & McMonagle, S. (2019). Translanguaging challenges in multilingual classrooms: Scholar, teacher and student perspectives. *International Journal of Multilingualism, 18*(3), 491–514. DOI: 10.1080/14790718.2019.1686002

Troudi, S., & Jendli, A. (2011). Emirati students' experiences of English as a medium of instruction. In Al-Issa, A., & Dahan, L. S. (Eds.), *Global English and Arabic: Issues of language, culture, and identity* (pp. 23–48). Peter Lang.

Tupas, R. (2019). Entanglements of colonialism, social class, and unequal Englishes. *Journal of Sociolinguistics, 23*(5), 529–542. https://doi-org.uaeu.idm.oclc.org/10.1111/josl.12384. DOI: 10.1111/josl.12384

UNESCO. (2023). *Declaration on the common agenda for education and climate change at COP28.* https://www.unesco.org/sites/default/files/medias/fichiers/2023/12/Declaration-on-education-and-climate-change-en.pdf

United Arab Emirates Government. (2017). *UAE and 2030 Agenda for Sustainable Development: Excellence in Implementation.* Federal Competitiveness and Statistics Authority. https://fcsc.gov.ae/wp-content/uploads/2026/03/2018-UAE-and-the-2030-Agenda-for-Sustainable-Development-Excellence-in-Implementation.pdf

United Nations. (2015) Transforming Our World. The 2030 Agenda for Sustainable Development. https://sdgs.un.org/2030agenda

Van De Wetering, J., Leijten, P., Spitzer, J., & Thomaes, S. (2022). Does environmental education benefit environmental outcomes in children and adolescents? A meta-analysis. *Journal of Environmental Psychology, 81*, 101782. DOI: 10.1016/j.jenvp.2022.101782

Vander Ark, T. (2018, May 9). Get kids in the community and change the world. *Forbes.* https://www.forbes.com/sites/tomvanderark/2018/05/09/get-kids-in-the-community-and-change-the-world/

Vogel, S., & García, O. (2017). *Translanguaging*. Oxford Research Encyclopedia of Education., DOI: 10.1093/acrefore/9780190264093.013.181

WAM. (2024, November 3). *MoE, MOHESR cancel EmSAT exam for Grade 12 students*. Retrieved from https://www.wam.ae/en/article/b60yota-moe-mohesr-cancel-emsat-exam-for-grade-students

World Commission on Environment and Development. (1987). *Our common future*. Oxford University Press.

Yilmaz, T. (2021). Translanguaging as a pedagogy for equity of language minoritized students. *International Journal of Multilingualism, 18*(3), 435–454. https://doi-org.uaeu.idm.oclc.org/10.1080/14790718.2019.1640705. DOI: 10.1080/14790718.2019.1640705

Zoghbor, W. S. (2023). Negotiating the Arabic and English space in UAE higher education: The ownership of English in the English-medium instruction context. In *English as a Medium of Instruction on the Arabian Peninsula* (pp. 89–111). Routledge. DOI: 10.4324/9781003183594-8

Chapter 10
Investigating the Challenges Faced by Social Studies Teachers to Implement GCED in Primary Schools in Bangladesh

Tanjima Rahman
Institute of Education and Research, University of Dhaka, Bangladesh

Arifur Rahman
http://orcid.org/0000-0002-8828-4372
Institute of Education and Research, University of Dhaka, Bangladesh

Khairul Islam
http://orcid.org/0000-0003-0209-2896
Institute of Education and Research, University of Dhaka, Bangladesh

Raju Ahammed Fakir
http://orcid.org/0009-0009-9122-140X
Institute of Education and Research, University of Dhaka, Bangladesh

ABSTRACT

The study aimed to examine the challenges social studies teachers face in implementing global citizenship education in two primary schools in Bangladesh. This study employed a qualitative case study approach, focusing on two elementary schools:

DOI: 10.4018/979-8-3373-7650-9.ch010

one in an urban area and the other in a rural area. The primary data for this study were acquired through in-depth interviews with four social studies teachers and two head teachers from two schools. Additionally, four classroom observations and two focus group sessions with students were conducted to gain more clarity. This study analyzed data thematically. This study identifies seven core challenges regarding knowledge of GC and GCED, integrating GCED into facilities and infrastructure, classroom management and teaching skills, training and professional development, assessment of GCED, and a heavy workload. So, policymakers and school authorities need to focus on solving these challenges to make students appropriate global citizens.

INTRODUCTION

Global citizenship means being a member of a broader human society and recognizing the interconnectedness and interdependence of local, national, and global political, economic, social, and cultural dynamics (UNESCO, 2015). According to Oxfam (2015, p. 5), a global citizen is someone who respects people and their backgrounds, is dedicated to social justice, knows how the wider world works, connects with others locally and globally, works with others to make the world a fairer place, and carries responsibility for their actions. Accordingly, global citizenship education (GCED) is an important part of current education designed to empower global citizens who are knowledgeable about international issues and can engage in global affairs (Liu, 2020). GCED is the top priority for the United Nations and UNESCO. As a result, it is emphasized in the Education 2030 Agenda and Framework for Action. Specifically, it is related to SDG 4.7, which states that human rights, the rights of women, gender equality, peace and pacifism, the value of practical education, sustainable lifestyles, globalization, and the need for acceptance of diversity, as well as cultural engagement in its role in sustainable development and necessity for sustainable development by 2030. As a result, countries worldwide are prioritizing global citizenship education in their education policies and curricula (Browes, 2017).

In developed and developing countries, implementing GCED faces implicit challenges due to differences in socio-economic conditions and education systemsIssues in some advanced economies are attributable to longstanding nationalistic views prioritizing local citizenship rather than a global outlook. There may also be concerns about the reception of GCED in the political landscape, where some actors may perceive it as a threat to national identity or sovereignty (Starkey, 2022). On the other hand, infrastructure deficiencies and resource limitations are frequently linked to GCED hurdles in developing countries. Most educational establishments need more resources and qualified personnel to instruct GCED. Cultural considerations and regional political agendas may also hamper the adoption of a curriculum that encourages global connection; communities may still decide to prioritize pressing domestic or regional issues above global ones (Sant & González-Valencia, 2018; Browes, 2017). Bangladesh is an emerging country that started to focus on global citizenship education in the Education Policy 2010, the 2012 curriculum, and the new curricular framework of 2021 through integrating concepts of global citizenship, including global fraternity, awareness of human rights, sense of justice, and soft skills for global citizenship (NCTB, 2021).

However, through an extensive search, only one study was found on global citizenship education in Bangladesh, which examined the national curriculum framework 2021 for its embeddedness of GCED components in the curriculum (Ahsan et al., 2024), suggesting that nationwide policy efforts on GCED are yet to be prioritized into school–level implementation in the context of Bangladesh. Due to the status of Bangladesh as a growing country, it is extremely important to incorporate GCED in primary schools to prepare young people who are socially responsible, culturally aware, and globally engaged. Given Bangladesh's diverse society and its increasing importance in the global economy, GCED encourages the cultivation of critical skills in young people, such as empathy, tolerance, and citizenship (Ahsan et al., 2024; Rahman et al., 2025). GCED, particularly through social studies in primary school, is important to accomplish such a target. However, issues of teaching global citizenship education in primary school are hardly researched, especially the challenges teachers face in carrying it out. As it is a relatively new concept in Bangladesh, some challenges must be explored to solve them. So, this study aims to answer the following research question: How do Bangladeshi social studies teachers at primary school's experience challenges in implementing global citizenship education? This study will fill the research gap and contribute to literature by exploring the answer to this research question. Besides, understanding these challenges might be essential for policymakers and educators to formulate targeted measures that support teachers and facilitate the significant incorporation of GCED. This research seeks to illuminate these challenges, contributing to the continuing discourse on GCED in Bangladesh and promoting a more inclusive and supportive environment for primary educators dedicated to nurturing global citizens.

LITERATURE REVIEW

Global citizenship is an educational approach that helps learners understand global issues, respect diversity, think critically, and act responsibly for a more peaceful, just, and sustainable world. GCED develops learners' knowledge, skills, values, and action for a more just, peaceful, and sustainable world (UNESCO, 2015). GCED is also a multidimensional approach that helps learners achieve the quality of civic engagement and broaden participatory learning and social responsibility for sustainable change. Teachers are one of the main stakeholders in implementing GCED in school. However, studies show that they face multiple challenges in implementing GCED in the classroom and they need help with different obstacles in the practice and implementation of GCED (McGregor, 2023). Mairi et al. (2023) also argued that teachers have trouble integrating GCED and multilingual competencies into their planning, highlighting the necessity for further tools and professional development opportunities. As a result, this literature review section organizes the relevant literature thematically and highlights the key barriers that negatively influence GCED practice in schools.

Lack of Curriculum Integration and Policy Positioning

There is no specific curriculum on GCED in Bangladesh. This creates real problem in practicing GCED in schools because there lacks a clear guideline for the teachers to follow and implement GCED. Moreover, as GCED is not integrated into the curricula, it has posed significant challenges for the teachers who are working toward developing global awareness and social responsibility among students. Most of the curricula are driven by core courses and standardized assessments, leaving little room for emphasizing complicated, value-oriented themes central to GCED. GCED's absence from the national curricula means that teachers wishing to incorporate GCED often take it upon themselves in isolation, without structured support, resources, or time in the school day, which may discourage them from prioritizing it (Browes, 2017). When only the teachers are responsible for GCED practices, it becomes a burden for them, and such dependence creates uneven practice across schools and classrooms. Thus, some teachers may deliberately practice it, while others may practice it only fragmentarily or avoid it entirely. The literature therefore suggests that a specific curriculum, along with policy reform, is required to implement GCED in schools.

Lack of Teachers' Knowledge and Pedagogical Preparedness

Teachers' lack of GCED knowledge including human rights, global interconnection, cultural diversity, and sustainability, make it difficult for many teachers to implement them. GCED and sustainability are interlinked because it develops the knowledge, values, and competencies needed to address social inequality, cultural diversity, human rights, and environmental responsibility. This interconnection provides a pedagogical pathway through which sustainability can move from policy into classroom practice (McGregor, 2023; Mairi et al., 2023). However, this outcome depends on teachers' preparedness, curricular support, and institutional conditions in primary schools. GCED remains only partially connected to sustainability in practice where these are weak. Thus, it hinders the educational approach for sustainable development (Adaspayeva & Parkes, 2021; Bourn et al., 2017).

The multidisciplinary nature of GCED drew upon concepts from political science, geography, history, ethics, and environmental studies— all disciplines teachers may lack formal training or deep expertise. Teachers without a solid grounding in these subjects might think they cannot facilitate conversations that require critical thinking and nuance or engage with complex global issues. Such a difference may lead to superficial GCED implementation, where topics are approached but not adequately addressed, and limiting students' ability to engage with global issues. Moreover, GCED lessons may not relate to students or current issues due to weak content knowledge of GCED themes among teachers, which reduces the relevance and effectiveness of GCED. Building teachers' content knowledge in these areas is essential if teachers are to feel confident and effective in delivering GCED in their classrooms through dedicated professional development and resource availability (Bourn et al., 2017).

Lack of Assessment and Accountability

In addition, many educational systems have rigid and standardized curricula and focus on core subjects with high-pressure assessments. Standardized tests remain the most important measure of performance in many educational systems. However, these competencies' complex, multifaceted, and inherently subjective nature means that teachers face significant challenges in assessing students' global citizenship skills and knowledge. Unlike traditional subjects that rely heavily on quantitative evaluation and standardized tests, GCED aims to foster attitudes, values, and skills, including empathy, critical thinking, cultural sensitivity, and social responsibility, which are challenging, if not impossible, to measure through conventional assessment methods. Due to the nature of these competencies, teachers may struggle to create adequate instruments to assess their ability to demonstrate these proficiencies, especially since most standardized testing mechanisms do not consider GCED or prioritize it. These structural hurdles are implementing GCED in the classroom and the school. Moreover, evaluating students' understanding of global issues requires that teachers be aware of different viewpoints and experiences, making it difficult to develop fair assessments that account for disparities and then have the ability to achieve that understanding. Moreover, GCED-specific rubrics or guidelines are not readily available to teachers, forcing them to develop their evaluative criteria, leading to inconsistency and misalignment among different classrooms or institutes. This vagueness makes it difficult to monitor students' development concerning global citizenship and may put teachers off spending time on GCED due to the lack of clear evaluative goals. Teachers can better measure student progress in global citizenship through flexible and comprehensive assessment systems. Therefore, this literature suggests reformation in assessment to move GCED from policy to classroom practice.

Lack of Classroom Management and Interactive Pedagogy

Classroom management is a key skill that the teachers require to implement GCED. Because teachers with poor classroom management skills can seriously affect the effective implementation of Global Citizenship Education (GCED). It often requires facilitating open discussions of complex, controversial topics that can inspire passionate feelings and differing opinions among students. While addressing classroom management issues, teachers having difficulty managing classes may struggle to maintain a respectful and inclusive environment, leading to potential disruptions, hegemony by vocal students, or reluctance from others to speak (Assalihee & Boonsuk, 2023). In addition, GCED often requires collaborative work, group discussion, and problem-solving activities, all of which depend on strong classroom management skills. The teachers who are unable to manage classroom effectively, often struggle to ensure interactive pedagogies in the classroom that eventually hinders GCED implementation. Therefore, this study suggests that teachers require classroom management skills to implement GCED.

Lack of Professional Development and Teacher Support

Lack of training and professional development significantly affects the ability of teachers to implement Global Citizenship Education (GCED) effectively. Many teacher preparation programs and professional development initiatives ignore GCED and focus primarily on traditional academic subjects and instructional methods. This leads to teachers missing the indispensable knowledge and refinement of skills adapted to dealing appropriately with complex global challenges such as human rights, sustainability, and intercultural understanding (Norman, 2021). Teachers have few in-service training opportunities for Global Citizenship Education (GCED), particularly in under-resourced contexts, where they often do not have access to workshops, materials, and networks that enable them to develop their GCED competencies. Without sustained, practical, and even task-focused professional development opportunities in Global Citizenship Education (GCED) content areas—like cultivating sensitive dialogues in class spaces, implementing inclusive global curricula, and using interactive pedagogical approaches—teachers often feel unprepared and unsupported in their GCED endeavors (Parker et al., 2023). The literature suggests that one-off training sessions are insufficient; instead, teachers require continuous, practical, and contextually relevant professional development that helps them develop both the conceptual understanding and pedagogical skills necessary for GCED implementation.

Infrastructural Constraints

Infrastructural barriers are also a significant challenge for teachers trying to trace out Global Citizenship Education (GCED), as the lack of resources, poor facilities, and low access to technology could hinder the smooth course of teaching global issues. Many schools, particularly in underdeveloped regions, are challenged with overcrowded classrooms, lack of instructional materials, and restricted access to digital tools, which should provide an path for connecting students with global perspectives and interactive learning resources. Without the appropriate technology, teachers cannot use multimedia materials, interact with classrooms in other countries, and be informed about current events worldwide—all essential components of a fully realized Global Citizenship Education. Additionally, a lack of classroom space and high student-teacher ratios create barriers to interactive or group activities necessary for construing critical thinking, collaborative action, or discussions of global issues (Browes, 2017). School venues often have no designated space or resources for GCED, such as libraries and classroom resources for studies of the world. These infrastructural drawbacks limit teachers' ability to effectively deliver GCED while constraining students' exposure to diverse viewpoints and active learning opportunities fundamental to understanding and engaging with their surroundings (Parker et al., 2023).

Therefore, the literature suggests that teachers need specific training in GCED and sufficient infrastructure support to implement it in the classroom. This literature shows that GCED implementation is not hindered by only one obstacle. Rather, it is a set of interconnected challenges. Curriculum integration, teacher knowledge, professional development, assessment systems, classroom management, and infrastructure all interact to influence how GCED is understood and implemented in schools. While the global literature has identified many barriers to GCED, few studies have examined how primary school teachers in Bangladesh experience these challenges in practice. A study by Rahman et al. (2025) focused on social science teachers' perceptions of global citizenship in primary school but did not extensively examine the challenges of implementing GCED in primary schools. There is limited research on the challenges teachers face in implementing GCED in Bangladesh. As none of these addresses the primary education sector alone, this study seeks to address a strong gap. This study contributes a context-specific, more integrated understanding of GCED implementation and directly responds to the research question guiding the study. Furthermore, this study extends the GCED literature by providing an integrated theoretical explanation of how teacher cognition, pedagogical capacity, and systemic constraints interact to shape classroom implementation in a Global South context.

CONCEPTUAL FRAMEWORK OF THE STUDY

The conceptual framework guides this study by showing that the implementation of GCED is a multidimensional and theoretically informed process that includes teacher-level and system-level factors. Figure 1 (see below) illustrates that teachers' difficulties in integrating GCED are an outcome of six interrelated challenges: curriculum integration, subject knowledge, classroom management capacity, professional development, assessment practices, and infrastructural conditions. This framework functions as an umbrella as it includes all the challenges under one figure. The explanation is the following:

Content knowledge is important for implementing GCED in classroom. For example, Shulman's pedagogical content knowledge (PCK) theory emphasizes that GCED implementation requires both mastery of subject knowledge and the ability to transform that knowledge into pedagogically meaningful practices (Shulman, 1986). Based on literature, the framework shows that insufficient subject knowledge and pedagogical skills constrain teachers' ability to transform GCED concepts into effective classroom practices. In addition, Bandura's self-efficacy theory explains that teachers' beliefs about their capabilities influence their willingness to adopt innovative practices (Bandura, 1977). As a result, the Self-efficacy theory by Bandura (1977) helps explain how limited training and institutional support may reduce teachers' confidence in addressing complex global issues, thereby affecting implementation. Moreover, Ajzen's theory of planned behavior (TPB) further demonstrates that attitude, subjective norms, and perceived behavioral control shape teachers' intentions and actions (Ajzen, 1991).

This framework aligns with Theory of Planned Behavior by suggesting that external constraints such as curriculum rigidity, assessment pressure, and infrastructural limitations shape teachers perceived behavioral control over GCED practices. Thus, they constrain GCED implementation. More significantly, this framework does not involve only one obstacle; rather it shows how the combinations of all the challenges negatively affect GCED implementation in the primary schools in Bangladesh. For example, weak curriculum integration may limit opportunities for professional development, while inadequate infrastructure may restrict the use of participatory pedagogies, reinforcing classroom management challenges. These interdependencies highlight that GCED implementation is not solely determined by individual teacher capacity; rather, it is structurally embedded together to make the challenges more intense within broader educational systems. By integrating these challenges into a single visual model, the framework extends existing literature that often examines GCED challenges in isolation. This framework provides a context-sensitive lens for analyzing how multiple constraints converge in primary schools in Bangladesh. It also shows a more comprehensive understanding of the challenges that hinder GCED implementation in the primary education in Bangladesh.

Figure 1. Conceptual framework of the study

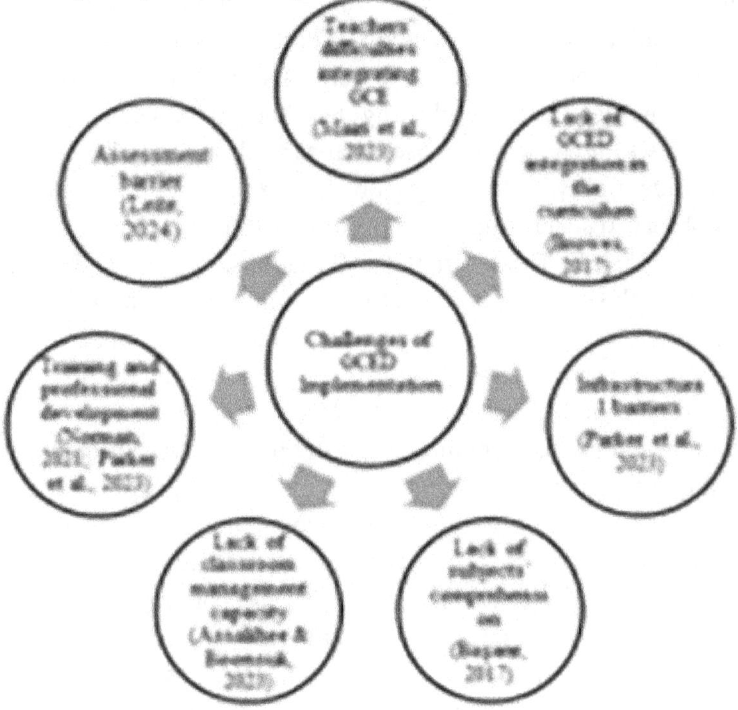

Note. This figure has been adopted from (Mairi et al., 2023; Browes, 2017; Parker et al., 2023; Assalihee & Boonsuk, 2023; Norman, 2021)

METHODOLOGY

A qualitative case study is suitable for examining the challenges encountered by social studies teachers in the implementation of Global Citizenship Education (GCED) in primary schools in Bangladesh, since it facilitates a comprehensive, context-specific analysis of the issues. This study employed a qualitative approach to reveal teachers' perspectives, experiences, and perceptions, offering insights into the deep and diverse realities they face within the Bangladeshi educational and cultural setting. Two primary schools were selected purposively considering urban and rural settings in Bangladesh. This intentional selection allows the study to encompass varied perspectives and underscore disparities in GCED implementation across these contexts, thereby offering a more thorough comprehension of the distinct and common obstacles encountered by social studies teachers in diverse socio-geographical environments.

In-depth interviews were conducted with two head teachers from rural and urban locations and two social science teachers from each school. Two FGDs of class five students were conducted from two schools. Additionally, four classrooms of social studies teacher from these two schools were observed. Each in-depth interview lasted 40 to 50 minutes. Before collecting data, the researcher invited participants to attend interviews and provided consent paper pledging for adhering ethical principles and confidentiality. After getting consent, researcher conducted interviews. Researchers observed social studies classrooms with the permission of head teachers and teachers. Each FGD needed 35 to 40 minutes. Following data collection, a comprehensive analysis stage was used for the qualitative data to elicit meaningful information about global citizenship education from the participants. Since Bangla is the mother tongue and accurately represents the issue, the original transcription of the recorded interviews was done in Bangla.

Following the data transcription, a cleansing process was performed to identify and address any errors or inconsistencies. A systematic coding process was followed. Every text segment was given a descriptive code, which effectively captures the essence of the content and makes systematic classification and retrieval possible during the analytical process. Recurring themes and patterns in the data were discovered during the coding process. As the study developed, codes were arranged into larger theme groups, making it possible to examine the underlying issues and points of view in greater detail. Redundant or superfluous data were eliminated in order to prioritize significant discoveries. Several strategies were employed to strengthen the analysis's validity and credibility, such as triangulation and debriefing. These steps helped and increased the credibility of the analysis, resulting in reliable findings that accurately represent teachers' actual experiences and challenges they face to implement GCED.

Table 1. Participant Demographics

Teachers Name	School Role	Teaching Experience (years)	Educational Qualification	Gender	Region Type
Refat Ara Begum (HU)	Primary School (Headteacher)	20 years	Postgraduate	Female	Urban
Khadiza Sultana Etu (TU1)	Primary School (Assistant teacher)	15 years	Graduate	Female	Urban
Rohima Khamon (TU2)	Primary School (Assistant teacher	6years	Postgraduate	Female	Urban
MD Sofiq Ahmed (HR)	Primary School (Headteacher)	23	Postgraduate	Male	Rural
Nitu Rani Majhi (TR1)	Primary School (Assistant teacher)	7	Graduate	Female	Rural
Sathi Akter	Primary School (Assistant teacher)	13 years	Graduate	Female	Rural
Two FGDs from Class Five students (1 Urban and 1 rural school). Each FGD consists of 6 students.					
*** Pseudonyms were used due to ethical considerations. HU= Head teacher of Urban School, TU= Teacher of Urban School. HR=Head teacher of Rural School, TR= Teacher of Rural School.					

FINDINGS

This study found that the teachers encounter multiple challenges in implementing GCED in the schools. The empirical data of the study demonstrated that because of the intense obstacles altogether, teachers often lose their motivation and thus fail to apply the pedagogical approachs that strengthen GCED implementation in broader educational contexts. The data were analyzed thematically to find the pattern of the challenges. As a result, the data analysis revealed seven themes: knowledge about GC and GCED, integrating GCED, facilities and infrastructural support, Classroom management, and teaching skills, Training and professional development, Assessment of GCED, and a load of tasks. These themes are based on the empirical data from the participants that require further explanation and analysis. Therefore, the themes are separately analyzed below.

Knowledge About GC and GCED

In-depth knowledge about global citizenship (GC) and global citizenship education (GCED) is crucial for teachers to implement in the classroom. However, this study finding indicated that four class teachers, two rural, one urban teacher, and a head teacher from a rural primary school, needed more knowledge about global citizenship and global citizenship education. These terms were new for them as they started to be popularized under the new curriculum framework in 2021. They had only a basic idea of global citizenship. They consider global citizenship to prepare a person to be skilled enough to compete with other people worldwide or feel sympathy and empathy for other people when other countries inflict them. A class teacher at a rural school said,

"To be frank, I do not know what global citizenship means. I have a concept about citizenship. In the new curriculum, 5 days training, it is discussed briefly. As a result, I am not clear about this term." –TR1.

The head teacher at a rural school needed to have a clear idea about global citizenship and global citizenship education but tried to explain it based on assumptions. This urban class teacher said,

"Generally, we are considered global citizens for living in this same world. But I think to be a global citizen, anyone must be competent enough to compete with the people of other countries to get a job and access any country easily." HR

However, a class teacher at an urban primary school clearly understands global citizenship terms. He learned about this issue through the internet as he was interested in learning this as a social studies teacher. He said,

"We can consider world brotherhood feelings as global citizenship. We are all connected and responsible for making this world peaceful. Human rights, people's voice, respect of diversity are indicators of global citizenship and global citizenship education"- TU1.

The head teacher at the urban school also had in-depth knowledge, and he admitted that the lack of knowledge about global citizenship was the main challenge in implementing this. During FGDs, it was found that teachers hardly discussed global issues, and they focused on only Bangladesh-related topics. As a result, students do not know the term global citizen or global citizenship. Teachers' lack of knowledge and practice about global citizenship issues was also found through classroom observations. From this case study, primary school head teachers and teachers' lack of knowledge about global citizenship is a major challenge to implementing global citizenship education.

Integrating GCED

Global citizenship education implementation relies on proper integration into the teaching-learning process, curriculum, and textbooks. The teachers and head teachers, both urban and rural, admitted the challenges of integrating global citizenship education in their schools. The indicators and concepts of global citizenship were explained to teachers who lacked knowledge by the researchers. After understanding, they all raised questions about implementing it due to the precise focus in the curriculum and textbook. The headteacher of a rural school said,

> *"Global citizenship is a new concept to us. The new curriculum framework is discussed briefly, and there is not much topic in the textbook. Through social studies subject, it can be implemented. So, Global citizenship issues need to be properly integrated into the curriculum and textbook."- HR.*

A rural school teacher was concerned about the limited time to complete the syllabus. Global citizenship is not directly included in the textbook, so it is difficult to discuss this issue with students intensely. Teacher said,

> *"I do not know where GCED is in the curriculum. We need to complete the syllabus within due time. As a result, we mainly focus on topics that adhere to the textbook. Sometimes, we can teach global citizenship issues while teaching related topics with examples. But time-constrained is the barrier." TR2*

An urban teacher raised concerns about implementing global citizenship, considering the weakness of teaching approaches.

> *"I do not know how I can effectively teach global citizenship education. First, I need to know how to teach global citizenship issues in the class." TU2*

The head teachers at the urban schools considered the lack of teachers' commitment as a challenge to integrating GCED in school. The headteacher of an urban school said,

> *"Unfortunately, our school teachers lack motivation to apply new things to their teaching. They do not take any initiative willingly for the betterment of the students. They mainly focus on completing their syllabus."- HU.*

In classroom observation, the lack of motivation of teachers to teach students about GCED is noticed. They could easily discuss global citizenship issues and actively engage students through the content they taught. Through FGDs, it is found that teachers do not integrate global citizenship issues into their teaching, which makes students unaware of global citizenship. So, global citizenship education integration in curriculum and textbook, lack of knowledge about how to teach it, and teachers' lack of motivation are the barriers to implementing GCED in primary schools.

Facilities and Infrastructural Support

Global citizenship education implementation in primary schools needs facilities and infrastructural support. The findings demonstrated that facilities and infrastructural barriers exist in schools. Teachers need school administration support to implement global citizenship education, but the study found that they need more facilities from school. A teacher from an urban school who had a clear idea about global citizenship education blamed the school authority for not providing support to execute the global citizenship issue. She said with a sorrowful tone,

"One day, I want to take my students to the local slum to make them realize people's sufferings and grow sympathy and empathy for them. But my head teacher did not permit me. You can not ignore these types of challenges."- TU1.

A teacher from a rural school indicated that the classroom size and teacher-student ratio are the barriers to implementing GCED. A teacher from a rural school mentioned,

"There are many students in the small classroom. I cannot pay attention to everyone. I cannot even teach something work-based if I want."- TR1.

A head teacher from a rural school said another fundamental challenge to implementing GCED. She said with grief,

"Our teachers get a minimal salary. They are suffering with their life. As a result, they can not focus on students' betterment wholeheartedly. If they get a sufficient salary, they could concentrate on students' betterment, including global citizenship education." HR

The head teacher of an urban school expressed frustration about the insufficient need for more projectors, multimedia, and other technology that support implementing GCED. Though teachers need different instructional materials, visual aids, equipment, and manuals, the head teacher could not provide these due to insufficient financial allotment and support from the government. He said,

"Our school has multimedia, but it is not working. I provided a requisition to the government office, but there has been no response for a long time. If our multimedia works, my teachers can show the video about global issues, which will be more effective"- HU.

From classroom observation, it is found that class size and student ratio is higher, which is a barrier to activity-based teaching. It is tough to implement GCED effectively, and activity-based teaching is crucial. Besides, teachers took classes through lectures; they did not use visual aids and did not engage students, which was necessary. From FGDs, it is also clear that teachers generally do not use any technology or visual aids that can enhance students' understanding of global citizenship.

Classroom Management and Teaching Skills

Teachers' classroom management capacity and teaching skills are essential to implementing global citizenship education. The study's findings revealed that all teachers from rural and urban schools practice religious and cultural diversity in their classes. All of them admitted that they did not consider who is from other cultures and religions and their socio-economic background during teaching. They even inspired students not to discriminate against other students. A teacher from a rural school said,

"In my class, there is a student from a Dalit family. One day, I noticed some students bullied him and insulted him. I made all students understand their misdeeds. I explained to them that every man is equal in this world. "- TR2.

This study found that teachers lack classroom management and teaching skills. All teachers from rural and urban schools used only the lecture method, which is not sufficient for global citizenship education. Teachers also lack pedagogical knowledge and skills. Besides, due to small classrooms, high teacher-student ratios, and a lack of other facilities, teachers rely on the lecture method solely.

A teacher from an urban school said,

"In foundation training, I learned a variety of teaching approaches. But due to sitting organization in the classroom, excessive students, and time constraint to complete syllabus, I can not apply other teaching approaches rather than lecture."- TU2

The study's findings showed that teachers tried to control the classroom effectively but could not do it properly because of students' disruptive behavior. For example, one teacher mentioned

"I try to control the students differently in the classroom, but some naughty students obey first and then disturb again. Because of this, I sometimes try to control them by treating them."- TR1

Both head teachers admitted the need for teaching skills and classroom management capacity. They expressed concern about teachers' reluctance to use various teaching methods in their classes and teachers needing to appropriately manage courses, which would be the barrier to global citizenship education. A head teacher at a rural school said,

"Our teachers always use the lecture method and can not control chaos in the classroom. Sometimes I requested them to control class and use a different teaching method, but they do not pay heed."-HR.

The classroom observation found teachers lacking classroom management and teaching skills. Teachers used a teacher-centric approach in their social studies teaching and did not focus on student engagement, which is the barrier to implementing GCED. From FGDs, it is also evident that teachers need more classroom management and teaching skills. In the teacher's interview, though they said they practiced diversity regarding culture, religion, and socio-economic background, students expressed concern about this issue. They sometimes get humiliated and discriminated against by teachers.

Training and Professional Development

Teachers' training and professional development are essential to implement GCED. However, this study found that all the teachers and head teachers of rural and urban schools needed more training and professional development facilities for global citizenship education. Teachers get foundation training for ten months from PTI (Primary Teachers Training Institute), and sometimes they get subject-based training from URC (Upazila Resource Centre). URC-based training is needed for them. But no one receives GCED specialized training.

A teacher from an urban school mentioned,

"Within 2-3 years, I get training, luckily. However, most of the training sessions are about basic teaching strategies. I never get any training on GCED. So I do not know how to teach global citizenship education effectively."-TU2.

A teacher from a rural school teacher said,

"In my whole life, I have received foundation training and only three training sessions, including new curriculum implementation training. But no training focuses on global topics or global citizenship education."-TR1.

Both head teachers also admitted the need for more training and professional development regarding global citizenship education. Though global citizenship education is connected with social studies, teachers did not get any training about this issue. Headteacher from an urban school mentioned,

"Teachers need more training to understand Bangladesh's new curriculum. Besides, they need to get specialized training on global citizenship. Without training and workshops on this issue, it is not possible to implement it properly."- HU.

From classroom observation and FGDs, it is evident that teachers lack the skills to implement GCED in the classroom. Lack of training and professional development is one of the major reasons for their unskilled classroom teaching and management.

Assessment of GCED

Assessment of global citizenship is crucial to implementing global citizenship education effectively. This study finding indicated that all teachers and head teachers had little knowledge about the evaluation of GCED. They did not know about specific tools for measuring the global citizenship skills of students. Traditionally, they take written tests to measure student competency. As a result, they are concerned about the global citizenship skill assessment process. A teacher from a rural school indicated,

"We use paper and pencil tests to evaluate our student performance. But I realized now that assessment of global citizenship needs a different approach which I do not know."-TR2.

Another teacher from an urban school said,

"In our country, all subject teachers follow almost the same system of assessment to measure student performance. However, GCED is related to the psychomotor and affective domains. So, the assessment of GCED will be based on hands-on experience and activities, though we do not have clear guidelines for assessing global citizenship."- TU2.

The headteacher of a rural school mentioned,

"There is a separate assessment guideline from the previous curriculum in the new curriculum. The new curriculum has no instructions on assessing our student's global citizenship skills, which I think is a significant barrier to implementing GCED''-HR.

Classroom observation and FGDs also confirmed teachers' assessment barriers to global citizenship, which indicated that teachers and students need clear guidance on assessing global citizenship education.

Task Load of the Teachers

Teachers need sufficient time to prepare for class. This study indicated insufficient teachers in urban and rural schools, contributing to increased task loads. Teachers take classes and do all types of administrative work because no clerk has been recruited in the primary schools in Bangladesh. A teacher at an urban school mentioned

"We not only take our assigned classes but also engage with all administrative works, including student admission, communicating with the education office and documentation, and meeting with guardians and other stakeholders. Besides, we must engage with government campaigns and activities, which is responsible for not taking class properly."-TU1

A teacher from a rural school said,

"In our school, we have only three teachers. I take classes in Bangla, social studies, and Islamic studies. Multiple subjects teaching and overload administrative work compelled me to stick to the syllabus. There is little time left for content not directly related to the syllabus. ``- TR2

Headteachers of rural and urban schools admitted to the pressure of teachers. GCED will be an extra burden if it is not directly included in the subject matter.

A head teacher of an urban school admitted,

"My teachers are overburdened with their assigned subject and administrative work. As a result, they lacked the time and energy to give their classes with full attention. To implement global citizenship education, I think the teacher load needs to be reduced."-HU

From classroom observation and FGDs, it is found that teachers are exhausted from taking five to six classes a day. As a result, they do not prepare lesson plans and do not discuss anything beyond the fixed topic. Overloading work is one of the barriers to implementing GCED in the primary schools of Bangladesh.

DISCUSSION

The results showed that a shortage of space prevented teachers from implementing activity-based teaching and learning strategies. The findings aligned with Rahman (2024), who argued that there were several challenges, such as teachers' insufficient administrative competencies, limited infrastructure, and excessive class numbers. This limitation refrains teachers from implementing GCED in primary schools. This challenge hinders the implementation of GCED in primary schools. To overcome this, Ovcharova (2023) suggests that teachers must employ efficient classroom management techniques and establish a compelling learning environment to foster involvement. Aligning the study of Mamaile & Omodan (2023) and Xu et al. (2023), who also noted that overcrowded classrooms, inadequate parental engagement, and linguistic obstacles in educational settings might impede the effectiveness of management initiatives. Therefore, teachers should learn more about classroom management strategies to implement GCED. Infrastructure is crucial for implementing GCED. A study in Indonesia conducted by Sabilah (2024) found that educational institutions in Indonesia also challenges related to inadequate infrastructure and financial resources, which adversely affected learning conditions and student motivation (Sabilah, 2024). Likewise, the study's results unveiled several obstacles to obtaining and implementing GCED in Bangladesh. Like in Bangladesh, comparable challenges are evident in Liberia, where restricted infrastructure, unsuitable educational programs, and insufficient teaching personnel impede the availability of high-quality education (Yele & Specioza, 2024). The study of Khairunnisa et al. (2024) also aligns with these findings, as they mentioned that the lack of adequate infrastructure and financial resources has a detrimental effect on learning environments.

The study found that the absence of professional development and training among teachers impeded the implementation of GCED, while inadequate training and instruction undermined teachers' professional development. The study found that teachers in Bangladesh, Zambia, and Ghana considered professional development advantageous for improving their pedagogical knowledge and teaching abilities (Muyunda et al., 2023; Borna et al., 2023). Zindi and Sibanda (2023) argued that implementing well-planned professional development programs aligned with curricular goals and targeted toward specific requirements is a practical approach to tackling these issues. Therefore, teachers require sufficient training to implement GCED in the class. This present study revealed substantial differences in rural and urban students' social and economic circumstances, resulting in inequities. Vadivel et al. (2023) and Mulyani et al. (2024) argued that children from economically disadvantaged households frequently face academic challenges and prematurely join the labor market. This underscores the necessity for free vocational education and initiatives to raise parental awareness. Adverse economic conditions harm children's health, education, and general development, especially in impoverished urban regions. Therefore, the socio-economic conditions should be developed to implement GCED.

The study's results indicated a shortage of teachers in urban and rural schools, leading to a rise in workload. Both rural and urban schools displayed an inadequate student-teacher ratio, giving teachers limited time to deliver lectures. Mukuka and Alex (2024) argued that the disparity in the ratio of teachers to students gives rise to apprehensions over the caliber of education and the possible proliferation of extremism. Research findings also indicated areas for improvement in the geometric expertise of student teachers. Therefore, the ratio of teachers to students should be balanced in order to implement GCED. The study found that one of the most important challenges was insufficient research. The research findings indicate that policymakers should have considered the stakeholders' opinions appropriately during the policy-making process. Aligning to the study, Mulyani et al. (2024) noted that implementing Global Citizenship Education (GCED) encounters challenges like insufficient comprehension, curricular limitations, cultural hindrances, and resource shortages. Therefore, it is recommended that the issues of GCED should be investigated.

Finally, this study demonstrates that the challenges of GCED implementation in Bangladesh are not simply operational or logistical but are deeply theoretical and systemic in nature. By integrating TPB, PCK, and self-efficacy perspectives, the study provides a more comprehensive explanation of how and why these challenges remain, and how they may be addressed through coordinated interventions at the policy, institutional, and classroom levels. This theoretical discussion extends existing research by moving beyond descriptions of barriers and offering an explanatory framework that links teacher cognition, pedagogical capacity, and systemic structures in shaping GCED practice (UNESCO, 2015; NCTB, 2021).

Limitations

The researcher purposefully and conveniently selected the study region (including the site and schools) based on considerations of financial resources, time constraints, and appropriate authorization. The researcher selected schools that were both willing to provide empirical evidence and readily accessible. The study exclusively included government primary schools located in urban and rural locations. The researcher's failure to account for variability in primary schools resulted in restricted findings. Furthermore, this study did not investigate the views and approaches of teachers in other disciplines. Notwithstanding this, the classroom observation data was not captured due to permission issues. Ultimately, due to the qualitative nature of the study, it was not feasible to generalize the findings. Further quantitative study could be conducted in a large-scale sample to generalize the findings. This would strengthen the implications of the study.

Implications

There is no specific policy on GCED in Bangladesh. As informing a new policy requires research findings to understand the gap and to recommend initiatives, this study contributes to broader policy and systematic implications. The policy makers would find this study useful to address the gap in policy and practice. Thus, analysis of research findings in the study would provide valuable information to guide and shape the development of new plans and policies. Furthermore, the results would contribute to enhancing the attitudes and practices of social science teachers in the classroom, as they showed a deficiency in comprehending global citizenship education (GCED). Furthermore, the research undertaken would contribute to implementing the National Curriculum Framework 2021 (NCTB, 2021) and facilitate the attainment of Sustainable Development Goal 4.7, as announced by the United Nations in 2015.

CONCLUSION

Global citizenship education is essential for establishing an impartial and equitable global society. The study's findings underscore social science teachers' challenges when integrating global citizenship education (GCED) into classes and schools in Bangladesh. Despite the importance of a comprehensive understanding of global citizenship education (GCED), the expertise of teachers in this field is limited. Implementing classroom activities and practices associated with global citizenship education (GCED) was impeded by the absence of instructional resources. Teachers were instrumental in implementing GCED in classrooms and institutions; however, they required sufficient training. As a result, the school's lack of diversity practice and the insufficient training of teachers presented substantial challenges. In order to safeguard their liberties, the students were required to engage in co-curricular activities. Providing teachers with comprehensive professional development and practical legislative guidelines would assist in the removal of the challenges associated with global citizenship education (GCED).

REFERENCES

Adaspayeva, N., & Parkes, S. (2021). Global Citizenship Education conceptualisation in curriculum guidelines of the New Zealand Curriculum. *The New Zealand Annual Review of Education*, *25*, 57–75. DOI: 10.26686/nzaroe.v25.6935

Ahsan, S., Islam, M. K., & Khan, M. R. H. (2024). Scope of Global Citizenship Education in the National Curriculum Framework 2021 of Bangladesh: A critical Review. *Teacher's. World Journal of Educational Research (Los Angeles, Calif.)*, *49*(1), 49–76. DOI: 10.3329/twjer.v49i1.70238

Ajzen, I. (1991). The theory of planned behavior. *Organizational Behavior and Human Decision Processes*, *50*(2), 179–211. DOI: 10.1016/0749-5978(91)90020-T

Assalihee, M., & Boonsuk, Y. (2023). Teaching Management Strategies on 21st Century Islamic Education for Southernmost Thai Private Islamic Schools. *Anatolian Journal of Education*, *8*(1), 13–28. DOI: 10.29333/aje.2023.812a

Bandura, A. (1977). Self-efficacy: Toward a unifying theory of behavioral change. *Psychological Review*, *84*(2), 191–215. DOI: 10.1037/0033-295X.84.2.191 PMID: 847061

Borna, C. S., Oppon-Wusu, M., Bedu-Addo, P. K. A., Dabone, K. T., Kwarteng-Nantwi, E., Churcher, K. A., Aninganigu, K., Owusu-Sekyere, F., & Kissi-Abrokwah, B. (2023). Professional development of mathematics teachers: Its impact on their classroom delivery in Ghana. *East African Journal of Education Studies*, *6*(1), 320–337. DOI: 10.37284/eajes.6.1.1152

Bourn, D., Hunt, F., & Bamber, P. (2017). *A review of education for sustainable development and global citizenship education in teacher education: Background paper prepared for the 2017/8 Global Education Monitoring Report, Accountability in education: Meeting our commitments*. UNESCO. Retrieved from https://discovery.ucl.ac.uk/id/eprint/10030831/1/bournhuntbamber.pdf

Browes, N. (2017). *Global citizenship concepts in the curricula of four countries*. UNESCO International Bureau of Education.

Khairunnisa, K., Dongoran, I. Y., Harahap, K. G., Lubis, N. A., Azhar, Y. A., & Haerunisa, A. (2024). Problematika Sarana dan Prasarana dalam Proses Belajar Mengajar di SD Negeri 104202 Bandar Setia. *Journal of Education*, *6*(4), 20586–20591. DOI: 10.31004/joe.v6i4.6112

Liu, H. (2020). Globalization and world Citizenship education. *International Journal of Learning and Teaching*, *94–97*, 94–97. Advance online publication. DOI: 10.18178/ijlt.6.2.94-97

Mairi, S., Gruber, J., Mercer, S., Schartner, A., Ybema, J., Young, T., & Van Der Meer, C. (2023). Teacher educators' perspectives on global citizenship education and multilingual competences. *Journal of Multilingual and Multicultural Development*, *46*(9), 2631–2647. DOI: 10.1080/01434632.2023.2170388

Mamaile, D., & Omodan, B. I. (2023). Exploring challenges hindering teachers' implementation of classroom management strategies in Gauteng High Schools, South Africa. *Research in Educational Policy and Management*, *5*(2), 245–262. DOI: 10.46303/repam.2023.22

McGregor, I. M. (2023). Teaching for global citizenship: A case study on the challenges of implementing a human rights course. *Compare: A Journal of Comparative Education*, •••, 1–18. DOI: 10.1080/03057925.2023.2254220

Mukuka, A., & Alex, J. K. (2024). Student Teachers' Knowledge of School-level Geometry: Implications for Teaching and learning. *European Journal of Educational Research*, *volume–13–2024*(volume–13–issue–3–july–2024), 1375–1389. https://doi.org/DOI: 10.12973/eu-jer.13.3.1375

Mulyani, H., Komalasari, K., Permatasari, M., Bribin, M. L., & Suriaman, S. (2024). Transformasi Pendidikan Kewarganegaraan Global di Era Abad 21: Analisis Implementasi dan Tantangan. *Jurnal Kewarganegaraan*, *21*(1), 88–101. DOI: 10.24114/jk.v21i1.55115

Muyunda, G., Yue, L., & Oranga, J. (2023). Teachers' professional development in Zambia: Perceptions and practices. [IJSL]. *International Journal of Social Learning*, *3*(2), 222–233. DOI: 10.47134/ijsl.v3i2.177

National Curriculum and Textbook Board [NCTB]. (2021). *National Curriculum Framework 2021.*: Dhaka Ministry of Education, Peoples' Republic of Bangladesh. Retrieved from https://nctb.gov.bd/pages/static-pages/6922dc5d933eb65569e0ff0f

Norman, E. J. (2021). Global Citizenship Education in the English Language Classroom: Theory and practice. *Advances in Social Science, Education and Humanities Research. Advances in Social Science, Education and Humanities Research.* Advance online publication. DOI: 10.2991/assehr.k.211224.033

Ovcharova, K. (2023). The role of the teacher in Classroom Management: Promoting active participation and engagement. *Педагогически Форум*, *11*(3), 45–54. DOI: 10.15547/PF.2023.018

Oxfam. (2015). *Education for Global Citizenship.* A guide for schools. Oxford, Oxfam.

Parker, R., Chainey, J., Goundar, P., Richardson, S., Dabrowski, A., Berry, A., & Scoular, C. (2023). *Being and becoming global citizens: Measuring progress toward SDG 4.7. Phase I: Monitoring teacher and school readiness to enact global citizenship in the Asia-Pacific region.* https://doi.org/DOI: 10.37517/978-1-74286-718-2

Rahman, M. M. (2024). Practice and perceptions of classroom management: A study at the tertiary level. *Asian Journal of Contemporary Education, 8*(2), 62–76. DOI: 10.55493/5052.v8i2.5098

Rahman, T., Islam, K., & Rahman, A. (2025). Social science teachers' perceptions and practices of global citizenship education: A case study of primary schools in Bangladesh. *Journal of Interdisciplinary Studies in Education, 14*(4), 49–72. DOI: 10.32674/c7dz0p23

Sabilah, A. I. (2024). Revitalisasi fasilitas PAUD Pelangi guna meningkatkan kualitas pendidikan usia dini. *Jurnal Sains Teknologi Dalam Pemberdayaan Masyarakat, 3*(1), 43–48. DOI: 10.31599/kg02s789

Sant, E., & González-Valencia, G. (2018). Global Citizenship Education in Latin America. In The Palgrave Handbook of Global Citizenship and Education (pp. 67–82). London: Palgrave Macmillan. Retrieved from https://www.researchgate.net/publication/325270842_Sant_E_Gonzalez-Valencia_G_2018_Global_Citizenship_Education_in_Latin_America_In_The_Palgrave_Handbook_of_Global_Citizenship_and_Education_pp_67-82_London_Palgrave_Macmillan

Shulman, L. S. (1986). Those who understand: Knowledge growth in teaching. *Educational Researcher, 15*(2), 4–14. DOI: 10.3102/0013189X015002004

Starkey, H. (2022). Challenges to global citizenship education: Nationalism and cosmopolitanism. In *Global Citizenship in Foreign Language Education* (pp. 62–78). Routledge. DOI: 10.4324/9781003183839-5

UNESCO. (2015). *Global Citizenship Education: Topics and Learning Objectives.* UNESCO.

Vadivel, B., Alam, S., Nikpoo, I., & Ajanil, B. (2023). The impact of low socioeconomic background on a child's educational achievements. *Education Research International, 2023*, 1–11. DOI: 10.1155/2023/6565088

Xu, C., Zhu, K., & Liu, S. (2023). Classroom management Strategies in secondary language education. *Journal of Education Humanities and Social Sciences, 8*, 1649–1654. DOI: 10.54097/ehss.v8i.4539

Yele, W. B., & Specioza, A. (2024). Examining the access, quality, and relevance gaps in Liberia's educational policy environment. *International Journal for Multidisciplinary Research*, *6*(3), 20482. Advance online publication. DOI: 10.36948/ijfmr.2024.v06i03.20482

Zindi, B., & Sibanda, M. M. (2023). Challenges impeding the implementation of effective continuous teacher professional development programs in Eastern Cape Department of Education. *Business Ethics and Leadership*, *7*(4), 46–61. DOI: 10.61093/bel.7(4).46-61.2023

Chapter 11
Reimagining Chemistry Education Through Place-Based and Sustainability-Oriented Learning

Antuni Wiyarsi
http://orcid.org/0000-0001-5573-9345
Universitas Negeri Yogyakarta, Indonesia

ABSTRACT

This chapter explores how chemistry education can become more meaningful by connecting learning to students' local environments, cultures, and real-life experiences. It highlights place-based education (PBE) as an approach that encourages students to engage with community issues and sustainability challenges. By integrating socio-scientific issues and ethno-chemistry, the chapter shows how traditional knowledge and cultural practices can be linked to scientific concepts. It also presents a community-engaged, project-based learning model where students collaborate with local communities to address real problems. This approach enhances understanding, strengthens relevance, and fosters responsible, reflective, and environmentally aware learners

A. REDEFINING PLACE-BASED EDUCATION (PBE) IN SUSTAINABILITY-ORIENTED CHEMISTRY LEARNING

Place-based education (PBE) is a pedagogical umbrella that prioritizes experiential learning, community, and ecological contexts to strengthen the connectivity between students and their local environment, culture, and context (Gruenewald,

DOI: 10.4018/979-8-3373-7650-9.ch011

2003a; Smith, 2002; Sobel, 2004; Orr, 2013). This approach focuses on actively transforming students into proactive researchers of the various dimensions of the environment—whether social, political, economic, or cultural—in which they grow and interact (Nichols et al., 2016). Thus, PBE facilitates students' active exploration of the sociopolitical and ecological realities of their community, thereby fostering sensitivity to local culture and the environment. Within the framework of PBE, the concept of 'place' goes beyond mere geographic markers or physical settings. PBE discourse emphasizes that place is a meaningful lived space, with a particular focus on the importance of shared cultural heritage (cultural commons) in shaping learning experiences. Place is viewed as an entity that provides political, social, and ethical dimensions to the learning process (Israel, 2012). This depth of meaning is reflected in the PBE framework, which encompasses four main aspects (Ardoin, 2006; Ardoin et al., 2012), covers:

1. Biophysical: Refers to the basic physical context and ecological characteristics of a location.
2. Psychological: Emphasizes the unique experiences and emotional bonds each individual forms within that physical environment.
3. Socio-Cultural: Views individuals as part of a particular society and culture that builds and maintains a collective relationship with that place.
4. Political-Economic: Highlights the various political processes and economic forces that shape a place and influence people's attitudes toward it (Qablan & Al-Qaderi, 2009).

Practically, the implementation of PBE is manifested in four integrative dimensions proposed by Granit-Dgani (2021), which are learning in place, study of place, learning from place, and for the sake of the place. These four dimensions provide a progressive framework for understanding the relationship between students and their environments:

1. Learning in place

This first dimension focuses on the shift in the physical learning space from the classroom to an outdoor environment. In this context, the essence of the curriculum or lesson plan may not change significantly; the main change lies in the arrangement of the learning environment or setting to provide contextual refreshment. For example, in inquiry learning, teachers take students to observe (one steps of inquiry-based learning) the different of many leaf shapes in the garden, then discuss together outdoors to extend their knowledge (Qablan, 2015). That means, students do observation in real situation, not only in the laboratory.

2. Study of place

The second dimension involves an in-depth investigation of a specific environment. Learners observe and analyse various processes—both natural and social—occurring in that location, making the place the primary object of study. There are efforts to build a sense of place between students and the specific environment around them. For instance, students collect water samples over several months to measure pH, dissolved oxygen, and water hardness in limestone areas. They map out the characteristics of the area's limestone soils, compare water content in various wells and streams, and analyse local industries and waste management processes arounds. This provides a comprehensive description of water quality, as well as the environmental elements and community activities that influence it.

3. Learning from place

At this stage, the environment and all its components are considered to have an active pedagogical role. Places are no longer mere objects but rather unique sources of knowledge that provide valuable lessons not found in theoretical texts, for both educators and learners. In this dimension, there is a shift from "learning about something" to "learning with something." For example, rather than treating environmental issues as abstract concepts, students investigate real-world phenomena, such as a local river changing colour or emitting odors—clear indicators of pollution within their own community. Students map local industries and their waste management, observe residents' behaviour in managing household waste, analyse clean water quality and how they 'maintain' the river. Students not only learn about scientific concepts, but begin to grasp the intricate link between human agency, public health, and ecological integrity in build environmental awareness and responsibility as citizens. In this context, place provides moral, emotional, and spiritual inspiration to students.

4. Learning for the sake of the place.

This final dimension is action- and advocacy-oriented. Based on the insights gained from the previous three dimensions, this dimension aims to encourage positive change and conservation efforts, so that education directly contributes to the sustainability and well-being of the place (Qablan & DeBaz, 2015). Students do not merely diagnose local issues/problems and their root, but they also provide solutions to these problems. For example, students collaborate to build and test low-cost, sustainable filtration systems using locally available materials (like charcoal and sand) in areas with high iron levels in the water. The students provide these designs to local community members, as a service dedicated to the health, safety, and collective well-being of the residents who inhabit that place.

Rooted in constructivist principles, PBE fosters students to actively participate in learning through engagement with their environment and community activities. The implementation of PBE enhances students' understanding, critical thinking, learning experiences, interest, and concern for sustainability (Ardoin et al., 2012; Badawy et al., 2024; Bascopé & Reiss, 2021; Hernandez Gonzalez, 2023; Wade-Lyles, 2016). PBE enables students to reflect on their ancestors' practice and relevance in different aspects of life, think critically about their own practices, and keep place-based knowledge in thoughts and action. This fortified learners to get connected with the people of the community, apply their learning concept to the real-life situation, and most of all appreciate the place-based knowledge of our ancestors and its contribution in the environmental aspect.

On the other hand, over the past few decades, global awareness of the urgency of environmental conservation has significantly escalated, along with the emergence of new scientific insights into the vulnerability of our planet's ecosystems. This phenomenon emphasizes that sustainability is no longer merely an option, but rather a fundamental prerequisite that integrates ecological, social, and economic aspects to ensure the future survival and health of the Earth. In this context, each emerging new understanding demands a paradigm shift in how humans interact with nature, in order to mitigate damage and maintain the balance of the increasingly fragile biosphere. The practical global application of the sustainability concept is currently guided by the 17 SDGs, which integrate peace, justice, and climate action (Tahat et al., 2024). Fundamentally, sustainability represents the complex challenges of harmonizing economic advancement with ecological preservation. It revolves around the strategic utilization of natural capital to drive growth while ensuring that these resources are neither degraded nor exhausted. At its essence, sustainability serves as a vital framework for intergenerational equity, aiming to safeguard environmental integrity while simultaneously navigating social and economic transitions. However, this balance is increasingly threatened by global pressures such as accelerated urbanization, climate change, and demographic shifts. Within this context, education emerges as a critical catalyst; without a profound public awareness of environmental systems and the consequences of human agency, progress toward a sustainable future remains significantly impeded. it implies the need for in-depth discussion of sustainability issues from school level that are integrated in school subject (such as Chemistry) and real action in activities outside of classroom.

PBE, that focuses on students' environment study, is the main pedagogy to reach Education for Sustainable Development (ESD) goals. Specially, study of Burmeister et al. (2012) suggests four basic models how to integrate chemistry learning with ESD, include (1) Model 1, that is integrating practices from Green and Sustainable

Chemistry (GSC) into school chemistry practical work; (2) Model 2, which is including GSC practices as content in the chemistry curriculum; (3) Model 3, that means using socio-scientific issues (SSI) that are related to sustainability issues and chemistry as starting points for learning chemistry and about the relevance of GSC for sustainable development, and (4) Model 4, that is making sustainability a driver for school reform and changing school life accordingly including aspects that are related to chemistry and its applications. This shows the reciprocal relationship between PBE and sustainability that can be applied in chemistry learning. For example, PBE brings global sustainability issues like climate change or the biodiversity crisis to the forefront, making them easily understandable for students. In PBE, students not only learn about global warming theoretically but also explore how changing seasonal patterns affect farmers' harvests in their area or how sea level rise affects coastal areas. Besides environmental aspects, two other pillars of sustainability, social/cultural and economic, are also a concern for PBE. The social/cultural pillar relates to the community as a place and resource for students to learn about local environmental issues, which will then benefit the community through the proposed solutions.

Furthermore, sustainability and SSI-based instruction overlap since they adopt an issue-based perspective, address multiple perspectives, and seek solutions through ongoing inquiry. So forth, the use of SSI related to sustainability issues as a starting point in chemistry learning is an effort to support the SGDs agenda in quality education. The multidimensional characteristics of SSI are very close to the 17 SGGs agendas. For example, SSI with an environmental dimension is closely related to the goals in the aspects of Clean and Affordable Energy, Climate Change Management, Marine Ecosystems, and Terrestrial Ecosystems. Meanwhile, SSI with a health dimension is related to the goals of Healthy and Prosperous Living, as well as Responsible Consumption and Production (Almomani et al., 2024). This is in line with research results showing that the chemistry-focused SSI topics most often integrated in chemistry learning are related to chemicals and environmental pollution, the use of fossil fuels, everyday chemicals and nutrition (Çalık & Wiyarsi, 2021).

B. ETHNO-CHEMISTRY: MAPPING CULTURAL ASSETS AS CHEMISTRY LEARNING CONTEXT

The meaning of culture itself is not simple, in anthropology, culture is defined as the whole complex of ideas and everything that humans have produced in their historical experience. Nature of culture could be described, such as; (1) Culture could be learned. This means that culture is not inherited biologically, but is learned through socialization from childhood within the family and community; (2) Culture is social. Culture is shared by members of a society. Culture cannot exist without a human group; (3) Culture is symbolic. Culture uses symbols such as language, traditional clothing, flags, and ceremonies that have specific meanings; (4) Culture is dynamic. Culture is always changing according to the times, technology, and interactions with other cultures; (5) Culture is integrated. Cultural elements such as religion, economics, education, and customs are interrelated and influence one another; (6) Culture is adaptive. Culture helps humans adapt to their natural and social environments, and (7) Culture is cumulative. Culture is passed down from generation to generation and continues to evolve over time.

Culture can also be described as a way of human behaviour, including ethics, language, eating habits, religious and moral beliefs, knowledge, attitudes, and values that are the result of human work such as various objects including tools/objects produced by technology. Furthermore, it is also stated that culture is a pattern of a society's way of life. Systematically, cultural forms are divided into three views, namely (1) culture as an idea, concept, value, norm, and outlook on life, (2) culture as activities and patterned actions of humans in society, and (3) culture as objects produced by humans/artifacts (Byram, 1995; Macionis, 2017).

1. Culture as an idea, concept, value, norm, and outlook on life.

Culture, by its very nature, has an abstract dimension, making it intangible or visually intangible, unlike physical objects. This dimension resides entirely within the minds or collective consciousness of the communities where the culture grows and develops. Despite its immaterial nature, this ideal culture can manifest itself in written form, such as essays and literature by local authors, often known in the Indonesian context as *'Adat Istiadat'*. The manifestation of this system of values and norms ultimately results in Indigenous Knowledge, a body of knowledge developed by indigenous communities through generations of empirical experience and in-depth interaction with their surroundings.

Indonesia's cultural diversity has given rise to a variety of local wisdoms that serve as the foundation for social interaction and environmental conservation. In Java, the value of *Gotong Royong* (mutual cooperation) is a pillar of community cooperation, aligning with the Balinese *Tri Hita Karana* philosophy, which emphasizes a balanced relationship between humans, God, and nature. This spirit of togetherness is also reflected in the *Huma Betang* philosophy of life in Kalimantan, which upholds tolerance in communal life. In addition to regulating interpersonal relationships, local wisdom also serves as an instrument for nature conservation, as seen in the *Sasi* tradition in Maluku, which imposes periodic bans on the harvesting of marine products to maintain the ecosystem. Similar principles are upheld by the *Baduy* people of Banten, who live in strict harmony to maintain the purity of nature.

Overall, all of these cultural practices are bound by norms of politeness that underlie ethical behaviour in social life throughout the archipelago.

2. Culture as the patterned activities and actions of humans in society

This form is known as a social system, which encompasses all the patterned actions of humans themselves in social life. This social system consists of various interactive activities, reciprocal relationships, and social interactions between individuals that occur continuously over time, always following certain patterns based on customary behaviour. As a series of real activities, the social system is concrete because it occurs around us in everyday life, so its existence can be observed, photographed, and documented. In practice, this form of culture is most often manifested through the implementation of traditional ceremonies or traditions that serve as physical representations of the social interactions of that society.

The diversity of traditions in Indonesia embodies a rich cultural identity reflected in various traditional ceremonies in each region. In Bali, there is the *Ngaben* Ceremony, a ritual of soul purification through cremation, while in Java, the *Sekaten* and *Nyadran* traditions symbolize religious acculturation and respect for ancestors. Cultural dynamics are also evident in the entertainment and sportsmanship aspects of *Karapan Sapi* in Madura, and the sacredness of final respects in the *Rambu Solo*' Ceremony in Toraja. Not to be missed, the Papuan people uphold togetherness and gratitude through the *Bakar Batu* tradition. All of these cultural manifestations are not merely ceremonies, but concrete manifestations of the social system of society that continues to be preserved from generation to generation.

3. Culture as objects/artifacts made by humans

The third form of culture is physical culture, which represents all the results of human activities, deeds, and material works in society. Compared to other forms, physical culture is the most concrete because it consists of objects or artifacts that can be directly touched, seen, and photographed.

The manifestations of this physical culture are extremely diverse, ranging from magnificent architecture like the *Rumah Gadang* in West Sumatra to textile masterpieces like *batik* and *ikat* weaving from East Nusa Tenggara. Furthermore, functional and symbolic objects such as regional musical instruments and *keris* (traditional weapon) and culinary riches (traditional Indonesian food) like *Gudeg* (Yogyakarta), Papeda (Maluku), and Rendang (Padang) are also inseparable parts of physical culture, serving as tangible evidence of a nation's creativity and civilization.

Meanwhile, ethno-chemistry studies the relationship between the cultural practices, traditions, and local wisdom of a community and the principles of chemistry.

The term is derived from the words 'ethno' (culture/nation) and 'chemistry'. Simply put, ethno-chemistry explores how traditional communities understand, utilize, and process the natural materials around them through chemical processes—though they may not use modern scientific terms to describe them. Furthermore, ethno-chemistry, as part of indigenous scientific knowledge in society, is a community's perception of a phenomenon and develops through a pattern of continuous transmission from generation to generation in an unstructured, informal, and local manner. It is contrast to formal science, or what is now called scientific science, it can be understood scientifically using structured scientific methods, thus being objective and accountable. Consequently, there is a need for the reconstruction of authentic science integrated into chemistry learning. The process of integrating culture-based chemistry education can be carried out through systematic content structuring steps to bridge local knowledge with formal science.

1. The first step begins with dissecting specific cultural practices, such as the fermentation process in making tempeh or the interaction of dyes in batik, to discover the underlying pure chemical phenomena.
2. Once these phenomena are discovered, the next step is to identify related chemical concepts—for example, reaction kinetics, thermochemistry, or chemical bonding
3. Explain the application of these concepts in depth within the context of the relevant cultural practices.
4. As a final step, all these findings are formulated to design relevant learning activities, so that students can learn chemistry through the lens of indigenous knowledge, or the knowledge they encounter in their daily lives.

These steps will make it easier for teachers to recognize the connection between culture and chemistry and then develop PBE with the community's culture as a 'place'. The following is an example of an ethno-chemistry study in the Indonesian context.

Jamasan pusaka (Legacy Collection/traditional weapon)

The *Jamasan Pusaka* ceremony is a traditional Javanese ritual performed periodically, especially during the month of *Suro*, as an effort to conserve cultural heritage objects and a manifestation of spiritual values. Technically, this ritual involves the process of cleaning metal from oxidation using natural ingredients such as lime juice containing citric acid to dissolve rust, and coating with *warangan* liquid (arsenic) to protect the blade and emphasize the *pamor* pattern (Haryoguritno et al., 2006). Behind these technical aspects, *jamasan* contains the philosophical meaning of *manunggaling rasa*, namely a symbol of purification of the human soul so that it is always clean from worldly stains (Suhardi, 2016). Recent research also shows that *Jamasan Pusaka* is now seen as a form of indigenous knowledge that is effective in maintaining the material integrity of historical artifacts while strengthening cultural identity amidst modernity (Prabowo et al., 2023). There some chemistry concept could be learn from this ceremony.

Lime fruits can be used to clean rust, as they contain citric acid. Citric acid has proven to be an efficient solution in addressing oxidation and rust problems on metal surfaces. Citric acid is an organic compound that is weak acidic, but effective enough to dissolve rust which is generally in the form of iron oxide (Fe_2O_3). Citric acid works by binding iron ions (Fe^{3+}) which causes rust to form. When citric acid comes into contact with rust, a chemical reaction occurs in the form of a redox reaction in which citric acid acts as a reducing agent that helps convert iron oxide into a compound that is more soluble in water. This process occurs because citric acid has a carboxyl functional group (-COOH) capable of bonding to iron ions, forming a stable and water-soluble complex. With the formation of this complex, iron ions no longer bind strongly to metals, so rust can be easily cleaned from metal surfaces such as keris or other heirlooms.

The reaction between citric acid ($C_6H_8O_7$) and rust ($Fe_2O_3 \cdot xH_2O$) is a redox reaction, in which citric acid acts as a reducing agent, helping to convert Fe^{3+} into a more soluble form. Reaction equation in general: $Fe_2O_3(s) + 6H^+(aq) \rightarrow 2Fe^{3+}(aq) + 3\ H_2O(l)$ Citric acid provides H^+ ions, which help dissolve Fe_2O_3 into Fe^{3+} in solution. Once Fe^{3+} ions are formed, citric acid can synthesize those ions, forming a more soluble iron-citrate complex: $Fe^{3+}(aq) + C_6H_8O_7(s) \rightarrow [Fe(C_6H_5O_7)]\ (aq) + 3\ H^+(aq)$ Fe^{3+} ions The dissolved ones will form a ferry-citrate complex, which is easier to clean with water.

The fragrances used in heirloom banquets contain natural chemical compounds, most of which belong to benzene derivatives. After the cleaning process, heirlooms such as keris are usually smeared with perfumed oïls such as sandalwood oil, jasmine oil, or clove oil. The chemical compounds in this fragrance give it a distinctive aroma and also play a role in maintaining the condition of heirloom metal. Some examples of chemical compounds in perfumes used in heirloom jewellery are (1) Eugenol, is found in clove oil, giving it a spicy and sweet aroma; (2) Benzyl acetate is found in jasmine oil, gives it a distinctive floral fragrance; (3) Vanillin, comes from cinnamon or vanilla, giving it a sweet and warm scent, and (4) Limonen is found in citrus oil, giving it a fresh and citrus aroma. These compounds are aromatic compounds that have a benzene (C_6H_6) structure or derivatives, resulting in a distinctive and long-lasting aroma.

Batik (traditional fabric)

The word batik in Javanese comes from the root word tik which refers to the work of hands that are delicate, soft, and detailed, and contain elements of beauty (art). The word tik also refers to the process of making fabric patterns by '"driping"' night (candle) with a tool called canting so that it forms a pattern consisting of an arrangement of stitches and strokes. The batik making process often uses wax, bleach, and synthetic chemical dyes, which can pollute the environment. Batik dyes are divided into two types: (1) Natural dyes, derived from plants, animals, and minerals, such as indigo, *soga*, and turmeric. It is more environmentally friendly, but the colour is limited and less durable; and (2) Synthetic dyes, such as azo, aniline, remazole, indigosol, and napthol. More varied and durable, yet contains harmful chemicals.

In the *batik* production process, *batik artisans* use synthetic textile dyes and other processes such as wax removal, soaking and rinsing will produce residual dyeing water substances. These residual substances produce residual waste rich in reactive dyes and chemicals that can pollute the environment. So that waste treatment is needed before the substance is simply dumped into the surrounding environment. Natural dyes are dyes obtained from natural sources such as plants, animals, and minerals, without going through complex chemical synthesis processes in the laboratory. The use of natural dyes such as Indigofera tinctoria in batik dyeing is in line with the principles of green chemistry, which emphasizes the use of environmentally friendly materials and more sustainable processes. Compared to synthetic dyes that can produce toxic waste and pollute the environment, natural dyes are safer for humans and ecosystems.

Additionally, natural ingredients such as Indigofera tinctoria can be renewed, reducing reliance on petroleum-based chemicals. In terms of chemical processes, these natural dyes work through oxidation reduction reactions. The leaves of Indigofera tinctoria contain the indican compound (indoxyl-β-D-glucoside), which are colourless. When the leaves are fermented in anaerobic conditions (without oxygen), the indican decomposes into indoxyl. After the fabric is dipped in this solution and exposed to air, the indoxyl oxidizes into indigo, which produces a distinctive blue colour, which following reaction:

enzim

$$C_{14}H_{17}NO_6\text{(indican)} + H_2O \rightarrow C_8H_7NO\text{ (indoxyl)} + C_6H_{12}O_6\text{(glucose)}$$

$$2C_8H_7NO\text{ (indoxyl)} + O_2 \rightarrow C_{16}H_{10}N_2O_2\text{ (indigo)} + 2H_2O$$

This process does not involve harmful chemicals, making it more environmentally friendly compared to synthetic dyes that often use heavy metals and other toxic substances. In terms of sustainability, the use of natural dyes also supports the local economy and cultural preservation. Colouring plants such as Indigofera tinctoria can be cultivated organically without harmful pesticides, maintaining the balance of the ecosystem. By applying the principles of green chemistry, the *batik* industry can reduce the impact of environmental pollution due to synthetic dye waste and produce products that are safer for human health and the environment (Almomani et al., 2022).

Furthermore, the batik industry, despite being a cultural and economic pillar, has significant ecological impacts due to the discharge of liquid waste without adequate treatment. *Batik* waste contains various hazardous chemicals, including synthetic dyes (such as azo compounds), heavy metals, and suspended solids with very high levels of Chemical Oxygen Demand (COD) and Biological Oxygen Demand (BOD). When released into water bodies, these substances inhibit sunlight penetration and reduce dissolved oxygen levels, drastically disrupting photosynthesis by aquatic plants and threatening the survival of aquatic organisms (Setiyandono et al., 2024). Recent research has shown that the accumulation of heavy metals from dyeing waste not only permanently damages river ecosystems but also poses long-term health risks to communities through contamination of groundwater and the food chain. Therefore, the implementation of sustainable waste treatment technologies is an urgent need to maintain a balance between preserving cultural heritage and protecting environmental integrity.

Wayang kulit (Shadow Puppets)

Shadow puppets is a traditional Indonesian performing art—especially developed in Java and Bali—that uses carved puppets (usually buffalo leather) as characters, whose shadows are projected onto a white cloth screen (*kelir*) using a lamp (*blencong*). Etymologically, the word "*Wayang*" comes from the word "Ma Hyang," meaning "toward the Almighty," or "*Wayangan*," meaning "reflection of life."

Before being used as a basic ingredient in making puppets shadow, cowhide must go through a series of processing processes to make it stronger, more flexible, and more durable. The first stage is skin cleansing, which aims to remove dirt, salts, and fat that still sticks. This process is usually done by soaking the skin in fairly strong running water, such as in a river or in a special washing bath. The torrential flow of water helps to remove the remnants of dirt as well as salt that was previously used to preserve raw skin. In addition, this washing also serves to re-soften the skin that may have hardened during storage.

After the cleaning stage is completed, the cowhide then enters the tanning process, which is an important stage that aims to prevent decay and increase the leather's resistance to the environment. This tanning process uses chemicals such as lime ($Ca(OH)_2$). As we know that the cowhide chosen for the material for making shadow puppets still has fur and fat tissue. Therefore, hair and fatty tissue must be removed with a lime solution ($Ca(OH)_2$) which is alkaline so that it can loosen the hair follicles and break down fat. The leather becomes cleaner and ready for the next stage of tanning. After soaking in a lime solution, the leather is washed to remove the remnants of the lime chemical.

Calcium hydroxide, or more commonly known as lime or betel lime, with the formula $Ca(OH)_2$. This compound is the result of the reaction of calcium oxide (CaO) with water: $CaO(s) + H_2O(l) \rightarrow Ca(OH)_2(s)$. Calcium hydroxide is weak bases and is often used in a variety of industrial processes, including tanning, building mortar, as well as an acidity control agent in soil and water. It plays an important role in the tanning process for puppet making. This compound helps remove hair and fat, opens up leather fiber, prevents decay, and prepares the leather to absorb tanning materials more easily.

The wick on the *blencong* (traditional oil lamps) is made of natural fiber materials such as cotton or fabric, which is a natural polymer of cellulose. Cellulose is a complex carbohydrate compound composed of many glucose chains (polysaccharides). When the wick is dipped in kerosene, the capillarity of the cellulose fiber allows the oil to rise to the top of the wick. When the wick is ignited, it works by absorbing kerosene through the capillaries, then burning slowly. The oil that has been absorbed undergoes a combustion reaction, producing fire and light. The wick itself does not burn completely because the fire burns more oil that continues to be absorbed, so the flame is maintained as long as there is still fuel. When the wick is lit, the cellulose in the cotton will be burned, producing carbon dioxide (CO_2) and water vapor (H_2O). $C_6H_{10}O(s) + 8O_2(g) \rightarrow 6\ CO_2(g) + 5\ H_2O(g) + \text{Energy}$. If combustion is imperfect, carbon monoxide (CO) and carbon (C) can be formed, so that the reaction becomes: $C_6H_{10}O(s) + 10\ O_2(g) \rightarrow CO_2(g) + 5\ H_2O(g) + 5\ CO(g) + 4\ C(s) + \text{Energy}$. Incomplete combustion occurs when in an enclosed space the amount of oxygen is not enough to burn the fuel completely into carbon dioxide gas and water vapor.

This culture can be used as a context in chemistry learning on topics such as reaction equations, combustion reactions, and acids and bases. In addition to enhancing the relevance of learning and for increasing student interest, this integration will develop students' awareness of local cultures that support sustainability principles.

Wedhang uwuh (traditional drink)

Wedang uwuh is a traditional drink from Yogyakarta, Indonesia. It is a blend of various spices, including ginger, sappanwood, cinnamon, lemongrass, and cloves. Literally translated, *wedang uwuh* means "garbage drink" due to the messy spice residue, but it boasts numerous health benefits, including boosting immunity, warming the body, and being rich in antioxidants. When *wedang uwuh* is left open for a long time, there are several chemical changes that affect its colour, aroma, and taste. One of the main reactions that occurs is oxidation of certain compounds, such as brazilin found in sappang wood. Brazil oxidizes into hematoxilin, which causes the color of the wheat to become more concentrated over time. This process is similar to the discoloration of sappan wood that is soaked or heated for a long time, with following reaction;

$$C_{16}H_{14}O_5 \text{ (brazillin)} + 1/2O_2 \rightarrow C_{16}H_{12}O_5 \text{ (brazilien)} + H_2O$$

Under acidic conditions ($pH < 6$), the color of brazilein tends to be yellow or orange (for example, if wedang uwuh is given lime). Conversely, under alkaline conditions ($pH > 7$), the color changes to purple or deep purplish red. Therefore, sappan wood can be used as a natural acid-base indicator.

Additionally, the components of aromatic compounds in spices such as ginger, cloves, and cinnamon can undergo volatilization or evaporation. Compounds such as gingerol (from ginger), eugenol (from cloves), and sinamaldehyde (from cinnamon) have a relatively low boiling point, making them volatile easily when exposed to air for long periods of time. As a result, the distinctive aroma of *wedang uwuh* will be reduced, and the taste can also change due to the loss of some of the components that make up the flavor. In addition to oxidation and evaporation, there may also be a degradation reaction of organic compounds that cause changes in the taste and quality of the drink. Therefore, in order for the quality of *wedang uwuh* to remain optimal, it is best to store *wedang uwuh* in a closed container or consume immediately after brewing.

C. COMMUNITY-ENGAGED PROJECT-BASED LEARNING FOR SUSTAINABILITY RELEVANCE

PBE enables students to reflect on their ancestors' practice and relevance in different aspects of life, think critically about their own practices, and keep place-based knowledge in thoughts and action. In the process of identifying the problems associated with pedagogical practices, teachers accepted that they have not been able to engage the students in their own learning, connect their learning in the real-life situation due to the implementation of decontextualized pedagogical practice (Maharjan et al., 2024). PBE is one of the pedagogies in overcoming the problem of relevance in chemistry learning, which is one of the main problems besides the problem of isolated facts, low transfer, excessive load and inappropriate emphasis on material in chemistry learning (Gilbert, 2006).

For over 40 years, science educators have debated how to create relevant and engaging science education for students. This debate has resulted in science-chemistry curriculum reform and/or movements (Fensham, 2004). Although the term 'relevance' in chemistry has had different dimensions and emphases over time, it has played a crucial role in achieving reform in chemistry learning and education (Stuckey et al., 2013). The concept of 'relevance' in science learning has evolved over the years, with chemistry learning considered relevant to students continuing to evolve, from simply using analogies, to involving students' daily life experiences (contextual), to perceiving the meaningfulness of students' lives across multiple dimensions. Furthermore, Stuckey et al. (2013) formulated the concept of science relevance more comprehensively, involving three dimensions (individual, societal, vocational) and four components (present, future, intrinsic, and extrinsic). Based on the development of this multidimensional relevance concept (Table 1), to make chemistry learning relevant for students, it must be designed to at least (a) prepare students for potential careers in science and engineering; (b) develop transferable skills in overcoming challenges in students' lives; and (c) encourage students to become socially responsible citizens.

Table 3. Key features of the multidimensional science relevance concept

Dimensions	**Components**			
	Present		**Future**	
	Intrinsic	**Extrinsic**	**Intrinsic**	**Extrinsic**
Individual	**Satisfying curiosity and interests**	**Getting good grades**	**Skills in dealing with personal life in the future**	**Acting responsibly and showing solidarity with others in the future**
Social	Finding one's place in society	Learning how to behave in society	Promoting one's own interests in public discourse	Behaving as a responsible citizen
Vocational	Orientation about potential careers	Passing qualifying exams for further education	Get a satisfying and high-paying job	Contribute to the economic growth of the community

Adapted from Stuckey et al. (2013)

Meanwhile, in line with the goals of global science education, sustainability becomes a relevant orientation of chemistry learning, especially related to the individual-future, social-future, and vocational-future (extrinsic) components. In line with Model 3 on how to integrate chemistry learning with ESD, proposed by

Burmeister et al. (2012), teachers are suggested to use SSI-related sustainability issues and chemistry as starting points for learning chemistry. SSI-related sustainability, which is a controversial/dilemmatic scientific issue in society, has great potential to be used in the context of PBE, especially in the dimensions of learning from place and learning for the sake of the place. Furthermore, local SSI, which is more familiar to students because it is in their living environment, is more optimal in developing students' scientific habits of mind as well as their scientific literacy (Wiyarsi et al., 2024), which is strengthens PBE implementation.

Additionally, as an educational approach, connections can be made between PBE, community-based education practices, and Indigenous on-country learning (Burgess et al., 2022), among other pedagogies (Gruenewald, 2003). Community can be represented by a cultural community with customs or norms, beliefs regarding traditional ceremonies, and unique cultural products. Furthermore, learning to develop sustainability competencies must emphasize active and practical student involvement. Relevant pedagogy for developing sustainability competencies involves cognitive, affective, and psychomotor activities (Sipos et al., 2018), particularly project-based learning/PjBL (Filho et al., 2016; Han et al., 2015; Haatainen & Aksela, 2021).

Implementation of Community-Engaged Project-Based Chemistry Learning: a Case of *Batik*

Community-Engaged Project-Based Chemistry Learning (CE-PjBChemL) is a learning strategy that is a manifestation of the PBE framework with local SSI representing the place of study, as well as learning from place and learning for the sake of the place represented by community culture, in this case the process of making batik, a traditional fabric from Yogyakarta, Indonesia that is full of philosophical content. CE-PjBChemL strengthens sustainability-relevance by developing critical thinking skills, argumentation on SSI, collaboration, awareness of sustainability and culture, and sustainability project competencies. CE-PjBChemL that integrates academic projects with real-world needs in the community is implemented through 7 steps.

1. Problem Identification and Community Mapping

The initial step is to identify the needs or challenges faced by the partner community (*Batik* artisan group). Students conduct observations, interviews, or focus group discussions (FGDs) with community leaders. This activity supports the relevance of science in the societal dimension, both present and future, where involvement with the community will facilitate students in learning how to behave in society and become responsible citizens in the future. Students also develop their interpersonal competence.

2. Determining the Driving Question

The teacher formulates alternative questions that challenges students to find solutions to identified community problems. This question must be open-ended, complex, and meaningful to both the students and the community. For instance, "How can we improve the waste management system for the *batik* industry in this area to prevent polluting the local river?". In this phase, students satisfying curiosity and interests with teachers' facilitation (individual relevance dimensions).

3. Project Design and Collaboration

Students plan a project strategy with input from community members. They assigned roles within the team, determining a timeline, and selecting a method (e.g., scientific research, tool development, or social campaign). Students must be active involvement from the community as partners, not just as objects of research. Dimensions of individual are facilitated in term enhance skills in dealing with personal life in the future. In the sustainability competence, students foster their anticipatory competence, that means students thinking in long term horizons (decision making, problem solving), to envision future scenarios to challenge the sustainability issues.

4. Guided Inquiry and Implementation

Students conduct in-depth investigations and begin working on practical solutions, i.e., data collection, laboratory experiments (if science-related), or prototype construction. Meanwhile, teachers provide technical guidance, while the community provides field data and practical feedback. During this process, students apply their strategic action competencies. Furthermore, the learning phase indicates societal and vocational relevance dimensions, which relate to the usefulness of students' learning processes in their future social and professional lives.

5. Reflection and Self-Evaluation

Throughout the process, students reflect on what they have learned and how their actions impact others. These activities are oriented to connect field experiences with theoretical concepts learned in class. Further, fostering empathy and social responsibility are represented social reflection during the process. That means, students develop their normative competence as part of sustainability competencies.

6. Product Presentation and Real Contribution (Public Product)

Project results are presented to a public audience, primarily members of the partner community. The activities cover exhibitions of work, program outreach, or distribution of tools to the community. This part provides a sense of accomplishment for students because their work is actually being used by others.

7. Impact Assessment & Sustainability

Evaluates the extent to which the project has had a positive impact on the community. Students collaborate with teachers assesses whether the proposed solution can continue to be implemented independently by the community after the school project ends (the principle of sustainability).

All the stages that emphasize the active involvement of students in constructing knowledge through problems that are relevant and familiar to students' lives, the implementation of CE-PjBChemL encourages better conceptual understanding, critical thinking, and scientific literacy in students. Argumentation and collaboration skills, as well as creative thinking, are also facilitated while working on sustainability-oriented projects to respond to SSI issues in existing communities. Therefore, this strategy has the potential to develop students' sustainability competencies.

To Sum up, the integration of PBE-ethnoscience is indeed a good alternative to improve the quality of chemistry learning, especially in accommodating sustainability's relevance and competence. However, implementation challenges can stem from teacher competency, language barriers in understanding local culture (availability of resources), misunderstandings of culture, cultural complexity, and inadequate school facilities. Efforts to overcome these challenges, in the context of classroom implementation, can be done by exploring students' experiences related to culture, which will make it easier for students to express their ideas. This will also strengthen the emotional connection between teachers and students, because teachers care about students' backgrounds. During the learning process, teacher pay attention to the possibility of conflict arising from students' differing views on certain cultures, as well as gaps in understanding. Various other learning strategies and practices need to be explored further to strengthen the implementation of integrated PBE with ethnoscience.

REFERENCES

Almomani, E. Y., Jarrar, W., Alhadid, A., Hamadneh, L., Qablan, A., & Almomani, H. Y. (2022). Shedding light on pharmacists' knowledge of kidney stones' etiology and treatment. *Pharmacy Practice*, *20*(3), 2712. DOI: 10.18549/PharmPract.2022.3.2712 PMID: 36733510

Almomani, H., Almomani, E., Qablan, A., & Al-Momany, A. (2024). Pharmacists' perspectives and perceived barriers to counseling patients with kidney stones. *Research in Social & Administrative Pharmacy*, *20*(4), 1551–7411. DOI: 10.1016/j.sapharm.2024.02.018 PMID: 38462437

Ardoin, N. M. (2006). Toward an interdisciplinary understanding of place: Lessons for environmental education. [CJEE]. *Canadian Journal of Environmental Education*, *11*(1), 112–126.

Ardoin, N. M., Schuh, J. S., & Gould, R. K. (2012). Exploring the dimensions of place: A confirmatory factor analysis of data from three ecoregional sites. *Environmental Education Research*, *18*(5), 583–607. DOI: 10.1080/13504622.2011.640930

Badawy, H., Alkaabi, A., Qablan, A., Badawy, H., Almaazmi, K., & Ndlela, N. (2024). Nurturing Future Innovators Integrating Computational Thinking, AI, and STEM in Early Education. In *Empowering Early Education with Computational Thinking, AI, and STEM*. IGI Publishers. DOI: 10.4018/979-8-3693-6210-5.ch004

Bascopé, M., & Reiss, K. (2021). Place-Based STEM Education for Sustainability: A Path towards Socioecological Resilience. *Sustainability (Basel)*, *13*(15), 8414. DOI: 10.3390/su13158414

Burgess, C., Thorpe, K., Egan, S., & Harwood, V. (2022). Towards a conceptual framework for Country-centred teaching and learning. *Teachers and Teaching*, *28*(8), 925–942. DOI: 10.1080/13540602.2022.2137132

Burmeister, M., Rauch, F., & Eilks, I. (2012). Education for Sustainable Development (ESD) and Chemistry Education. *Chemistry Education Research and Practice*, *13*(2), 59–68. DOI: 10.1039/C1RP90060A

Byram, M. (1995). Acquiring Intercultural Competence: A Review of Learning Theories. In Sercu, L. (Ed.), *Intercultural Competence: The Secondary School* (pp. 45–56). Aalborg University Press.

Çalık, M., & Wiyarsi, A. (2021). A systematic review of the research papers on chemistry-focused socio-scientific issues. Journal of Baltic Science Education. al of Baltic Science Education. DOI: 10.33225/jbse/21.20.360

Fensham, P. J. (2004). *Defining an identity: The evolution of science education as a field of research.* Kluwer Academic Publishers. DOI: 10.1007/978-94-010-0175-5

Filho, W. L. Shiel, Chris dan Paco, Arminda. 2016. Implementing and Operationalising Integrative Approaches to Sustainability In Higher Education: The Role of Project-Oriented Learning. Journal of cleaner Production 133 (2016) 126-135, (online), (http://www.sciencedirect.com, access 2026)

Gilbert, J. K. (2006). On the nature of "Context" in chemical education. *International Journal of Science Education, 28*(9), 957–976. DOI: 10.1080/09500690600702470

Granit-Dgani, D. (2021). Paths to place-based education. The Mofet Institute. Hebrew.

Gruenewald, D. A. (2003a). Foundations of place: A multidisciplinary framework for place-conscious education. *American Educational Research Journal, 40*(3), 619–654. DOI: 10.3102/00028312040003619

Gruenewald, D. A. (2003b). The best of both worlds: A critical pedagogy of place. *Educational Researcher, 32*(4), 3–12. DOI: 10.3102/0013189X032004003

Haatainen, O. M., & Aksela, M. (2021). Project-based learning in integrated science education: Active teachers' perceptions and practices. LUMAT: International Journal on Math. *Science and Technology Education, 9*(1), 149–173. DOI: 10.31129/LUMAT.9.1.1392

Han, S., Yalvac, B., Capraro, M. M., & Capraro, R. M. (2015). In-service teachers' implementation and understanding of STEM project based learning. *Eurasia Journal of Mathematics, Science and Technology Education, 11*(1), 63–76. DOI: 10.12973/eurasia.2015.1306a

Hernandez Gonzalez, F. (2023). Exploring the Affordances of Place-Based Education for Advancing Sustainability Education: The Role of Cognitive, Socio-Emotional and Behavioural Learning. *Education Sciences, 13*(7), 676. DOI: 10.3390/educsci13070676

Israel, A. L. (2012). Putting geography education into place: What geography educators can learn from Place-Based Education, and vice versa. Journal of Geography, 111(2), 76– Ives,81. DOI: 10.1080/00221341.2011.583264

Macionis, J. J. (2017). *Sociology* (16th ed.). Pearson.

Maharjan, B., Pant, B. P., & Dahal, N. (2024). Participatory learning: Exploring place-based pedagogy for future teachers. *Pedagogical Research, 9*(4), em0222. Advance online publication. DOI: 10.29333/pr/15155

Nichols, J. B., Howson, P. H., Mulrey, B. C., Ackerman, A., & Gately, S. E. (2016). The promise of place: Using place-based education principles to enhance learning. *The International Journal of Pedagogy and Curriculum*, *23*(2), 27–41. DOI: 10.18848/2327-7963/CGP/v23i02/27-41

Orr, D. (2013). Place and pedagogy. *NAMTA Journal*, *38*(1), 183–188.

Qablan, A., & Al-Qaderi, S. (2009). How to Change University Faculty Members' Attitudes and Behavior in the Context of Education for Sustainable Development. *Applied Environmental Education and Communication*, *8*(3-4), 184–194. DOI: 10.1080/15330150903269407

Qablan, A., & DeBaz, T. (2015). Facilitating Elementary Science Teachers Implementation of Inquiry-Based Science Teaching. *Teacher Development*, *19*(1), 3–21. DOI: 10.1080/13664530.2014.959552

Qablan, A. M., & DeBaz, T. (2015). Facilitating elementary science teachers' implementation of inquiry-based science teaching. *Teacher Development*, *19*(1), 3–21. DOI: 10.1080/13664530.2014.959552

Smith, G. A. (2002). Place-based education: Learning to be where we are. *Phi Delta Kappan*, *83*(8), 584–594. DOI: 10.1177/003172170208300806

Sobel, D. (2004). Place-based education: Connecting classroom and community. *Nature & Listening*, *4*(1), 1–7.

Stuckey, M., Hofstein, A., Mamlok-Naaman, R., & Eilks, I. (2013). The meaning of 'relevance' in science education and its implications for the science curriculum. *Studies in Science Education*, *49*(1), 1–34. DOI: 10.1080/03057267.2013.802463

Tahat, K., Mansoori, A., Qablan, A., & Tahat, D. (2024). Analyzing the newspapers' coverage of sustainable development issues in the United Arab Emirates. *Online Journal of Communication and Media Technologies*. Advance online publication. DOI: 10.30935/ojcmt/14710

Wade-Lyles, T. A. (2016). Integration of place-based education into science classes from prekindergarten through grade 5 [Doctoral dissertation, Walden University].

Wiyarsi, A., Çalik, M., Priyambodo, E., & Dina, D. (2024). Indonesian prospective teachers' scientific habits of mind: A cross-grade study in the context of local and global socio-scientific issues. *Science & Education*, *33*(5), 1257–1283. DOI: 10.1007/s11191-023-00429-4 PMID: 37359261

Chapter 12
Make It Real, Student Voice, Conditional Trust, and Design Implications for Place-Based Sustainability Science in the UAE:
Ethical Considerations for Aligning With Indigenous Knowledge Agendas

Noor Al Haj Ibrahem
http://orcid.org/0009-0000-7689-1477
Mid Sweden University, Sweden

Magnus Oskarsson
Mid Sweden University, Sweden

ABSTRACT

This chapter explores how UAE lower secondary students perceive the relevance, purpose, and trustworthiness of school science, and what this means for place-based sustainability education. Using focus group data from Grade 9 and 10 students, the data show that science becomes meaningful when linked to everyday life, future

DOI: 10.4018/979-8-3373-7650-9.ch012

goals, and community needs. The findings also highlight the importance of interactive pedagogy and students' conditional trust in science for sustainability-oriented teaching.

INTRODUCTION

Across many education systems, sustainability is now treated not only as an environmental concern but also as an educational one. Schools are increasingly expected to help young people understand complex socio-scientific issues, evaluate evidence, and take part in collective problem-solving. In this context, Indigenous knowledge systems and place-based learning have gained more attention in sustainability education because they foreground relationships among land, culture, identity, and community responsibility.

Sustainability-oriented curricula often emphasise 'real-world relevance', but relevance does not speak for itself (Qablan & Al-Qaderi, 2009; Qablan & DeBaz, 2015). Students are more likely to engage when scientific knowledge connects to their everyday lives and imagined futures, and when science is seen as a credible way to address societal challenges. Research on student attitudes toward science has repeatedly shown that perceived relevance is closely tied to sustained engagement. In this chapter, relevance refers to usefulness in present life, future goals, and social contribution. From an Expectancy-Value Theory (EVT) perspective (Wigfield & Eccles, 2020), this form of relevance aligns primarily with utility and attainment value. In other words, students are more likely to value science when it helps them achieve their goals and connects with what matters to them.

Relevance alone is not enough if students do not also see science as trustworthy, especially in the context of public controversy and uneven social outcomes. Recent scholarship argues that science education should foster a critical but balanced trust in science. This means helping learners understand not only how scientific knowledge is produced, but also how uncertainty, values, institutions, and social relationships shape scientific work (Erduran & Cheung, 2026).

These issues become especially visible in contexts where national policy strongly prioritises science education for development (Al-Zoubi et al., 2024). In the United Arab Emirates (UAE), science learning is increasingly tied to scientific literacy, 21st-century skills, and future workforce preparation, in line with broader national visions and reforms.

Regional research also suggests that students' everyday experiences in science classrooms, including teaching methods, assessment, and opportunities for agency, are still underexplored (Kayan-Fadlelmula et al., 2022). Large-scale assessments can show patterns of achievement, but they rarely reveal how students decide whether school science feels meaningful, credible, and worth investing in (Mansour & Alshamrani, 2015).

For sustainability education, this means that adding new content alone is unlikely to increase engagement if classroom routines stay the same.

This chapter examines how lower secondary students in the UAE talk about the relevance, purpose, and trustworthiness of school science, and how these perceptions shape opportunities for place-based sustainability education. It draws on focus group data from Grade 9 and 10 students in gender-segregated schools and brings together the perspectives of both boys and girls.

Edited volumes on Indigenous knowledge and place-based sustainability education often focus on community epistemologies and pedagogical change. They pay less attention to how students themselves make sense of science, and what this means for classroom design. Here, it is argued that students' 'relevance logics "-linked to everyday life, future aspirations, and community development-together with their conditional trust in science, offer practical guidance for designing place-based sustainability education in the UAE. Rather than documenting Indigenous epistemologies, the chapter uses student voices to identify classroom conditions under which community and heritage knowledge might be respectfully brought into sustainability science, with attention to attribution, verification, and reciprocity, and without making claims beyond the dataset.

Here, "place-based" learning is not treated as simply adding local examples or using local knowledge as a motivational hook. Rather, place is understood as relational and interpretive: it includes everyday practices, family narratives, community values, and national visions through which students come to see why science matters. This aligns with the volume's focus on local wisdom and learning ecologies, while also keeping a clear distinction between "local" and "Indigenous" knowledge claims.

Empirically, the chapter focuses on students' accounts of how school science can become meaningful, or fail to become meaningful, in their lived environment, and what this suggests for designing sustainability education that is both culturally responsive and educationally effective.

The chapter makes two main contributions. First, it offers an empirically grounded account of students' "relevance logics" in UAE school science: how they connect scientific content to everyday life, future aspirations, and societal contribution, and how these connections interact with beliefs about science as a way of addressing complex problems. Second, it translates these student-voice findings into practical design implications for place-based sustainability science teaching in Gulf contexts. It concludes by bringing these student-derived patterns together into three meta-principles for place-based sustainability-oriented science teaching in the UAE lower secondary: (1) local anchoring with explicit relevance routines, (2) participation and inquiry routines that normalise uncertainty, and (3) socio-scientific ethics routines that cultivate conditional trust.

The chapter begins with a conceptual framing that links place-based sustainability education with students' perceived relevance and conditional trust in science. It then describes the focus group datasets and analytic approach, presents integrated themes across boys' and girls' accounts, and closes by synthesising these patterns into three meta-principles and implementable routines for UAE lower secondary science teaching.

CONCEPTUAL FRAMING

This chapter brings together three strands: place-based approaches to sustainability education, student motivation through perceived value and relevance, and trust in science as a socially situated problem-solving practice.

It examines how students understand relevance and trust, and how these shape place-based sustainability learning.

Place, Local Knowledge, and Sustainability Education

Place-based learning starts from the idea that education becomes more meaningful when it connects learners to their local surroundings, community histories, and everyday activities (Gruenewald, 2003). In the call for chapters, place-based education is linked to an "ecology of learning," including the relationship among knowledge, land, and identity, as well as the tension between "global policies" and "local realities" in sustainability and Education for Sustainable Development (ESD).

From this viewpoint, place is not just something added onto the curriculum. It shapes which problems are treated as worth solving, which kinds of knowledge are accepted as valid, and how learners imagine their roles in community and societal futures.

This also connects to epistemic justice in science education: whose knowledge is treated as legitimate, whose experiences count as evidence, and how credibility is distributed in the classroom. (Fricker, 2007; Tahat et al., 2024). This matters in sustainability education because learners face complex issues that require them to weigh scientific claims, personal experience, and community priorities, rather than simply receive information.

In sustainability education, this often appears as a shift away from abstract environmental content toward more relational and situated understandings of socio-ecological systems.

In this chapter, place-based learning is understood through students' accounts of when science becomes meaningful to them: when it connects to daily life, imagined futures, and perceived societal needs, and when teaching practices either support or limit those connections. This is consistent with recent place-based sustainability work that emphasises locally grounded inquiry and community relevance.

What "Indigenous/Heritage/Community Knowledge" can Mean in UAE/Gulf schooling, and What This Chapter can (and Cannot) Claim

In the Gulf/UAE context, school discussions of "local knowledge" often overlap with heritage practices, family and community ways of knowing, and locally trusted knowledge sources such as municipal guidance, public health messaging, environmental agencies, and intergenerational practices in homes and communities. In this chapter, these are treated as community and heritage knowledge resources that may inform place-based sustainability pedagogy, while recognising that Indigenous knowledge, as discussed in decolonising science education, is not the same as local examples or cultural references.

Accordingly, this chapter does not claim to document Indigenous epistemologies from Indigenous communities. Instead, it offers a design bridge by using student voice to identify conditions for bringing community and heritage knowledge into sustainability-oriented science in respectful ways that reduce tokenism, strengthen reciprocity and attribution, and keep the epistemic work of evidence, explanation, and ethical reasoning visible. In other words, students' own meaning-making is treated both as a resource, in what they already connect to place, and as a limit, in what classroom routines still need to change, for ethically grounded place-based sustainability science.

Relevance as a Motivational Mechanism: A Value-Oriented Lens

To explain why place-based approaches matter for student engagement, this chapter draws on motivational research that links engagement to students' expectations and task values. Drawing on EVT-informed work, perceived relevance is treated here as closely linked to utility and attainment value: students' sense that science is useful in everyday life or for future goals, and meaningful in relation to identity and personal values.

Pedagogically, the point is straightforward: if students do not see school science as connected to their lives and futures, they are less likely to invest effort, persist through difficulty, or view science as worth engaging in.

In the UAE, students often emphasise the importance of science for future careers, its practical use in daily life, and its role in society. For instance, girls often link different scientific fields to future careers, such as physics to engineering and biology to medicine, while also highlighting science's relevance to national progress and infrastructure development.

These accounts point to a situated sense of utility: science gains value when students see it as connected to locally relevant futures. This matters for sustainability education because sustainability-related skills are unlikely to develop through decontextualised teaching alone (Wiek et al., 2011). They require students to see why scientific knowledge matters for decisions, practices, and collective futures in their own environment.

Trust in Science: From Deference to Informed Epistemic Confidence

Alongside relevance, this chapter treats trust in science as a second key factor shaping students' engagement with sustainability-focused science learning. This aligns with the view that science education should cultivate informed epistemic trust: the ability to assess scientific claims and sources while remaining confident that scientific methods can help address real-world problems.

In the UAE context, trust often takes the form of a nuanced, conditional stance. Students may feel confident that science can help address health, climate, or social issues, while also recognising its limits, the time required for such work, and the role of human decisions in shaping outcomes.

For example, girls often connect science to major problems such as climate change and to broader questions of well-being. However, their reasoning often goes beyond technical optimism. It includes moral and social concerns such as responsibility, effects on human lives, and the wider influence of science on societal change.

These accounts matter in sustainability education because sustainability challenges are classic socio-scientific problems. They involve uncertainty, conflicting values, and contested options for action (Sadler, 2011). Sustainability-focused science education, therefore, requires both relevance and a capacity to live with uncertainty and ethical complexity.

From Student Meaning-Making to Pedagogical Design

The call for chapters highlights the need for pedagogical innovation and for a move "from knowledge transmission to action competence."

Student accounts are, therefore, important evidence of when science feels meaningful, credible, and worth the effort.

The analysis begins from the assumption that students' descriptions of what they find "memorable," "boring," or "useful" are not just personal preferences. They also indicate whether classroom activities create meaningful links among scientific knowledge, everyday experience, and societal goals.

The analytic framework used here defines place-based sustainability pedagogy through three key components. First, it uses locally recognisable problems and practices as entry points into scientific concepts. Second, it encourages participatory inquiry and discussion in ways that support agency. Third, it makes visible how scientific knowledge is produced and used, thereby supporting informed trust. This framework fits the book's interest in connecting local knowledge traditions with contemporary educational practice and in supporting sustainability learning that is equitable, culturally rooted, and action-oriented.

The next section builds on this framing by outlining how relevance and trust are analysed as key factors in sustainability learning.

CONTEXT

In the UAE, science education is closely tied to national development goals and workforce preparation. This focus, together with pressure from international benchmarking, places greater emphasis on student engagement and the quality of classroom experience (Badawy et al., 2025). Under these conditions, student voice becomes especially important for designing sustainability-oriented pedagogy.

At the same time, research on science education in Gulf contexts has noted a lack of qualitative evidence on how students themselves interpret school science, including whether they connect classroom learning to life outside school. This gap matters for sustainability-oriented education because ESD requires more than coverage of environmental topics. It also depends on pedagogical conditions that support meaning-making, agency, and students' ability to relate scientific ideas to local realities and societal challenges (Ibrahim et al., 2023). This emphasis is central to the book's focus on place-based pedagogies and "transformative approaches to sustainability education".

Empirically, the chapter draws on two complementary focus group datasets with lower secondary students in gender-segregated UAE school contexts. Full methodological details are provided in the Methods section.

METHODS

Data Sources and Participants

This chapter draws on two complementary focus group datasets from Grade 9 and 10 students (approximately 14–15 years old) in the UAE. The first dataset comprises a semi-structured focus group with nine boys in a lower secondary school in Ras Al Khaimah. The second dataset comprises a focus group with 10 Grade 10 girls in a gender-segregated schooling context. Together, these datasets support an integrated analysis of how young people in the UAE evaluate school science in relation to relevance, purpose, and trust.

Dataset comparability and limits of integration. The two focus groups are comparable in grade level, broad prompts (interest/boring, purpose, relevance, and trust), and shared schooling structures, including gender-segregated contexts and exam-oriented expectations. They differ in emirate and school setting, transcript conventions, and the original purposes of the two studies, which may have shaped how participants developed their accounts. For this reason, the synthesis is used to identify cross-cutting patterns in how students made meaning, rather than to make claims about prevalence or treat differences as stable gender effects.

Data Generation

The boys' focus group was conducted in Arabic to allow greater expressive nuance and reduce language barriers. It lasted about 60 minutes and was audio-recorded. The prompts encouraged students to discuss whether they found science interesting or dull, what made science lessons memorable, why they studied science, and whether they could trust science to help solve real-world problems. The girls' focus group addressed similar topics. It highlighted students' varied views on biology, chemistry, and physics, as well as how the value of these subjects related to future plans and everyday usefulness.

Analytic Approach and Ethical Considerations

The analysis was guided by an inductive thematic approach that emphasised students' own language and interpretations (Clarke & Braun, 2013). Interpretation was organised around three key concepts: relevance, purpose, and trust. Each dataset was first analysed within its own case to identify important patterns (Almomani et al., 2024; Almomani et al., 2022). The two datasets were then combined to develop overarching themes and identify differences. Examples were selected to show both similarities and variations in students' responses. Ethical approval was secured for both the boys' and girls' studies, with permission obtained from schools and parents. Across all datasets, participants' identities were protected by using pseudonyms and removing identifiable information. In reporting the findings, students' perspectives are treated as shaped by schooling structures and expectations rather than as direct indicators of teaching quality.

Researcher Positionality and Interpretation

Interpretation drew on analytic work with both Arabic and English transcripts, with particular care taken to avoid cultural tokenism in claims related to place-based sustainability. Theme labels, excerpt choices, and synthesis statements were checked against the original transcripts to ensure that they reflected participants' intended meanings and did not make cultural or Indigenous claims beyond what students explicitly expressed.

FINDINGS: INTEGRATED THEMES ACROSS BOYS' AND GIRLS' ACCOUNTS

Across both datasets, five themes emerged in students' accounts of school science and its relevance to place-based sustainability learning: (A) relevance is anchored in everyday life and community; (B) students are motivated by future usefulness and practical value; (C) pedagogy acts as a gatekeeper between rote learning and applied or participatory learning; (D) trust in science is conditional and shaped by time, constraints, and human ethics; and (E) classroom discourse and opportunities for agency shape participation and operate as a hidden curriculum. The themes below are illustrated with quotations from both boys and girls and interpreted through a place-based sustainability lens.

Theme A. Relevance Is Place-Based: Everyday Life and Community as Interpretive Anchors

Students in both groups did not describe relevance as an abstract quality of science. They described it as a link between what was taught and what they recognised in daily life. They most often described relevance through health, household routines, familiar phenomena, and community needs such as infrastructure and societal progress.

The girls' focus group gave especially detailed examples from everyday life. Biology was often seen as useful because it helped students make sense of bodily sensations: "In biology, maybe if I feel stomach pain and I studied this before, I could understand what's wrong with me." For chemistry, relevance was tied to household safety and daily routines. One student described chemistry's presence in cooking and cleaning, highlighting its connection to toxic risks: "When you're cooking, you mix things together. And we learned that some chemicals shouldn't be mixed because they can be toxic… The same thing when cleaning. I wouldn't mix certain products together." Another student drew a parallel between home activities like baking and classroom learning, saying: "I bake and cook a lot, so I know which ingredients go together. That helped me a lot in chemistry last year."

The girls also linked relevance more explicitly to place and community. For example, a student explained the importance of physics as socially meaningful through its role in infrastructure and national operation: "It helps with infrastructure, like roads and buildings. If no one studied physics, it would affect the country." Peers immediately challenged this view, demonstrating that personal goals and divisions of labour also influence place-based relevance: "I feel like it's important only to the people who are going to pursue it in their life." These exchanges show that community and personal relevance can coexist, but they do not always align in students' reasoning.

The boys also linked relevance to recognisable real-world applications. One student emphasised how science is engaging because of its practical usefulness: "We learn things we can use later, like in hospitals." When students described memorable lessons, they usually pointed to situations in which science felt meaningful because they could observe it and connect it to familiar phenomena. For instance, they compared "physical and chemical changes" (like boiling water versus a rotting apple) to understand daily life processes. Even topics often seen as decontextualised, such as the periodic table, became more meaningful when taught in practical ways—for example, by discussing which elements are "dangerous or good, and how we can use them to make medicines."

Interpretation for Place-Based Sustainability Learning

In both groups, relevance was strongest when scientific concepts were linked to students' immediate environment, including health, household routines, tangible objects, and community infrastructure. This matters for sustainability education, where scientific literacy depends in part on connecting knowledge to daily choices and local impacts. Students' accounts suggest that sustainability-focused science is more likely to feel meaningful when it begins with familiar local practices such as household chemical safety, health, and the built environment. It can also help learners see how small-scale actions connect to wider social and ecological effects.

One practical implication is to begin sustainability education with household and community phenomena, and then connect these to wider systems such as waste streams, indoor air quality, water use, and energy demand, rather than starting with distant global concepts. This follows from the way students themselves linked relevance to household chemical practices, health experiences, and community infrastructure.

Theme B. Futures and Usefulness Drive Engagement: Aspirations, Responsibility, and National Development

Across both datasets, students linked science learning to their future plans. They often evaluated the value of school science in relation to careers, economic security, and social impact. Their sense of the future extended beyond personal advancement to include responsibility toward family, community, and national progress. Science was therefore seen as a way of shaping the future at personal, social, and sometimes national levels.

The girls often evaluated science subjects in relation to their future education and personal goals. This appeared both in the way they distinguished between disciplines and in how they thought about future study pathways. For instance, one student said: "Physics... I don't really find it useful unless it relates to my future major." Another discussed the value of subjects by linking specific careers to certain disciplines: "If someone wants to be an engineer, they need physics. But if they want to be a doctor, they need biology." Several also associated science with enhancing lives through healthcare-focused careers: "Studying the human body is important because it helps with the future, like medicine or healthcare. It helps improve people's lives."

Family role models also added another layer of meaning; for example, a student related her father's medical work to ethical responsibility ("you are dealing with people's lives"), subtly connecting scientific study with service and moral significance.

Beyond personal career goals, some girls also spoke about national and societal futures, especially in relation to infrastructure and access to technology. One student specifically linked scientific and technological progress to national strength and public good, citing issues such as limited access to medical technologies and the need to travel for healthcare. She concluded that scientific advancement "would help people a lot… science is important for the country."

The boys' group also framed science in future-oriented terms, often distinguishing between personal motivations and societal ones. On the personal side, one student described science as a path to stability and family support: "the goal is to get a job… to have a good position and salary… to secure my life and help my family." Another summarised this as a future safety goal: "We study to make our future safe. To have a profession that gives us money and respect." On the societal side, students described science as a way to combat disease and suffering: "I think we study science to find cures for diseases," and as a way to create inventions that "help people." One student also linked science learning to national development and sustainable technologies: "We develop our country like in the UAE, we make new technologies, solar energy, and robots… to benefit the nation." Students also explicitly saw personal and societal goals as compatible, stating that "I think both, because if you help society, you will also get money."

Interpretation for Place-Based Sustainability Learning

These accounts suggest that sustainability may be especially motivating when it is framed as future-oriented and connected to multiple kinds of value: personal goals such as careers and identity, responsibilities toward family and community, and collective visions of national progress. The prominence of "solar energy" and "technologies" in the boys' discussions also suggests that sustainability can be framed not only through environmental protection, but also through innovation, public benefit, and national aspiration. For these framings to resonate, they still need to be tied to local needs and ethical concerns.

Theme C. Pedagogy as the Gatekeeper: Rote Instruction Versus Experiential and Applied Learning

Across both groups, students' sense of "interest" or "boredom" depended less on the subject matter itself than on teaching style and opportunities for participation. Students repeatedly contrasted abstract, memorisation-focused teaching with more hands-on forms of learning involving application, activity, and interaction. In their accounts, pedagogy often determined whether science felt meaningful.

The boys' accounts clearly identified rote learning as a key source of disengagement. One student summed this up clearly: "Science class is boring because it is all about memorising." Students also described lessons as repetitive, "the lessons are all the same," and mainly involving copying, with comments like "no experiments, no practical side," and "just writing, copying, memorising, no action." By contrast, they associated engagement with hands-on activity and discovery. Science was enjoyable, as one student put it, "because we do nice things sometimes, like experiments," and because "experiments and discovering new things make science fun." Their most memorable experiences were consistently practical, such as the "volcano project." They also described learning as meaningful when they could "see it with my own eyes," which points to the importance of "learning by doing."

The girls raised similar concerns, especially about classroom pacing, opportunities to ask questions, and the emphasis on "numbers" across subjects. They mentioned that physics and chemistry can feel overwhelming because of the dense formulas and laws, saying, "There are so many formulas… so many laws," and called for a more balanced curriculum. Several comments linked the difficulty to teaching methods that limit interaction: "When the teacher explains, we can't ask questions until the end… and when we do, the explanations are not as thorough." They also noted emotional barriers to participating, such as: "Sometimes we're scared to ask questions because we think maybe it's a bad question or something we should already know."

The girls were also quite specific about the kinds of classroom conditions they wanted. They emphasised application and activity, saying, "First, we should apply what we learn. Not just listen, but actually use the knowledge in the lesson," and they dismissed lesson formats focused heavily on writing: "Something that doesn't make us sit the whole lesson writing. We should move, do activities, not just write." They were not asking for constant experimentation, but for a more predictable rhythm of interaction and clearer explanations of relevance: teachers "should try to do something interactive at least once a week," and should "always explain how what we learn connects to real life."

Interpretation for Place-Based Sustainability Learning

These findings suggest that place-based sustainability education cannot simply mean adding "local examples" to otherwise traditional lessons. Students were not only critiquing the content of lessons, but the teaching approach itself. They valued opportunities to explore, observe, discuss, and engage actively with knowledge. This matters for sustainability education, which depends on inquiry, systems thinking, and deliberation around socio-scientific issues. Students' repeated calls for weekly interaction and real-world connection suggest that sustainability learning should be treated as a regular pedagogical practice, not as a one-off topic. In both groups, boredom was tied less to the difficulty of the topic than to the lesson format itself: copying, memorising, and limited opportunity for questioning.

Theme D. Trust Is Conditional: Science as Powerful but Dependent on Time, Constraints, and Human Use

Students generally believed that science could help solve major problems, but their confidence was usually qualified. They saw trust as depending on factors such as time, institutional limits, and, above all, how people use science. Their accounts point more toward informed trust than uncritical optimism (Cobern et al., 2022).

The boys clearly distinguished between science as a form of capability and science as it is actually put into practice. They noted that scientific work faces restrictions, such as experiments being disallowed or requiring years of approval. They repeatedly suggested that the problem was not science itself, but how it was applied. Their strongest expression of conditional trust was about ethical use: "Maybe science can solve problems, but it depends on how people use it," and they also acknowledged that "sometimes people use science for bad things too." Students linked responsible use to sustainability-related goals such as "saving energy" and "finding new cures," which suggests that their trust in science was closely tied to moral concerns.

Girls' accounts. The girls also recognised science's potential across different fields, but they grounded their claims in specific issues such as climate change and health. For instance: "I think chemistry is very important for problems like climate change. Understanding substances and chemicals that affect the environment can help us create solutions." Others emphasised health-related challenges and moral duty: "biology is especially important because human lives are literally in our hands," and personal illness experiences motivated goals to "fix cancer." They also expanded the idea of "science solving problems" to include psychosocial well-being, linking scientific and psychological knowledge to adolescent emotional development and social norms.

Interpretation for Place-Based Sustainability Learning

Conditional trust appears to be an important resource for sustainability education because sustainability challenges are inherently socio-scientific. They involve uncertainty, trade-offs, and ethical implications. Students' emphasis on human use and constraint suggests they are ready for pedagogies that foreground the relationships among scientific knowledge, social decision-making, and responsibility. These findings support the inclusion of explicit ethical and civic dimensions within place-based sustainability science, where learners examine not only what science can do, but also how scientific knowledge is mobilised, governed, and used in particular places." Students' statements about trust routinely included qualifiers about time, approval, and how people use science. This helps explain why SSI-style routines fit the logic already present in their reasoning.

Theme E. Classroom Discourse and Agency: Participation Conditions as a Hidden Curriculum

Especially in girls' accounts, students identified classroom discourse norms and opportunities for agency as consequential for learning. These conditions influenced students' willingness to participate, their perceptions of difficulty, and their sense of relevance. They also sparked debates about choosing and timing specialisation, such as whether physics should be postponed or optional, highlighting concerns about agency, curricular choices, and identity exploration. These discussions were central, serving as a meta-commentary on how schooling systems allocate legitimacy, confidence, and opportunities for understanding.

Although less detailed in the boys' write-up, students also emphasised the importance of diverse teaching methods, such as group work and discussions. They agreed that applying concepts during labs could boost enjoyment, with comments like "If we had more lab time, everyone would like it" and "If we learn the lesson and then apply it in the lab, it will be much more enjoyable."

Interpretation for Place-Based Sustainability Learning

Place-based sustainability pedagogy relies on active participation, requiring learners to have a space to express local concerns, negotiate understanding, and discuss disputed issues. Students' accounts show that such participation cannot be assumed. It has to be built through classroom norms, routines, and structures that reduce the fear of asking questions and broaden the forms of contribution considered legitimate. Agency and discourse conditions should therefore be treated as part of the design of sustainability learning, rather than as incidental matters of classroom management.

DESIGN PRINCIPLES FOR PLACE-BASED SUSTAINABILITY-ORIENTED SCIENCE TEACHING (UAE LOWER SECONDARY)

The themes identified in students' accounts are treated here as design constraints, pointing to the classroom routines that need to shift if sustainability-oriented science is to feel relevant, participatory, and credible. To make the practical contribution clearer, these student-derived implications are brought together into three meta-principles that can guide lesson design and professional learning.

Meta-Principle 1: Local Anchoring With Explicit Relevance Routines

Students consistently found science meaningful when it was connected to familiar practices and observable local phenomena, but they disengaged when the purpose of learning remained implicit.

Implementable Routines

Start each unit with a locally recognisable phenomenon (5–10 minutes). Begin with a concrete practice students can easily name and picture (e.g., cleaning-product safety, mixtures in food preparation, hydration and bodily systems, or heat and cooling in buildings). Use it as an elicitation prompt before introducing disciplinary concepts: "What have you seen? What do you already do? What questions does this raise?"

Build a brief "relevance move" into every lesson (2–3 minutes). Explicitly identify: (a) where the concept appears in students' lived world, (b) what decisions or problems it informs, and (c) which community-level challenge it connects to (health, energy, or environment). The routine is deliberately brief to remain predictable and easy to sustain.

Use short "future-building" prompts as micro-bridges (1–2 minutes). Use concise prompts that connect concepts to practical pathways and local uses, such as "Which roles use this?" "Which UAE or Gulf technologies depend on these ideas?", and "Which community issue does it address?" Keep these links brief by using short career vignettes or role-model stories rather than lengthy digressions.

Ethically and pedagogically, this approach links everyday practices to broader socio-ecological outcomes, supporting sustainability literacy without reducing place to a decorative element.

Meta-Principle 2: Participation and Inquiry Routines That Normalise Uncertainty

Students highlighted obstacles to participation, including fear of making mistakes and limited opportunities to ask questions. They valued learning through doing, discussing, and investigating far more than copying and memorising.

Implementable Routines

Institutionalise "safe-question" channels through weekly low-risk participation structures such as anonymous question boxes (physical or digital), question quotas collected without evaluation, structured think-pair-share sessions, or brief "wonder time" periods in which questions are recorded before any evaluation of answers.

Establish a minimum viable inquiry routine each week. Make interactivity a regular expectation rather than a rare event by including one inquiry task, demonstration, micro-lab, simulation, structured discussion, or project checkpoint each week. Keep the format consistent so that students know what to expect, even as the content changes.

Use talk moves that normalise uncertainty. Replace deficit cues such as "you should already know" with routines that treat confusion as part of the learning process, such as "That's a strong question," "Let's test that idea," or "What evidence would settle this?" Treat student contributions as ideas to be developed rather than answers to be judged immediately.

This also reflects the ethical dimension of place-based sustainability, where issues are often contested and uncertain. Students need a safe environment for open discussion if they are to weigh evidence, trade-offs, and implications within specific contexts.

Meta-Principle 3: Socio-Scientific Ethics Routines That Cultivate Conditional Trust

Students generally saw science as a helpful problem-solving tool, but they also highlighted its limits, uncertainties, and the importance of human ethics. This reflects conditional trust rather than simple optimism.

Implementable Routines

Include one structured socio-scientific inquiry in each unit, using a case rooted in a local or regional context such as energy transitions, medical technologies, or chemical risks. Guide students through three questions: (a) what science can explain or enable, (b) what remains uncertain or constrained, and (c) how values, governance, and human decisions influence outcomes.

Introduce an "evidence uncertainty decision" triad. Make it standard practice to distinguish between the evidence supporting a position, what remains uncertain or unknown, and which decisions involve ethical or civic judgment beyond the evidence.

Avoid tokenism when drawing on local or heritage knowledge. When drawing on family practices or community stories, treat them as knowledge with ownership: cite sources, clarify how claims are being verified (for example, what counts as evidence and what requires expert input), and, where possible, use consultation or co-design so that classroom inquiry also serves community interests.

Ethically, conditional trust supports responsible sustainability citizenship by combining confidence in evidence-informed action with resistance to naïve techno-solutionism.

Concluding Synthesis

Together, these three meta-principles show how place-based sustainability-oriented science teaching can respond to students' expressed needs for relevance, participation, and credible problem-solving. The emphasis is not on choosing "more interesting" topics, but on establishing routines that connect disciplinary ideas to locally recognisable phenomena, normalise inquiry and uncertainty, and support ethical deliberation around socio-scientific problems. In this framing, student voice functions both as descriptive evidence and as a legitimate basis for pedagogical design aimed at sustainability-oriented scientific literacy in the UAE and comparable contexts.

DISCUSSION

This chapter examined how Grade 9 and 10 students in the UAE, both boys and girls, described the relevance, purpose, and trustworthiness of school science, and what these views suggest for place-based sustainability-oriented pedagogy. Across boys' and girls' accounts, engagement with science was shaped by three interrelated patterns: (a) relevance was tied to everyday practices and community imaginaries; (b) pedagogy acted as a gatekeeper by enabling or limiting application and participation; and (c) trust in science was generally positive, but clearly conditional on human use, constraints, and ethical considerations.

First, students' accounts make clear that relevance is context-dependent. Rather than treating relevance as a general preference for "real-world examples," students tied it to locally recognisable practices and phenomena such as health, household chemicals, and visible material changes. At times, they also linked relevance to community functioning, including infrastructure and national development. This resonates with place-based education scholarship, which treats place as both relational and epistemic, thereby shaping what counts as a meaningful problem and what knowledge is considered legitimate. For sustainability education, the implication is fairly direct: sustainability literacy and competence are more likely to develop when instruction begins with familiar practices and then expands outward toward systems thinking and collective consequences. This is consistent with ESD's emphasis on contextually grounded competencies.

Second, the findings suggest that pedagogy acts as a gatekeeper for whether place-based meaning can emerge: students criticised routines centred on copying and memorisation, and instead described "memorable" learning in terms of applying, interacting, and investigating. This aligns with science education research showing that engagement is more likely when students participate in inquiry and build explanations, rather than passively receive information. For sustainability education, the point is clear: if the goal is action competence and socio-scientific reasoning, transmissive routines will work against it. Learners need repeated opportunities to evaluate evidence, weigh trade-offs, and connect knowledge to decisions. In practice, sustainability cannot simply be added as new content without changing the dominant classroom routine. It requires regular routines of application and discussion.

Third, students' conditional trust in science appears constructive for sustainability education because it invites ethical reflection rather than simple obedience. When combined with awareness of limits, time, and ethical use, it aligns with what epistemic cognition researchers describe as informed trust rather than naive deference. Students viewed science as a powerful tool, but not as something automatically beneficial. Their accounts repeatedly pointed to the need for responsible governance and human judgment. This supports sustainability education because sustainability issues are inherently socio-scientific. They are uncertain, value-laden, and shaped by how knowledge is used in social life. Teaching that clearly distinguishes scientific evidence from societal decision-making can build on students' existing reasoning while avoiding both techno-optimism and cynical scepticism.

Although the combined findings show strong agreement between boys' and girls' accounts, especially in their criticism of memorisation-heavy instruction, the differences in emphasis are better treated as design considerations than as inherent group characteristics. Girls more often differentiated the usefulness of scientific disciplines in relation to imagined futures and were more explicit about the risks of participation associated with questioning. Boys, by contrast, more often linked science to national development narratives, including sustainability-relevant technologies. These patterns point to the importance of offering multiple entry points into relevance, including everyday life, futures, and societal contribution, while also designing participation structures that widen access to explanation and inquiry for all learners.

Taken together, the findings support the design principles outlined in the previous section: begin with locally recognisable phenomena, make relevance explicit, sequence learning through application and reflection, build predictable forms of interactivity, and teach conditional trust through socio-scientific problem framing. In the context of this edited volume, the chapter's contribution lies in showing how student voice can clarify the conditions under which place-based sustainability education becomes meaningful at the classroom level, and in translating those conditions into pedagogical commitments suited to UAE lower secondary schooling.

Contribution to the edited volume. The chapter contributes to the Indigenous and place-based sustainability agenda in three main ways: (1) Theoretically, it treats place as an interpretive anchor through students' "relevance logics," adding an empirical dimension to discussions of ecology of learning and the relationship between knowledge, place, and identity. (2) Empirically, it draws on students' accounts from gender-segregated UAE contexts to show convergent patterns in how pedagogy, value, and trust shape the meaning of sustainability-related science learning. (3) Practically, it translates these patterns into implementable design routines that show how community and heritage knowledge can be integrated ethically, with attention to attribution, verification, and reciprocity, without making Indigenous claims beyond the data.

REFERENCES

Al-Zoubi, Z., AlKaabi, A., Qablan, A., Bataineh, O., & Bany Issa, H. (2024). The impact of work pressure on decision-making effectiveness among department heads in faculties of educational sciences. *PLoS One*, *19*(8), e0304584. DOI: 10.1371/journal.pone.0304584 PMID: 39088449

Almomani, E. Y., Jarrar, W., Alhadid, A., Hamadneh, L., Qablan, A., & Almomani, H. Y. (2022). Shedding light on pharmacists' knowledge of kidney stones' etiology and treatment. *Pharmacy Practice*, *20*(3), 2712. DOI: 10.18549/PharmPract.2022.3.2712 PMID: 36733510

Almomani, H. Y., Almomani, E. Y., Qablan, A., & Al-Momany, A. (2024). Pharmacists' perspectives and perceived barriers to counselling patients with kidney stones. *Research in Social & Administrative Pharmacy*, *20*(6), 123–133. DOI: 10.1016/j.sapharm.2024.02.018 PMID: 38462437

Badawy, H. R., Alkaabi, A. M., Qablan, A. M., Badawy, H. R., Almaazmi, K. M. M. H., & Ndlela, N. P. (2025). Nurturing future innovators integrating computational thinking, AI, and STEM in early education. In *Empowering early education with computational thinking, AI, and STEM* (pp. 97-134). IGI Global Scientific Publishing.

Clarke, V., & Braun, V. (2013). *Successful Qualitative Research: A Practical Guide for Beginners*.

Cobern, W. W., Adams, B. A. J., Pleasants, B. A. S., Bentley, A., & Kagumba, R. (2022). Do We Have a Trust Problem? Exploring Undergraduate Student Views on the Tentativeness and Trustworthiness of Science. *Science & Education*, *31*(5), 1209–1238. DOI: 10.1007/s11191-021-00292-1 PMID: 35035099

Erduran, S., & Cheung, K. K. C. (2026). How science education research journals address (and neglect) trust in science. *Current Opinion in Psychology*, *68*, 102218. https://doi.org/https://doi.org/10.1016/j.copsyc.2025.102218. DOI: 10.1016/j.copsyc.2025.102218 PMID: 41352167

Fricker, M. (2007). *Epistemic Injustice: Power and the Ethics of Knowing*. Oxford University Press., DOI: 10.1093/acprof:oso/9780198237907.001.0001

Gruenewald, D. A. (2003). The Best of Both Worlds: A Critical Pedagogy of Place. *Educational Researcher*, *32*(4), 3–12. DOI: 10.3102/0013189X032004003

Ibrahim, H. R., Alghfeli, A. H., Alnuaimi, F. S., Alshamsi, N. N., & Alkaabi, A. M. (2023). STEM and Leadership in the Future: A Path to Innovation, Sustainability, and Entrepreneurship. In Abdallah, A., & Alkaabi, A. (Eds.), *Restructuring Leadership for School Improvement and Reform* (pp. 420–439). IGI Global., DOI: 10.4018/978-1-6684-7818-9.ch021

Kayan-Fadlelmula, F., Sellami, A., Abdelkader, N., & Umer, S. (2022). A systematic review of STEM education research in the GCC countries: Trends, gaps and barriers. *International Journal of STEM Education*, *9*(1), 2. DOI: 10.1186/s40594-021-00319-7

Mansour, N., & Alshamrani, S. (2015). *Science Education in the Arab Gulf States: Visions*. Sociocultural Contexts and Challenges., DOI: 10.1007/978-94-6300-049-9

Qablan, A., & Al-Qaderi, S. (2009). How to change university faculty members' attitudes and behavior in the context of education for sustainable development. *Applied Environmental Education and Communication*, *8*(3-4), 184–194. DOI: 10.1080/15330150903269407

Qablan, A. M., & DeBaz, T. (2015). Facilitating elementary science teachers' implementation of inquiry-based science teaching. *Teacher Development*, *19*(1), 3–21. DOI: 10.1080/13664530.2014.959552

Sadler, T. D. (2011). Situating Socio-scientific Issues in Classrooms as a Means of Achieving Goals of Science Education. In Sadler, T. D. (Ed.), *Socio-scientific Issues in the Classroom: Teaching, Learning and Research* (pp. 1–9). Springer Netherlands., DOI: 10.1007/978-94-007-1159-4_1

Tahat, K., Mansoori, A., Tahat, D., Qablan, A., & Haddad, I. (2024). Analyzing the newspapers' coverage of sustainable development goals in UAE. *Online Journal of Communication and Media Technologies*, *14*(3), e202439. DOI: 10.30935/ojcmt/14710

Wiek, A., Withycombe, L., & Redman, C. L. (2011). Key competencies in sustainability: A reference framework for academic program development. *Sustainability Science*, *6*(2), 203–218. DOI: 10.1007/s11625-011-0132-6

Wigfield, A., & Eccles, J. S. (2020). Chapter Five - 35 years of research on students' subjective task values and motivation: A look back and a look forward. In Elliot, A. J. (Ed.), Vol. 7, pp. 161–198). Advances in Motivation Science. Elsevier., https://doi.org/https://doi.org/10.1016/bs.adms.2019.05.002

Compilation of References

Aarthi, S., Ravikumar, R. N., Khudayberganov, S., Ismailov, A., & Khujaniyozova, O. (2025). Using Intelligent Tutoring Systems to enhance student engagement. In *Advances in Computational Intelligence and Robotics* (pp. 155–188). IGI Global. DOI: 10.4018/979-8-3373-1827-1.ch008

Aarthi, S., Ravikumar, R. N., Sobirova, K., & Mamadiyarov, Z. (2025). Empowering a sustainable future through digital learning and the SDGs. In *Harnessing E-Learning to Create a Sustainable Future* (pp. 145–172). IGI Global Scientific Publishing., DOI: 10.4018/979-8-3373-0112-9.ch006

Abdallah, A. K., Ismail, L. S., & Alkaabi, A. M. (2025). Green careers: Educating for the future of sustainability. In *Legal frameworks and educational strategies for sustainable development* (pp. 337–366). IGI Global Scientific Publishing. https://doi.org/10.4018/979-8-3693-2987-0.ch017

Abdel Qader, M. M. (2001). *Developing the listening skills of Arab learners of English as a foreign language in the UAE* (Master of Philosophy thesis, University of Glasgow).

Abdullah, W., Tarihoran, N., Supardi, S., Saenong, F., Wasehudin, W., Syarifuddin, E., Fauzi, A., & Mamur, I. (2026). Advancing sustainable Islamic management education through education for sustainable development. *Discover Sustainability*, *7*(1), 257. Advance online publication. DOI: 10.1007/s43621-025-02571-1

Abo-Khalil, A. G. (2024). Integrating sustainability into higher education challenges and opportunities for universities worldwide. *Heliyon, A Cell Press Journal, 10*(9), 1–13. DOI: 10.1016/j.heliyon.2024.e29946

Abu Dhabi Department of Education and Knowledge (ADEK). (2025). *School policy on sustainability* (Version 1.2). https://adek.gov.ae/-/media/Project/TAMM/ADEK/Policies/School-Policies/Health-safety-and-wellbeing/ADEK_S_Sustainability-Policy_EN-12.pdf

Adaspayeva, N., & Parkes, S. (2021). Global Citizenship Education conceptualisation in curriculum guidelines of the New Zealand Curriculum. *The New Zealand Annual Review of Education, 25*, 57–75. DOI: 10.26686/nzaroe.v25.6935

ADEK. (n.d.). School Performance Plaques. https://www.adek.gov.ae/en/Programmes-and-Initiatives/School-Performance-Plaques#:~:text=About%20National%20Identity%20Mark,belonging%2C%20volunteering%2C%20and%20conservation

Ahenakew, C. (2016). Grafting Indigenous Ways of Knowing Onto Non-Indigenous Ways of Being: The (Underestimated) Challenges of a Decolonial Imagination. *International Journal of Qualitative Research, 9*(3), 323–340. https://www.jstor.org/stable/26372211. DOI: 10.1525/irqr.2016.9.3.323

Ahmed, G., Al Amiri, N., & Abudaqa, A. (2024). Strategic leadership and economic transformation: The United Arab Emirates (UAE) model. *Journal of Global Business Research and Practice, 1*(1), 60–77. DOI: 10.70273/RSFK3414

Ahsan, S., Islam, M. K., & Khan, M. R. H. (2024). Scope of Global Citizenship Education in the National Curriculum Framework 2021 of Bangladesh: A critical Review. *Teacher's. World Journal of Educational Research (Los Angeles, Calif.), 49*(1), 49–76. DOI: 10.3329/twjer.v49i1.70238

Aiello, S., Aguayo, C., Wilkinson, N., & Govender, K. (2021). Developing culturally responsive practice using mixed reality (XR) simulation in Paramedicine Education. *Pacific Journal of Technology Enhanced Learning, 3*(1), 15–16. DOI: 10.24135/pjtel.v3i1.89

Ajzen, I. (1991). The theory of planned behavior. *Organizational Behavior and Human Decision Processes, 50*(2), 179–211. DOI: 10.1016/0749-5978(91)90020-T

Al Suwaiyan, L. A. (2018). Diglossia in the Arabic language. *International Journal of Language and Linguistics, 5*(3), 228–238.

Alam, M. (2025). Beyond Eurocentrism: Toward a philosophy of indigenous Fijian knowledge. *Cogent Arts & Humanities, 12*(1), 2588469. Advance online publication. DOI: 10.1080/23311983.2025.2588469

Al-Bataineh, A. (2021). Language policy in higher education in the United Arab Emirates: Proficiency, choices and the future of Arabic. *Language Policy, 20*(2), 215–236. DOI: 10.1007/s10993-020-09548-y

Al-Bataineh, A., & Gallagher, K. (2021). Attitudes towards translanguaging: How future teachers perceive the meshing of Arabic and English in children's storybooks. *International Journal of Bilingual Education and Bilingualism, 24*(3), 386–400. DOI: 10.1080/13670050.2018.1471039

Alkhaldi, M., Moonesar, I. A., Issa, S. T., Ghach, W., Okasha, A., Albada, M., Chelli, S., & Takshe, A. A. (2023). Analysis of the United Arab Emirates' contribution to the sustainable development goals with a focus on global health and climate change. *International Journal of Health Governance, 28*(4), 357–367. DOI: 10.1108/IJHG-04-2023-0040

Almomani, E. Y., Jarrar, W., Alhadid, A., Hamadneh, L., Qablan, A., & Almomani, H. Y. (2022). Shedding light on pharmacists' knowledge of kidney stones' etiology and treatment. *Pharmacy Practice, 20*(3), 2712. DOI: 10.18549/PharmPract.2022.3.2712 PMID: 36733510

Almomani, H., Almomani, E., Qablan, A., & Al-Momany, A. (2024). Pharmacists' perspectives and perceived barriers to counseling patients with kidney stones. *Research in Social & Administrative Pharmacy, 20*(4), 1551–7411. DOI: 10.1016/j.sapharm.2024.02.018 PMID: 38462437

Al-Taneiji, S., & McLeod, L. (2008). Towards decentralized management in United Arab Emirate (UAE) schools. *School Effectiveness and School Improvement, 19*(3), 275–291. DOI: 10.1080/09243450802246384

Alzaim, M. A., AlAli, M., Mattar, Y., & Samara, F. (2024). Integrated framework for enhancing liveability and ecological sustainability in UAE communities. *Sustainability (Basel), 16*(22), 9872. Advance online publication. DOI: 10.3390/su16229872

Al-Zoubi, Z., AlKaabi, A., Qablan, A., Bataineh, O., & Bany Issa, H. (2024). The impact of work pressure on decision-making effectiveness among department heads in faculties of educational sciences. *PLoS One, 19*(8), e0304584. DOI: 10.1371/journal.pone.0304584 PMID: 39088449

Ardoin, N. M. (2006). Toward an interdisciplinary understanding of place: Lessons for environmental education. [CJEE]. *Canadian Journal of Environmental Education, 11*(1), 112–126.

Ardoin, N. M., Schuh, J. S., & Gould, R. K. (2012). Exploring the dimensions of place: A confirmatory factor analysis of data from three ecoregional sites. *Environmental Education Research, 18*(5), 583–607. DOI: 10.1080/13504622.2011.640930

Areepattamannil, S. (2024, December). Building a knowledge economy: higher education as a catalyst for the United Arab Emirates' visionary growth. In *Frontiers in Education* (Vol. 9, p. 1510421). Frontiers Media SA. https://doi.org/DOI: 10.3389/feduc.2024.1510421

Arias-Gutierrez, R. I., & Minoia, P. (2023). Decoloniality and Critical Interculturality in Higher Education: Experiences and Challenges in Ecuadorian Amazonia. *Forum for Development Studies*, *50*(1), 11–34. DOI: 10.1080/08039410.2023.2177562

Aseron, J., Wilde, S., Miller, A., & Kelly, S. (2013). Indigenous student participation in higher education: Emergent themes and linkages. *Contemporary Issues in Education Research*, *6*(4), 417. DOI: 10.19030/cier.v6i4.8110

Assalihee, M., & Boonsuk, Y. (2023). Teaching Management Strategies on 21st Century Islamic Education for Southernmost Thai Private Islamic Schools. *Anatolian Journal of Education*, *8*(1), 13–28. DOI: 10.29333/aje.2023.812a

Assouline, S. G., & Whiteman, C. S. (2011). Twice-exceptionality: Implications for school psychologists in the post–IDEA 2004 era. *Journal of Applied School Psychology*, *27*(4), 380–402. DOI: 10.1080/15377903.2011.616576

Athayde, S., Silva-Lugo, J., Schmink, M., Kaiabi, A., & Heckenberger, M. (2017). Reconnecting art and science for sustainability: learning from indigenous knowledge through participatory action-research in the Amazon. *Ecology and Society: A Journal of Integrative Science for Resilience and Sustainability, 22*(2). DOI: 10.5751/ES-09323-220236

Atmaca, F., & Baloglu, M. (2022). Executive function profiles in gifted students with autism spectrum disorder. *Journal of Autism and Developmental Disorders*, *52*(8), 3472–3485. DOI: 10.1007/s10803-021-05229-5

Bada, S. O., & Olusegun, S. (2015). Constructivism learning theory: A paradigm for teaching and learning. *Journal of Research & Method in Education*, *5*(6), 66–70.

Badawy, H. R., Alkaabi, A. M., Qablan, A. M., Badawy, H. R., Almaazmi, K. M. M. H., & Ndlela, N. P. (2025). Nurturing future innovators integrating computational thinking, AI, and STEM in early education. In *Empowering early education with computational thinking, AI, and STEM* (pp. 97-134). IGI Global Scientific Publishing.

Badawy, H. R., Alkaabi, A. M., Qablan, A. M., Badawy, H. R., Almaazmi, K. M. M. H., & Ndlela, N. P. (2025). Nurturing future innovators integrating computational thinking, AI, and STEM in early education. In *Empowering early education with computational thinking, AI, and STEM* (pp. 97–134). IGI Global Scientific Publishing.

Badawy, H., Alkaabi, A., Qablan, A., Badawy, H., Almaazmi, K., & Ndlela, N. (2024). Nurturing Future Innovators Integrating Computational Thinking, AI, and STEM in Early Education. In *Empowering Early Education with Computational Thinking, AI, and STEM*. IGI Publishers. DOI: 10.4018/979-8-3693-6210-5.ch004

Badry, F., & Willoughby, J. (2016). Higher education revolutions in the gulf. In *Higher education revolutions: Short-term success Versus long-term viability?* (pp. 204–236). Routledge.

Baker, F. S. (2016). National pride and the new school model: English language education in Abu Dhabi, UAE. In *English language education policy in the Middle East and North Africa* (pp. 279–300). Springer International Publishing.

Baldwin, L., Omdal, S. N., & Pereles, D. A. (2015). Beyond stereotypes: Understanding, recognizing, and working with twice-exceptional learners. *Teaching Exceptional Children, 47*(4), 216–225. DOI: 10.1177/0040059915569361

Ball, S. J., Maguire, M., & Braun, A. (2012). *How schools do policy: Policy enactments in secondary schools.* Routledge.

Bandura, A. (1977). Self-efficacy: Toward a unifying theory of behavioral change. *Psychological Review, 84*(2), 191–215. DOI: 10.1037/0033-295X.84.2.191 PMID: 847061

Bang, M., Warren, B., Rosebery, A., & Medin, D. L. (2013). Desettling Expectations in Science Education. *Human Development, 55*(5-6), 302–318. DOI: 10.1159/000345322

Barnhardt, R. A. Y., & Kawagley, O. A. (2005). Indigenous knowledge systems and Alaska Native ways of knowing. *Anthropology & Education Quarterly, 36*(1), 8–23. DOI: 10.1525/aeq.2005.36.1.008

Barnhardt, R., & Kawagley, A. O. (2008). Indigenous knowledge systems and education. *Teachers College Record, 110*(13), 223–241. DOI: 10.1177/016146810811001318

Baron-Cohen, S. (2002). The extreme male brain theory of autism. *Trends in Cognitive Sciences, 6*(6), 248–254. DOI: 10.1016/S1364-6613(02)01904-6 PMID: 12039606

Baroudi, S., & Abi Haidar, H. (2025). Sustainability in the United Arab Emirates secondary schools: A policy practice analysis. *Sustainability (Basel), 17*(7), 3129. Advance online publication. DOI: 10.3390/su17073129

Barron, B. (2006). Interest and self-sustained learning as catalysts of development: A learning ecology perspective. *Human Development, 49*(4), 193–224. DOI: 10.1159/000094368

Bartlett, C., Marshall, M., & Marshall, A. (2012). Two-eyed seeing and other lessons learned within a co-learning journey of bringing together indigenous and mainstream knowledges and ways of knowing. *Journal of Environmental Studies and Sciences, 2*(4), 331–340. DOI: 10.1007/s13412-012-0086-8

Bascopé, M., Perasso, P., & Reiss, K. (2019). Systematic Review of Education for Sustainable Development at an Early Stage: Cornerstones and Pedagogical Approaches for Teacher Professional Development. *Sustainability (Basel), 11*(3), 719. DOI: 10.3390/su11030719

Bascopé, M., & Reiss, K. (2021). Place-Based STEM Education for Sustainability: A Path towards Socioecological Resilience. *Sustainability (Basel), 13*(15), 8414. DOI: 10.3390/su13158414

Battiste, M. (2013). Indigenous knowledge and indigenous peoples' education. In United Nations (Ed.), *Traditional knowledge in policy and practice* (pp. 31–51)., DOI: 10.18356/aa0ced95-

Battiste, M. (2019). *Decolonizing education.* Association of Canadian University Presses. DOI: 10.59962/9781895830972

Bebbington, J., & Unerman, J. (2018). Achieving the United Nations Sustainable Development Goals: An enabling role for accounting research. *Accounting, Auditing & Accountability Journal, 31*(1), 2–24. DOI: 10.1108/AAAJ-05-2017-2929

Bednarz, S. (1999). Fieldwork in K-12 geography in the United States. *International Research in Geographical and Environmental Education, 2*(8), 164–170. DOI: 10.1080/10382049908667604

Ben Jaafar, S., Alzouebi, K., & Bodolica, V. (2022). Accountability and quality assurance for leadership and governance in Dubai-based educational marketplace. *International Journal of Educational Management, 36*(5), 641–660. DOI: 10.1108/IJEM-11-2021-0439

Berkes, F. (2017). *Sacred ecology*. Routledge., DOI: 10.4324/9781315114644

Besnoy, K. D. (2018). The twice-exceptional dilemma: A synthesis of the literature. *Gifted Child Today, 41*(4), 195–203. DOI: 10.1177/1076217518786986

Bexell, S. M., Jarrett, O. S., & Ping, X. (2013). The effects of a summer camp program in China on children's knowledge, attitudes, and behaviors toward animals: A model for conservation education. *Visitor Studies, 16*(1), 59–81. DOI: 10.1080/10645578.2013.768072

Bianco, M., & Leech, N. L. (2010). Twice-exceptional learners: Effects of teacher preparation and disability labels on gifted referrals. *Teacher Education and Special Education, 33*(4), 319–334. DOI: 10.1177/0888406409356392

Bibri, S. E., & Huang, J. (2025). Generative AI of things for sustainable smart cities: Synergizing cognitive augmentation, resource efficiency, network traffic, cybersecurity, and anomaly detection for environmental performance. *Sustainable Cities and Society, 133*, 106826. Advance online publication. DOI: 10.1016/j.scs.2025.106826

Biermann, F., Abbott, K., Andresen, S., Bäckstrand, K., Bernstein, S., Betsill, M., Bulkeley, H., Cashore, B., Clapp, J., Folke, C., Gupta, A., Gupta, J., Haas, P. M., Jordan, A., Kanie, N., Kluvankova-Oravska, T., Lebel, L., Liverman, D., Meadowcroft, J., & Zondervan, R. (2012). Navigating the anthropocene: Improving earth system governance. *Science, 335*(6074), 1306–1307. DOI: 10.1126/science.1217255 PMID: 22422966

Bin Rashed, F. M., Aljanahi, M. H., & Al Qenai, R. M. (2025). Promoting Translanguaging in the ESL Classroom: A Closer Look into Kuwait and the United Arab Emirates. In *Intercultural Communication Education and Research in the Middle East and North Africa* (pp. 104-121). Routledge.

Bishop, R. (2005). Changing Power Relations in Education: Kaupapa Mäori Messages for "Mainstream" Education in Aotearoa/New Zealand. *Counterpoints, 275*, 253–269. https://www.jstor.org/stable/42978789

Blaik Hourani, R., & Litz, D. (2016). Perceptions of the school self-evaluation process: The case of Abu Dhabi. *School Leadership & Management, 36*(3), 247–270. DOI: 10.1080/13632434.2016.1247046

Bodkin-Andrews, G., Whittaker, A., Harrison, N., Craven, R., Parker, P., Trudgett, M., & Page, S. (2017). Exposing the patterns of statistical blindness: Centring Indigenous standpoints on student identity, motivation, and future aspirations. *Australian Journal of Education, 61*(3), 225–249. DOI: 10.1177/0004944117731360

Boling, E. (2010). The need for design cases: Disseminating design knowledge. *International Journal of Designs for Learning, 1*(1), 1–8. DOI: 10.14434/ijdl.v1i1.919

Bonnett, M. (2007). Environmental education and the issue of nature. *Journal of Curriculum Studies, 39*(6), 707-721.

Bonnett, M. (2002). Education for sustainability as a frame of mind. *Environmental Education Research, 8*(1), 9–20. DOI: 10.1080/13504620120109619

Borna, C. S., Oppon-Wusu, M., Bedu-Addo, P. K. A., Dabone, K. T., Kwarteng-Nantwi, E., Churcher, K. A., Aninganigu, K., Owusu-Sekyere, F., & Kissi-Abrokwah, B. (2023). Professional development of mathematics teachers: Its impact on their classroom delivery in Ghana. *East African Journal of Education Studies, 6*(1), 320–337. DOI: 10.37284/eajes.6.1.1152

Bourn, D., Hunt, F., & Bamber, P. (2017). *A review of education for sustainable development and global citizenship education in teacher education: Background paper prepared for the 2017/8 Global Education Monitoring Report, Accountability in education: Meeting our commitments*. UNESCO. Retrieved from https://discovery.ucl.ac.uk/id/eprint/10030831/1/bournhuntbamber.pdf

Brayboy, B. M. J., Solyom, J. A., & Castagno, A. E. (2014). Looking into the hearts of Native peoples: Nation building as an institutional orientation for graduate education. *American Journal of Education, 120*(4), 575–596. DOI: 10.1086/676908

Bray, M. (1996). *Decentralization of education: Community financing* (Vol. 36). World Bank Publication. DOI: 10.1596/0-8213-3724-6

Brett, P., Gurung, G., Shah, R., Yogi, K. K., & Dhungana, J. (2023). Enacting the Sustainable Development Goals in Nepalese Schools and Teacher Education. In Beasy, K., Smith, C., & Watson, J. (Eds.), *Education and the UN Sustainable Development Goals. Education for Sustainability* (Vol. 7). Springer., DOI: 10.1007/978-981-99-3802-5_10

Bridi, R. M., & Al Hosani, N. (2022). An examination of the status of higher education in the United Arab Emirates: From humble beginnings to future challenges. *Eurasian Journal of Educational Research, 2022*(99), 26–44.

Browes, N. (2017). *Global citizenship concepts in the curricula of four countries*. UNESCO International Bureau of Education.

Brown, P. R. (2019). Understanding how deficit discourses work against implementing participatory approaches in Australian Indigenous policy. *The Australian Journal of Social Issues, 54*(4), 401–417. DOI: 10.1002/ajs4.78

Burden-Leahy, S. M. (2009). Globalisation and education in the postcolonial world: The conundrum of the higher education system of the United Arab Emirates. *Comparative Education, 45*(4), 525–544. DOI: 10.1080/03050060903391578

Burgess, C., Thorpe, K., Egan, S., & Harwood, V. (2022). Towards a conceptual framework for Country-centred teaching and learning. *Teachers and Teaching, 28*(8), 925–942. DOI: 10.1080/13540602.2022.2137132

Burmeister, M., Rauch, F., & Eilks, I. (2012). Education for Sustainable Development (ESD) and Chemistry Education. *Chemistry Education Research and Practice, 13*(2), 59–68. DOI: 10.1039/C1RP90060A

Burns, T., Köster, F., & Fuster, M. (2016), Education Governance in Action: Lessons from Case Studies. Educational Research and Innovation, OECD Publishing, Paris. http://dx.doi.org/DOI: 10.1787/9789264262829-en

Byram, M. (1995). Acquiring Intercultural Competence: A Review of Learning Theories. In Sercu, L. (Ed.), *Intercultural Competence: The Secondary School* (pp. 45–56). Aalborg University Press.

Byram, M. (1997). *Teaching and assessing intercultural communicative competence*. Multilingual Matters.

Cajete, G. (1999). *Native science: Natural laws of interdependence*. Clear Light Pub.

Cajete, G. A. (2016). Indigenous education and the development of indigenous community leaders. *Leadership, 12*(3), 364–376. DOI: 10.1177/1742715015610412

Cajete, G. A., & Pueblo, S. C. (2010). Contemporary Indigenous education: A nature-centered American Indian philosophy for a 21st century world. *Futures, 42*(10), 1126–1132. DOI: 10.1016/j.futures.2010.08.013

Çalık, M., & Wiyarsi, A. (2021). A systematic review of the research papers on chemistry-focused socio-scientific issues. Journal of Baltic Science Education. al of Baltic Science Education. DOI: 10.33225/jbse/21.20.360

Canagarajah, S. (2011). Codemeshing in Academic Writing: Identifying Teachable Strategies of Translanguaging. *Modern Language Journal, 95*(3), 401–417. https://doi-org.uaeu.idm.oclc.org/10.1111/j.1540-4781.2011.01207.x. DOI: 10.1111/j.1540-4781.2011.01207.x

Canagarajah, S. (2011). Translanguaging in the classroom: Emerging issues for research and pedagogy. *Applied Linguistics Review, 2*(2011), 1–28. DOI: 10.1515/9783110239331.1

Carm, E. (2014). Inclusion of indigenous knowledge system (IKS) – A precondition for Sustainable Development and an integral part of environmental studies. *The Journal of Educational Research, 4*(1), 58–76. DOI: 10.3126/jer.v4i1.10726

Carney, S. (2022). Reimagining our futures together: A new social contract for education. *Comparative Education, 58*(4), 568–569. DOI: 10.1080/03050068.2022.2102326

CAST. (2018). *Universal Design for Learning guidelines version 2.2*. http://udlguidelines.cast.org

Castagno, A. E., & Brayboy, B. M. J. (2008). Culturally responsive schooling for Indigenous youth: A review of the literature. *Review of Educational Research, 78*(4), 941–993. DOI: 10.3102/0034654308323036

Cenoz, J., & Gorter, D. (2017). Minority languages and sustainable translanguaging: Threat or opportunity? *Journal of Multilingual and Multicultural Development, 38*(10), 901–912. DOI: 10.1080/01434632.2017.1284855

Cenoz, J., & Gorter, D. (2020). Teaching English through pedagogical translanguaging. *World Englishes*, *39*(2), 300–311. https://doi-org.uaeu.idm.oclc.org/10.1111/weng.12462. DOI: 10.1111/weng.12462

Cenoz, J., & Gorter, D. (2021). *Pedagogical translanguaging*. Cambridge University Press.

Cenoz, J., & Gorter, D. (2022). Pedagogical translanguaging and its application to language classes. *RELC Journal*, *53*(2), 342–354. DOI: 10.1177/00336882221082751

Chandra, D. V. (2014). Re-examining the importance of indigenous perspectives in the western environmental education for sustainability: "From tribal to mainstream education". *Journal of Teacher Education for Sustainability*, *16*(1), 117–127. DOI: 10.2478/jtes-2014-0007

Chang, S. E., & Kuo, M.-Y. (2021). A Place-Based Pedagogical Action Study to Enrich Rural Sustainability: Knowledge Ties of National Taiwan University's 10-Year Partnership with Pinglin. *Sustainability (Basel)*, *13*(5), 2916. DOI: 10.3390/su13052916

Cheeseman, A., & Wright, T. (2019). Examining environmental learning experiences at an earth education summer camp. *Environmental Education Research*, *25*(3), 375–387. DOI: 10.1080/13504622.2018.1509301

Cherubini, L. (2009). "taking Haig-Brown seriously": Implications of Indigenous thought on Ontario educators. *Journal of the Canadian Association for Curriculum Studies*, *7*(1), 6–23. DOI: 10.25071/1916-4467.18136

Chheang, S., & Huot, S. (2026). Safeguarding digital classrooms: Cybersecurity implications of AI-driven tools in Cambodian education. [JEILT]. *Journal of Education Innovation & Language Teaching*, *2*(1), 23–45. DOI: 10.63381/jeilt.v2i1.68

Chinn, P. W. U. (2007). Decolonizing methodologies and indigenous knowledge: The role of culture, place and personal experience in professional development. *Journal of Research in Science Teaching*, *44*(9), 1247–1268. DOI: 10.1002/tea.20192

Cirkony, C., Kenny, J., & Zandvliet, D. (2023). A two-eyed seeing teaching and learning framework for science education. *Canadian Journal of Science, Mathematics and Technology Education = Revue Canadienne de l'Enseignement des Sciences, des Mathématiques et de la Technologie*, *23*(2), 340–364. DOI: 10.1007/s42330-023-00276-z

Clarke, V., & Braun, V. (2013). *Successful Qualitative Research: A Practical Guide for Beginners*.

Cobern, W. W., Adams, B. A. J., Pleasants, B. A. S., Bentley, A., & Kagumba, R. (2022). Do We Have a Trust Problem? Exploring Undergraduate Student Views on the Tentativeness and Trustworthiness of Science. *Science & Education, 31*(5), 1209–1238. DOI: 10.1007/s11191-021-00292-1 PMID: 35035099

Coelho, D., Khalil, N., & Shankar, D. D. (2024). Teachers' perceptions and 'invisible'uses of plurilingual pedagogy in UAE K-12 schools. *International Journal of Multilingualism, 21*(2), 670–688. DOI: 10.1080/14790718.2022.2099400

Cohen, W. M., & Levinthal, D. A. (1990). Absorptive capacity: A new perspective on learning and innovation. *Administrative Science Quarterly, 35*(1), 128–152. DOI: 10.2307/2393553

Cook, V. A., Phillips, D., & Holden, J. (2006). Geography fieldwork in a 'risk society'. *Area, 38*(4), 413–420. DOI: 10.1111/j.1475-4762.2006.00707.x

Corntassel, J. (2008). Toward sustainable self-determination: Rethinking the contemporary Indigenous-rights discourse. *Alternatives, 33*(1), 105–132. DOI: 10.1177/030437540803300106

Cruz, A. R. (2020). Land Education. In *Oxford Research Encyclopedia of Education*. Oxford University Press., DOI: 10.1093/acrefore/9780190264093.013.682

Dabrowski, K. (2016). *Positive disintegration* (Revised ed.). Maurice Bassett. (Original work published 1964)

Dartevelle, O. (2012). State of the Art of Environmental Sustainability in Academic Curricula and Conditions for Registration.

Das, T. M., Medhi, M., & Banerjee, D. S. (2025). Indigenous knowledge systems and sustainable learning: A comprehensive regional analysis of Northeast India. *International Journal of Environmental Sciences*, •••, 1999–2012. DOI: 10.64252/vcp0c468

Datta, R. (2018). Decolonizing both researcher and research and its effectiveness in Indigenous research. *Research Ethics, 14*(2), 1–24. DOI: 10.1177/1747016117733296

Datta, R. (2024). Relationality in Indigenous climate change education research: A learning journey from Indigenous communities in Bangladesh. *Australian Journal of Environmental Education, 40*(2), 128–142. DOI: 10.1017/aee.2024.13

David, S. A., & Abukari, A. (2020). Perspectives of teachers' on the selection and the development of the school leaders in the United Arab Emirates. *International Journal of Educational Management, 34*(1), 56–69. DOI: 10.1108/IJEM-02-2019-0057

Davidson, C. M. (2009). The United Arab Emirates: Prospects for Political Reform. *The Brown Journal of World Affairs, 15*(2), 117–127. https://www.jstor.org/stable/24590846

Day, K., & McPherson, L. (2025). Resilient epistemologies: Integrating Crotty's framework with Indigenous yarning protocols for decolonial research. *International Journal of Qualitative Methods, 24*(16094069251404324), 16094069251404323. Advance online publication. DOI: 10.1177/16094069251404323

De la Fuente, M. J. (2021). Introduction: toward education for sustainable development (ESD) in foreign languages. In *Education for sustainable development in foreign language learning* (pp. 1-13). Routledge.

Degyansky, M. (2024). Relational, ecological cooperation with and as part of more-than-human world(s). In Warren, J. N., Biggiero, L., Hübner, J., & Ogunyemi, K. (Eds.), *The Routledge handbook of cooperative economics and management* (pp. 433–442). Routledge., DOI: 10.4324/9781003449850-34

Dei, D.-G. J., Anane-Donkor, L., Dzandza, P. E., Peasah, T., & Puttick, C. P. (2025). Synthesizing the characteristics and applications of indigenous knowledge for sustainable development: A systematic review. *SAGE Open, 15*(3), 21582440251383843. Advance online publication. DOI: 10.1177/21582440251383843

Dei, G. J. S. (2011). *Indigenous philosophies and critical education*. Peter Lang Group AG. DOI: 10.3726/978-1-4539-0131-1

Demakova, I. D., Valeeva, R. A., & Shipova, A. V. (2016). Socialization of Adolescents: Cultural Practices in Children's Summer Camp. *International Journal of Environmental and Science Education, 11*(7), 1631–1640.

Dillon, J., Rickinson, M., & Teamey, K. (2016). The value of outdoor learning: Evidence from research in the UK and elsewhere. in *Towards a convergence between science and environmental education* (ss. 193-200). Routledge.

Dodgson, J. E. (2019). Reflexivity in qualitative research. *Journal of Human Lactation : Official Journal of International Lactation Consultant Association, 35*(2), 220–222. DOI: 10.1177/0890334419830990 PMID: 30849272

Dorji, T., Rinchen, K., Morrison-Saunders, A., Blake, D., Banham, V., & Pelden, S. (2024). Understanding how indigenous knowledge contributes to climate change adaptation and resilience: A systematic literature review. *Environmental Management, 74*(6), 1101–1123. DOI: 10.1007/s00267-024-02032-x PMID: 39215837

Duarte, J. (2020). Translanguaging in the context of mainstream multilingual education. *International Journal of Multilingualism*, *17*(2), 232–247. https://doi-org.uaeu.idm.oclc.org/10.1080/14790718.2018.1512607. DOI: 10.1080/14790718.2018.1512607

Dubai Development Authority. (2019). *Master planning guidelines*. https://dda.gov.ae/-/media/5F89FD14770B491BB843C84BC8031C40.pdf

Duit, A., & Galaz, V. (2008). Governance and Complexity—Emerging Issues for Governance Theory. *Governance: An International Journal of Policy, Administration and Institutions*, *21*(3), 311–335. DOI: 10.1111/j.1468-0491.2008.00402.x

Dyment, J. E., & Potter, T. G. (2015). Is outdoor education a discipline? Provocations and possibilities. *Journal of Adventure Education and Outdoor Learning*, *15*(3), 193–208. DOI: 10.1080/14729679.2014.949808

Ehren, M. C. M., & Visscher, A. J. (2006). TOWARDS A THEORY ON THE IMPACT OF SCHOOL INSPECTIONS. *British Journal of Educational Studies*, *54*(1), 51–72. DOI: 10.1111/j.1467-8527.2006.00333.x

Environment Agency—Abu Dhabi. (2026). *Sustainable Schools*. https://sustainableschools.ead.ae/sis/ssiWebsite/

Erduran, S., & Cheung, K. K. C. (2026). How science education research journals address (and neglect) trust in science. *Current Opinion in Psychology*, *68*, 102218. https://doi.org/https://doi.org/10.1016/j.copsyc.2025.102218. DOI: 10.1016/j.copsyc.2025.102218 PMID: 41352167

Escobar, A. (2001). Culture sits in places: Reflections on globalism and subaltern strategies of localization. *Political Geography*, *20*(2), 139–174. DOI: 10.1016/S0962-6298(00)00064-0

Evans, P. B. (1995). *Embedded autonomy: States and industrial transformation*. Princeton University Press. DOI: 10.1515/9781400821723

Expo 2020 Dubai. (2020). *Sustainability report 2020: Connecting minds, creating the future*. https://www.expo2020dubai.com/-/media/expo2020/sustainability/expo-designed-report-digital-version-v12.pdf

Fang, W. T., Hassan, A. A., & LePage, B. A. (2022). Outdoor education. In *The living environmental education: Sound science toward a cleaner, safer, and healthier future* (pp. 229–260). Springer Nature Singapore.

Fatima, Y., Cleary, A., King, S., Solomon, S., McDaid, L., Hasan, M. M., Al Mamun, A., & Baxter, J. (2022). Cultural identity and social and emotional wellbeing in Aboriginal and Torres Strait Islander children. In *Family Dynamics over the Life Course* (pp. 57–70). Springer International Publishing., DOI: 10.1007/978-3-031-12224-8_4

Fensham, P. J. (2004). *Defining an identity: The evolution of science education as a field of research.* Kluwer Academic Publishers. DOI: 10.1007/978-94-010-0175-5

Ferdinan, A., Abdillah, , Rahman, A., & Bahtiar, A. R. (2026). Eco Islamic education and its contribution to sustainable development goals in Muhammadiyah boarding schools of South Sulawesi. *Discover Sustainability*, *7*(1), 310. DOI: 10.1007/s43621-026-02647-6

Fernández-Llamazares, Á., & Cabeza, M. (2017). Rediscovering the potential of indigenous storytelling for conservation practice. *Conservation Letters*, *11*(3), 1–12. DOI: 10.1111/conl.12398

Filho, W. L. Shiel, Chris dan Paco, Arminda. 2016. Implementing and Operationalising Integrative Approaches to Sustainability In Higher Education: The Role of Project-Oriented Learning. Journal of cleaner Production 133 (2016) 126-135, (online), (http://www.sciencedirect.com, access 2026)

Fleming, T. (2018). Mezirow and the theory of transformative learning. In Wang, V. (Ed.), *Critical theory and transformative learning* (pp. 120–136). IGI Global Publishing., DOI: 10.4018/978-1-5225-6086-9.ch009

Flores, N., & Schissel, J. L. (2014). Dynamic Bilingualism as the Norm: Envisioning a Heteroglossic Approach to Standards-Based Reform. *TESOL Quarterly*, *48*(3), 454–479. DOI: 10.1002/tesq.182

Foley-Nicpon, M. (2018). The social and emotional development of twice-exceptional children. In Pfeiffer, S. I. (Ed.), *Handbook of giftedness in children* (2nd ed., pp. 197–213). Springer., DOI: 10.1007/978-3-319-77004-8_12

Foley-Nicpon, M., & Assouline, S. G. (2020). High-ability students with coexisting disabilities: Implications for school psychological practice. *Psychology in the Schools*, *57*(10), 1615–1626. DOI: 10.1002/pits.22342

Foltz, R. (2000). Is there an Islamic environmentalism? *Environmental Ethics*, *22*(1), 63–72. DOI: 10.5840/enviroethics200022149

Ford, P. (1986). Outdoor Education: Definition and Philosophy.

France, D., & Haigh, M. (2018). Fieldwork@40: Fieldwork in geography higher education. *Journal of Geography in Higher Education*, *42*(4), 1–17. DOI: 10.1080/03098265.2018.1515187

Freire, P. (1970). *Pedagogy of the oppressed* (M. B. Ramos, Trans.). *Continuum.*

Fricker, M. (2007). *Epistemic Injustice: Power and the Ethics of Knowing*. Oxford University Press., DOI: 10.1093/acprof:oso/9780198237907.001.0001

Gagné, F. (2021). Differentiating giftedness from talent: The DMGT perspective on talent development. In Cross, T. L., & Cross, J. R. (Eds.), *Handbook for counselors serving students with gifts and talents* (2nd ed., pp. 43–74). Prufrock Press. DOI: 10.4324/9781003235415-3

Gainsford, A., & Evans, M. (2021). Integrating andragogical philosophy with Indigenous teaching and learning. *Management Learning*, *52*(5), 559–580. DOI: 10.1177/1350507620972528

Gallagher, K. (2011). Bilingual education in the UAE: Factors, variables and critical questions. *Education, Business and Society*, *4*(1), 62–79. DOI: 10.1108/17537981111111274

Gallay, E., Marckini-Polk, L., Schroeder, B., & Flanagan, C. (2016). Place-based stewardship education: Nurturing aspirations to protect the rural commons. *Peabody Journal of Education*, *91*(2), 155–175. DOI: 10.1080/0161956X.2016.1151736

García, O., Ibarra Johnson, S., & Seltzer, K. (2017). *The Translanguaging Classroom. Philadelphia.* Caslon.

García, O., & Torres-Guevara, R. (2010). Monoglossic ideologies and language policies in the education of Latinas/os. In Murillo, E., Villenas, S., Galván, R. T., Muñoz, J. S., Martínez, C., & Machado-Casas, M. (Eds.), *Handbook of Latinos and education: Research, theory and practice* (pp. 182–193). Routledge.

Gerelkhuu, S., Fiel'ardh, K., Fujii, H., Yembuu, B., & Dembereldorj, U. (2026). Towards place-responsive climate change education: Mongolian primary teachers' pedagogical judgement across urban and rural contexts. *Environmental Education Research*, •••, 1–22. DOI: 10.1080/13504622.2026.2635431

Gerlach, A. (2018). Thinking and researching relationally. *International Journal of Qualitative Methods*, *17*(1), 160940691877607. DOI: 10.1177/1609406918776075

Gilbert, J. K. (2006). On the nature of "Context" in chemical education. *International Journal of Science Education*, *28*(9), 957–976. DOI: 10.1080/09500690600702470

Global Center on Adaptation. (2024). *Educate to adapt: The interconnected paths of learning and climate.* https://gca.org/educate-to-adapt-the-interconnected-paths-of-learning-and-climate/

Global Environmental Education Partnership (GEEP). (n.d.). *Sustainable schools initiative* (Case study). https://thegeep.org/resources/case-studies/sustainable-schools-initiative

Goode, G. S., & MacGillivray, L. (2023). The construction of systems thinking pedagogy during a professional development institute. *Journal of Pedagogical Research*, *7*(4), 275–302. DOI: 10.33902/JPR.202318879

Government, U. A. E. (2017). *UAE Centennial 2071*. https://u.ae/en/about-the-uae/strategies-initiatives-and-awards/strategies-plans-and-visions/uae-centennial-2071

Government, U. A. E. (2021). *UAE Net Zero by 2050 Strategy*. The Official Platform of the UAE Government. https://u.ae/en/about-the-uae/strategies-initiatives-and-awards/strategies-plans-and-visions/environment-and-energy/the-uae-net-zero-2050-strategy

Government, U. A. E. (2025, August 4). *Regulatory authorities of K-12 education*. Retrieved March 8, 2026, from https://u.ae/en/information-and-services/education/school-education-k-12/regulatory-authorities-of-k-12-education

Govinda, R. (Ed.). (2002). *India education report: A profile of basic education*. Oxford University Press.

Granit-Dgani, D. (2021). Paths to place-based education. The Mofet Institute. Hebrew.

Greenwood, J., & Te Aika, L. H. (2010). *Hei tauira: Teaching and learning for success for Māori in tertiary settings*. Ako Aotearoa – National Centre for Tertiary Teaching Excellence.

Gruenewald, D. A. (2003). The best of both worlds: A critical pedagogy of place. *Educational Researcher*, *32*(4), 3–12. DOI: 10.3102/0013189X032004003

Gruenewald, D. A. (2003a). Foundations of place: A multidisciplinary framework for place-conscious education. *American Educational Research Journal*, *40*(3), 619–654. DOI: 10.3102/00028312040003619

Gunawan, K. D. H., & Indrawan, I. P. O. (2025). Decolonising Science Education: A Bibliometric Analysis of Indigenous Knowledge Integration in Global STEM. *Journal of Posthumanism*, *5*(5), 2598–2622. DOI: 10.63332/joph.v5i5.1651

Gwekwerere, Y. (2011). Bridging culture, indigenous and scientific ways of knowing. *Cultural and Pedagogical Inquiry*, *6*(2), 1–4. DOI: 10.18733/C3GW24

Haatainen, O. M., & Aksela, M. (2021). Project-based learning in integrated science education: Active teachers' perceptions and practices. LUMAT: International Journal on Math. *Science and Technology Education*, *9*(1), 149–173. DOI: 10.31129/LUMAT.9.1.1392

Hall, P. A., & Taylor, R. C. R. (1996). Political science and the three new institutionalisms. *Political Studies*, *44*(5), 936–957. DOI: 10.1111/j.1467-9248.1996.tb00343.x

Han, S., Yalvac, B., Capraro, M. M., & Capraro, R. M. (2015). In-service teachers' implementation and understanding of STEM project based learning. *Eurasia Journal of Mathematics, Science and Technology Education*, *11*(1), 63–76. DOI: 10.12973/eurasia.2015.1306a

Happé, F., & Frith, U. (2006). The weak coherence account: Detail-focused cognitive style in autism spectrum disorders. *Journal of Autism and Developmental Disorders*, *36*(1), 5–25. DOI: 10.1007/s10803-005-0039-0 PMID: 16450045

Harris, O., Taylor, S., & Mitchell, T. (2025). Telling the land: Aboriginal educational narratives and curriculum integration in Australian schools. *International Journal of Educational Narratives*, *3*(3), 218–229. DOI: 10.70177/ijen.v3i3.2211

Harushimana, I. (2022). Preparing educators who save lives: What can US schools of education do? *Journal for Multicultural Education*, *16*(3), 225–236. DOI: 10.1108/JME-11-2021-0198

Heard-Bey, F. (2005). The United Arab Emirates: Statehood and nation-building in a traditional society. *The Middle East Journal*, *59*(3), 357–375. https://www.jstor.org/stable/4330153. DOI: 10.3751/59.3.11

Hecht, M., & Crowley, K. (2019). Unpacking the learning ecosystems framework. *Journal of the Learning Sciences*, *28*(2), 264–284. DOI: 10.1080/10508406.2019.1693381

Hemara, W. (2000). *Māori pedagogies: A view from the literature*. New Zealand Council for Educational Research.

Herb, M. (1999). *All in the family: absolutism, revolution, and democracy in Middle Eastern monarchies*. State University of New York Press.

Hernandez Gonzalez, F. (2023). Exploring the Affordances of Place-Based Education for Advancing Sustainability Education: The Role of Cognitive, Socio-Emotional and Behavioural Learning. *Education Sciences*, *13*(7), 676. DOI: 10.3390/educsci13070676

Hertog, S. (2011). *Princes, brokers, and bureaucrats: Oil and the state in Saudi Arabia*. Cornell University Press. DOI: 10.7591/9780801458774

Higgins, B. J. (2026). Learning From Faculty Before Implementing a Culturally Responsive Teaching Program at a Community College. *New Directions for Community Colleges*, *2026*(213), 71–77. DOI: 10.1002/cc.70049

Hlalele, D. J. (2019). Indigenous knowledge systems and sustainable learning in rural South Africa. *Australian and International Journal of Rural Education*, *29*(1), 88–100. DOI: 10.47381/aijre.v29i1.187

Honig, M. I., & Hatch, T. C. (2004). Crafting Coherence: How Schools Strategically Manage Multiple, External Demands. *Educational Researcher*, *33*(8), 16–30. DOI: 10.3102/0013189X033008016

Hopkyns, S. (2022). Translanguaging in Gulf English-medium instruction higher education. In Al Rashdi, F., & Mehta, S. R. (Eds.), *Language and identity in the Arab world* (pp. 76–92). Routledge., DOI: 10.4324/9781003174981-6

Hopkyns, S., Zoghbor, W., & Hassall, P. J. (2021). The use of English and linguistic hybridity among Emirati millennials. *World Englishes*, *40*(2), 176–190. DOI: 10.1111/weng.12506

Hornberger, N. H. (2014). Bringing our languages home: Language revitalization for families. Ed. by Hinton Leanne. Berkeley, CA: Heyday Books, 2013. Pp. xx, 264. ISBN 9781597142007. $20. *Language*, *90*(2), 540–543. https://www.jstor.org/stable/24672189. DOI: 10.1353/lan.2014.0033

Housel, D. A., & Nikoui, M. (2024). Practical recommendations for accommodating students with limited or interrupted formal education in adult education programs in the United States. *Adult Learning*. Advance online publication. DOI: 10.1177/10451595241297660

Howard, C. D., Boling, E., Rowland, G., & Smith, K. M. (2012). Instructional design cases and why we need them. *Educational Technology*, •••, 34–38.

Hudson, A., & Vodden, K. (2020). Decolonizing pathways to sustainability: Lessons learned from three Inuit communities in NunatuKavut, Canada. *Sustainability (Basel)*, *12*(11), 4419. DOI: 10.3390/su12114419

Hughes, C. E. (2021). *Twice-exceptional gifted children: Understanding, teaching, and counseling gifted students* (2nd ed.). Routledge.

Hu, L., Cheng, J., Wu, C., Liu, Z., & Zheng, L. (2025). Influencing factors for sustainable high-quality implementation of comprehensive practical activity courses in rural areas. *Humanities & Social Sciences Communications*, *12*(1), 1937. Advance online publication. DOI: 10.1057/s41599-025-06226-4

Hunt (Nehiyaw), D. (2025). Inuit Qaujimajatuqangit (IQ) and the climate of colonialism: pedagogy, intellectual emancipation, and decolonial futures. *AlterNative (Auckland, N.Z. : 2005), 11771801251362692*. DOI: 10.1177/11771801251362691

Huot, S. (2026). Emerging voices from Cambodia: International student journeys and educational aspirations in a globalized academic landscape. In Irham, I., Miftahul, H., & Lalu Nurul, Y. (Eds.), *International student narratives in globalized higher education* (pp. 363–398). IGI Global Scientific Publishing., DOI: 10.4018/979-8-3373-4212-2.ch014

Huot, S., & Chheang, S. (2025). Unwriting empire: Decolonizing Khmer language, literature, and cultural memory. *The Context Journal, 12*(6), 280–293. https://doi.org/https://doi.org/10.5281/zenodo.17061523

Huot, S., & Em, S. (2025). The integration of English language teaching with environmental sustainability: A comprehensive review of pedagogical strategies and global impacts. [CJESS]. *Cambodian Journal of Educational and Social Sciences, 2*(2), 83–95. DOI: 10.69496/cjess.v2i2.68

Huot, S., & Hok, C. (2025a). Enhancing English language teaching for multilingual learners: Leveraging heritage languages in the classroom. *Journal of Social and Educational Research, 4*(1), 87–95. DOI: 10.5281/zenodo.15698963

Huot, S., & Hok, C. (2025b). Unpacking the struggle: Linguistic, pedagogical, and cognitive barriers in Cambodian students' English essay writing. *Context (Ibadan), 12*(4), 149–168. DOI: 10.5281/zenodo.15807401

Huot, S., Loch, S., & Rin, R. (2025). Land, life, and the sacred: Reimagining development through tribal worldviews. In Sayak, P., Susmita, H., & Nitesh, T. (Eds.), *Psychology of media influence: Shaping thoughts, behavior and society* (pp. 448–462). Mitra Press.

Hvidt, M. (2013). Economic diversification in GCC countries: Past record and future trends. *The London School of Economics Kuwait Programme on Development, Governance and Globalisation in the Gulf States*, 27.

IndigenTerra. (2026). https://sites.google.com/view/indigenterra/home

Institute of Education Sciences. (2020). *Supporting Pacific Island communities through place-based education*. U.S. Department of Education. https://ies.ed.gov/learn/blog/supporting-pacific-island-communities-through-place-based-education

Israel, A. L. (2012). Putting geography education into place: What geography educators can learn from Place-Based Education, and vice versa. Journal of Geography, 111(2), 76– Ives,81. DOI: 10.1080/00221341.2011.583264

Jacob, M. M., Sabzalian, L., Anderson, R. N., Muniz, H. R., Simmons, K., & Beavert, V. R. (2024). Indigenous genealogies of relational knowledge: Cedar tree and gray squirrel as important relatives and teachers. *Genealogy*, *8*(1), 19. DOI: 10.3390/genealogy8010019

Jacobson, A. M. (2015). Why Summer Camp Matters: The Benefits of Working at Summer Camp on Future Employment.

Jaiswal, D. A.Dr. Anita Jaiswal. (2025). Indigenous pedagogies teaching and learning practices rooted in local contexts. [NIJMS]. *Naveen International Journal of Multidisciplinary Sciences*, *1*(4), 97–104. DOI: 10.71126/nijms.v1i4.40

Jenkins, J., & Leung, C. (2019). From mythical 'standard' to standard reality: The need for alternatives to standardized English language tests. *Language Teaching*, *52*(1), 86–110. DOI: 10.1017/S0261444818000307

Jickling, B., & Wals, A. E. J. (2008). Globalization and environmental education: Looking beyond sustainable development. *Journal of Curriculum Studies*, *40*(1), 1–21. DOI: 10.1080/00220270701684667

Jones, T. J., Williams-Jones, G., & Nyce, H.Jr. (2025). Braiding Indigenous knowledge systems and Western science through co-creation and co-teaching. *Frontiers in Earth Science (Lausanne)*, *13*, 1587092. DOI: 10.3389/feart.2025.1587092

Jordan, A., & Lenschow, A. (2010). Environmental policy integration: A state of the art review. *Environmental Policy and Governance*, *20*(3), 147–158. DOI: 10.1002/eet.539

Kalidas, A. J. E. (2025). Challenges in implementing a culturally sustaining pedagogy in Indigenous formal education. *Journal of Indigenous Social Development*, *13*(3), 141–174. DOI: 10.55016/ojs/jisd.v13i3.81550

Kana'iaupuni, S. M., Ledward, B., & Keohokalole, K. (2011). New research on the impact of cultural influences in education on Native Hawaiian student outcomes. *AAPI Nexus Journal: Policy, Practice, and Community*, *9*(1–2), 211–229. DOI: 10.36650/nexus9.1-2_211-229_KanaiaupuniEtAl

Karlsen, G. E. (2000). Decentralized centralism: Framework for governance in education. *Journal of Education Policy*, *15*(5), 525–538. DOI: 10.1080/026809300750001676

Kato, D. S., & de Carvalho, L. M. (2026). Afro-Amerindian cosmopolitics in environmental education: A decolonial analysis of academic discourses involving the epistemological disputes around the term sustainability. *British Educational Research Journal*, *52*(1), 25–46. DOI: 10.1002/berj.70104

Kayan-Fadlelmula, F., Sellami, A., Abdelkader, N., & Umer, S. (2022). A systematic review of STEM education research in the GCC countries: Trends, gaps and barriers. *International Journal of STEM Education*, *9*(1), 2. DOI: 10.1186/s40594-021-00319-7

KDHA. February 21, 2025. Under the directive of HH Sheikh Hamdan bin Mohammed bin Rashid Al Maktoum, Dubai launches new policy for Arabic language education in early childhood education. https://web.khda.gov.ae/en/About-Us/News/2025/Under-the-directive-of-HH-Sheikh-Hamdan-bin-Mohamm#:~:text=The%20new%20policy%20encourages%20a,13%20in%20Dubai%20private%20schools

Khairunnisa, K., Dongoran, I. Y., Harahap, K. G., Lubis, N. A., Azhar, Y. A., & Haerunisa, A. (2024). Problematika Sarana dan Prasarana dalam Proses Belajar Mengajar di SD Negeri 104202 Bandar Setia. *Journal of Education*, *6*(4), 20586–20591. DOI: 10.31004/joe.v6i4.6112

Kimmerer, R. W. (2013). *Braiding Sweetgrass: Indigenous Wisdom, Scientific Knowledge and the Teachings of Plants* (1st ed.). Milkweed Editions.

Kimmerer, R. W. (2013). *Braiding sweetgrass: Indigenous wisdom, scientific knowledge and the teachings of plants*. Milkweed.

Kioupi, V., & Voulvoulis, N. (2019). Education for Sustainable Development: A systemic framework for connecting the SDGs to educational outcomes. *Sustainability (Basel)*, *11*(21), 6104. DOI: 10.3390/su11216104

Kirkness, V. J., & Barnhardt, R. (1991). First Nations and Higher Education: The Four R's—Respect, Relevance, Reciprocity, Responsibility. *Journal of American Indian Education*, *30*, 1–15.

Kolb, D. A. (1984). *Experiential learning: Experience as the source of learning and development*. FT press.

Kolb, D. A. (2015). *Experiential learning: Experience as the source of learning and development* (2nd ed.). Pearson Education.

Kovach, M. (2009). *Indigenous methodologies: Characteristics, conversations, and contexts*. University of Toronto Press.

Kramsch, C. J. (1998). *Language and culture*. Oxford University Press.

Krane, J. (2018). *Energy kingdoms: Oil and political survival in the Persian Gulf*. Columbia University Press. DOI: 10.7312/kran17930

Laver, Q., Andersen, O., Moore, S., Pottle, C., & Snook, J. (2024). Fostering fisheries futures: The promise of localized stewardship education in Nunatsiavut. *Frontiers in Environmental Science*, *12*(1409358), 1409358. Advance online publication. DOI: 10.3389/fenvs.2024.1409358

Leal Filho, W., Vargas, V. R., Salvia, A. L., Brandli, L. L., Pallant, E., Klavins, M., Ray, S., Moggi, S., Maruna, M., Conticelli, E., Ayanore, M. A., Radovic, V., Gupta, B., Sen, S., Paço, A., Michalopoulou, E., Saikim, F. H., Koh, H. L., Frankenberger, F., & Vaccari, M. (2019). The role of higher education institutions in sustainability initiatives at the local level. *Journal of Cleaner Production*, *233*, 1004–1015. DOI: 10.1016/j.jclepro.2019.06.059

Lee, E. (2021). Translingualism in the teaching of English: Theoretical considerations and pedagogical implications. In Vinogradova, P., & Shin, J. K. (Eds.), *Contemporary foundations for teaching English as an additional language* (pp. 121–130). Routledge.

Leimgruber, J. R., Al-Issa, A., Lorenz, E., & Siemund, P. (2024). Managing and investing in hybrid identities in the globalized United Arab Emirates. *Journal of Language, Identity, and Education*, *23*(6), 955–972. DOI: 10.1080/15348458.2022.2070849

Leithwood, K., Harris, A., & Hopkins, D. (2020). Seven strong claims about successful school leadership revisited. *School Leadership & Management*, *40*(1), 5–22. DOI: 10.1080/13632434.2019.1596077

Leonard, K., Buttigieg, P. L., Hudson, M., Paul, K., Pearlman, J., & Juniper, S. K. (2022). Two-eyed seeing: Embracing the power of indigenous knowledge for a healthy and sustainable Ocean. *PLoS Biology*, *20*(10), e3001876. Advance online publication. DOI: 10.1371/journal.pbio.3001876 PMID: 36269777

Levitan, J., & Johnson, K. M. (2020). salir adelante: Collaboratively developing culturally grounded curriculum with marginalized communities. *American Journal of Education*, *126*(2), 195–230. DOI: 10.1086/706921

Lin, J., Stoltz, A., Aruch, M., & Rappeport, A. (2021). Decolonization and transformation of higher education for sustainability. *Journal of Comparative & International Higher Education*, *13*(Summer), 134–156. DOI: 10.32674/jcihe.v13iSummer.3255

Liu, H. (2020). Globalization and world Citizenship education. *International Journal of Learning and Teaching*, *94–97*, 94–97. Advance online publication. DOI: 10.18178/ijlt.6.2.94-97

Liu, Y., & Fang, F. (2020). Translanguaging Theory and Practice: How Stakeholders Perceive Translanguaging as a Practical Theory of Language. *RELC Journal.* Advance online publication. DOI: 10.1177/0033688220939222

Lloyd, A., & Gray, T. (2014). Place-based outdoor learning and environmental sustainability within Australian primary schools. Journal of Sustainability Education, 41883.

Lotz Sisitka, H., Wals, A. E. J., Kronlid, D., & McGarry, D. (2015). Transformative, transgressive social learning: Rethinking higher education pedagogy in times of systemic global dysfunction. *Current Opinion in Environmental Sustainability, 16*, 73–80. DOI: 10.1016/j.cosust.2015.07.018

Lotz-Sisitka, H., Ali, M. B., Mphepo, G., Chaves, M., Macintyre, T., Pesanayi, T., Wals, A., Mukute, M., Kronlid, D., Tran, D. T., Joon, D., & McGarry, D. (2016). Co-designing research on transgressive learning in times of climate change. *Current Opinion in Environmental Sustainability, 20*, 50–55. DOI: 10.1016/j.cosust.2016.04.004

Lozano, R., Lukman, R., Lozano, F. J., Huisingh, D., & Lambrechts, W. (2013). Declarations for sustainability in higher education: Becoming better leaders, through addressing the university system. *Journal of Cleaner Production, 48*, 10–19. DOI: 10.1016/j.jclepro.2011.10.006

Luckin, R., Holmes, W., Griffiths, M., & Corcier, L. B. (2016). Pearson (Firm); University College, L. *Intelligence unleashed: An argument for AI in education.*

Macfarlane, A.H., Webber, M., Cookson-Cox, C., & McRae, H.S. (2014). Ka Awatea: An iwi case study of Māori students' success.

Macfarlane, A., Glynn, T., Cavanagh, T., & Bateman, S. (2007). Creating culturally safe schools for māori students. *Australian Journal of Indigenous Education, 36*(1), 65–76. https://doi.org/https://doi.org/10.1017/s1326011100004439. DOI: 10.1017/S1326011100004439

Macionis, J. J. (2017). *Sociology* (16th ed.). Pearson.

Magni, G. (2017). Indigenous knowledge and implications for the sustainable development agenda. *European Journal of Education, 52*(4), 437–447. DOI: 10.1111/ejed.12238

Maharjan, B., Pant, B. P., & Dahal, N. (2024). Participatory learning: Exploring place-based pedagogy for future teachers. *Pedagogical Research, 9*(4), em0222. Advance online publication. DOI: 10.29333/pr/15155

Mairi, S., Gruber, J., Mercer, S., Schartner, A., Ybema, J., Young, T., & Van Der Meer, C. (2023). Teacher educators' perspectives on global citizenship education and multilingual competences. *Journal of Multilingual and Multicultural Development*, *46*(9), 2631–2647. DOI: 10.1080/01434632.2023.2170388

Malcom, A. (2024). Integrating indigenous knowledge into Livestock Policy formulation. *International Journal of Livestock Policy*, *3*(1), 30–42. DOI: 10.47941/ijlp.1962

Mamaile, D., & Omodan, B. I. (2023). Exploring challenges hindering teachers' implementation of classroom management strategies in Gauteng High Schools, South Africa. *Research in Educational Policy and Management*, *5*(2), 245–262. DOI: 10.46303/repam.2023.22

Manns, M., Heiss, J., Morohashi, J., & Loupis, L. (2025). Harnessing indigenous knowledge for education for sustainable development. *Journal of Education for Sustainable Development*, *19*(2), 157–167. DOI: 10.1177/09734082251360875

Mansour, N., & Alshamrani, S. (2015). *Science Education in the Arab Gulf States: Visions*. Sociocultural Contexts and Challenges., DOI: 10.1007/978-94-6300-049-9

March, J. G., & Olsen, J. P. (1989). *Rediscovering institutions: The organizational basis of politics*. Free Press.

Marino, M. T., Black, A. C., Hayes, M. T., & Beecher, C. C. (2014). An analysis of factors that affect struggling readers' achievement using digital text and reading supports in middle school science classrooms. *Journal of Special Education Technology*, *29*(1), 1–16. DOI: 10.1177/016264341002500101

Martinez, N. (2023). Indigenous literacies: A look at pedagogies and policy in the Southwest United States. *Journal of Adolescent & Adult Literacy: A Journal From the International Reading Association, 66*(5), 282–289. DOI: 10.1002/jaal.1282

Massouti, A., Al-Rashaida, M., & Alhosani, M. (2024). A Qualitative Study on Dubai's Inclusive Education Policy from School Leaders' Perspectives. *Sustainability (Basel), 16*(3), 1252. DOI: 10.3390/su16031252

Mbah, M., Johnson, A. T., & Chipindi, F. M. (2021). Institutionalizing the intangible through research and engagement: Indigenous knowledge and higher education for sustainable development in Zambia. *International Journal of Educational Development*, *82*(102355), 102355. DOI: 10.1016/j.ijedudev.2021.102355

MBRF. (n.d.). Bil Arabi. https://mbrf.ae/en/reading-arabic/bil-arabi

McCarty, T., & Lee, T. (2014). Critical culturally sustaining/revitalizing pedagogy and Indigenous education sovereignty. *Harvard Educational Review*, *84*(1), 101–124. DOI: 10.17763/haer.84.1.q83746nl5pj34216

McCoach, D. B., & Siegle, D. (2003). Factors that differentiate underachieving gifted students from high-achieving gifted students. *Gifted Child Quarterly*, *47*(2), 144–154. DOI: 10.1177/001698620304700205

McGregor, D., Whitaker, S., & Sritharan, M. (2020). Indigenous environmental justice and sustainability. *Current Opinion in Environmental Sustainability*, *43*, 35–40. DOI: 10.1016/j.cosust.2020.01.007

McGregor, I. M. (2023). Teaching for global citizenship: A case study on the challenges of implementing a human rights course. *Compare: A Journal of Comparative Education*, •••, 1–18. DOI: 10.1080/03057925.2023.2254220

McLaren, P., & Giroux, H. (1990). Critical pedagogy and rural education: A challenge from Poland. *Peabody Journal of Education*, *67*(4), 154–165. DOI: 10.1080/01619569009538705

Mdletshe, L. N. (2025). Towards the application of inclusive indigenous pedagogies in tourism education in South Africa. In *Indigenous Teaching Disciplines and Perspectives for Higher Education* (pp. 171–196). IGI Global Scientific Publishing., DOI: 10.4018/979-8-3693-9296-6.ch009

Meadowcroft, J. (2007). National sustainable development strategies: Features, challenges and reflexivity. *European Environment*, *17*(3), 152–163. DOI: 10.1002/eet.450

Meier, S. (2022). Digital Storytelling: A Didactic Approach to Multimodal Coherence. *Frontiers in Communication*, *7*, 906268. DOI: 10.3389/fcomm.2022.906268

Meuleman, L. (2018). *Metagovernance for sustainability: A framework for implementing the sustainable development goals*. Routledge. DOI: 10.4324/9781351250603

Meuleman, L. (2021). Public administration and governance for the SDGs: Navigating between change and stability. *Sustainability (Basel)*, *13*(11), 5914. DOI: 10.3390/su13115914

Ministry of Climate Change and Environment (MOCCAE). (2017). *National climate change plan*. https://www.moccae.gov.ae/assets/1d66432d/Infosheet-%20NCCP.pdf.aspx

Ministry of Education (MoE). (2024). *Ministry of Education launches a national campaign to instil the values of "We the UAE 2031" vision among students.* https://www.moe.gov.ae/En/MediaCenter/News/Pages/Ministry-of-Education-launches-a-national-campaign-to-instil-the-values-of-We-the-UAE-2031-vision-among-students.aspx

Ministry of Education. (2017). The Emirates Standardized Test (EmSAT) to replace CEPA at universities. https://www.moe.gov.ae/En/MediaCenter/News/pages/emsat.aspx#:~:text=News,The%20Emirates%20Standardized%20Test%20(EmSAT)%20to%20replace%20CEPA%20at%20universities,the%20MOE%20English%20language%20curriculum

Ministry of Education. (2023). Green Education Partnership Roadmap. United A rab Emirates Ministry of Education. https://www.moe.gov.ae/En/MediaCenter/News/Pages/Green-Education.aspx

Miuka, N. K., & Tufu, J. (2025). Decolonizing the curriculum: Policy implications for Indigenous knowledge integration in national education systems. *International Journal of Post Axial: Futuristic Teaching and Learning*, 194–203. DOI: 10.59944/postaxial.v3i3.478

MoE. (2023). *UAE Ministry of Education launches Greening Education Hub themed 'Legacy from the Land of Zayed' ahead of COP28.* https://www.moe.gov.ae/En/MediaCenter/News/Pages/MOECOP28.aspx?utm_source=chatgpt.com

Morrison, W. F., & Rizza, M. G. (2007). Creating a toolkit for identifying twice-exceptional students. *Journal for the Education of the Gifted*, *31*(1), 57–76. DOI: 10.4219/jeg-2007-513

Mtitu, E. A. (2021). *Learner-centred teaching in Tanzania: Geography teachers' perceptions and experiences.* DOI: 10.26686/wgtn.17006401

Mudaly, R. (2018). Towards decolonising a module in the pre-service science teacher education curriculum: The role of indigenous knowledge systems in creating spaces for transforming the curriculum. *Journal of Education*, *74*(74). Advance online publication. DOI: 10.17159/2520-9868/i74a04

Mukuka, A., & Alex, J. K. (2024). Student Teachers' Knowledge of School-level Geometry: Implications for Teaching and learning. *European Journal of Educational Research, volume–13–2024*(volume–13–issue–3–july–2024), 1375–1389. https://doi.org/DOI: 10.12973/eu-jer.13.3.1375

Mulholland, E., & Berger, G. (2017). Multi-level governance and vertical policy integration: Implementation of the 2030 agenda for sustainable development at all levels of government. *ESDN quarterly report, 43*.

Mulyani, H., Komalasari, K., Permatasari, M., Bribin, M. L., & Suriaman, S. (2024). Transformasi Pendidikan Kewarganegaraan Global di Era Abad 21: Analisis Implementasi dan Tantangan. *Jurnal Kewarganegaraan*, *21*(1), 88–101. DOI: 10.24114/jk.v21i1.55115

Muyunda, G., Yue, L., & Oranga, J. (2023). Teachers' professional development in Zambia: Perceptions and practices. [IJSL]. *International Journal of Social Learning*, *3*(2), 222–233. DOI: 10.47134/ijsl.v3i2.177

Myende, P. E.P. E. (2019). Creating functional and sustainable School–Community Partnerships: Lessozns from three South African cases. *Educational Management Administration & Leadership*, *47*(6), 1001–1019. DOI: 10.1177/1741143218781070

National Committee on Sustainable Development Goals. (2023). *UAE and the 2030 Agenda for Sustainable Development*. https://mohap.gov.ae/documents/20117/1212145/the+2030+agenda+for+sdg.pdf/6e96b55c-73b6-d1d1-36cf-f1277da306f1?t=1739164957965

National Curriculum and Textbook Board [NCTB]. (2021). *National Curriculum Framework 2021.*: Dhaka Ministry of Education, Peoples' Republic of Bangladesh. Retrieved from https://nctb.gov.bd/pages/static-pages/6922dc5d933eb65569e0ff0f

National Education Policy. 2020. (2020). *Ministry of Education, Government of India.* https://www.education.gov.in/sites/upload_files/mhrd/files/NEP_Final_English_0.pdf\

Nesterova, Y. (2020). Rethinking environmental education with the help of indigenous ways of knowing and traditional ecological knowledge. *Journal of Philosophy of Education*, *54*(4), 1047–1052. DOI: 10.1111/1467-9752.12471

Nichols, J. B., Howson, P. H., Mulrey, B. C., Ackerman, A., & Gately, S. E. (2016). The promise of place: Using place-based education principles to enhance learning. *The International Journal of Pedagogy and Curriculum*, *23*(2), 27–41. DOI: 10.18848/2327-7963/CGP/v23i02/27-41

Norman, E. J. (2021). Global Citizenship Education in the English Language Classroom: Theory and practice. *Advances in Social Science, Education and Humanities Research. Advances in Social Science, Education and Humanities Research.* Advance online publication. DOI: 10.2991/assehr.k.211224.033

Nwigwe, E. O. (2026). Understanding place-based education in Nigeria: A structured narrative review of its status, barriers, and African comparative insights. *Social Sciences & Humanities Open*, *13*, 102409. Advance online publication. DOI: 10.1016/j.ssaho.2025.102409

OECD. (2015). *Education Policy Outlook 2015: Making Reforms Happen*. OECD Publishing., DOI: 10.1787/9789264225442-

OECD. (2016). *Policy coherence for sustainable development 2016: Towards sustainable and resilient societies*. OECD Publishing., DOI: 10.1787/9789264256996-

OECD. (2019). *Policy coherence for sustainable development 2019: Empowering people and ensuring inclusiveness and equality*. OECD Publishing., DOI: 10.1787/a90f851f-

OECD. (2023). *Driving policy coherence for sustainable development: Accelerating progress towards the SDGs*. OECD Publishing., DOI: 10.1787/a6cb4aa1-

Olszewski-Kubilius, P., & Subotnik, R. F. (2023). Talent development as a framework for gifted education. In Olszewski-Kubilius, P., & Subotnik, R. F. (Eds.), *Talent development as a framework for gifted education: Implications for best practices and applications in schools* (pp. 1–18). Prufrock Press.

Onesi-Ozigagun, O., Ololade, Y. J., Eyo-Udo, N. L., & Ogundipe, D. O. (2024). Revolutionizing education through AI: A comprehensive review of enhancing learning experiences. *International Journal of Applied Research in Social Sciences*, *6*(4), 589–607. DOI: 10.51594/ijarss.v6i4.1011

Opoku, M. J., & James, A. (2021). Pedagogical model for decolonising, indigenising and transforming science education curricula: A case of South Africa. *Journal of Baltic Science Education*, *20*(1), 93–107. DOI: 10.33225/jbse/21.20.93

Ormiston, N. T. (2014). Transforming stories and teachings into social work pedagogies. *Affilia*, *29*(3), 368–372. DOI: 10.1177/0886109914522630

Orr, D. (2013). Place and pedagogy. *NAMTA Journal*, *38*(1), 183–188.

Ovcharova, K. (2023). The role of the teacher in Classroom Management: Promoting active participation and engagement. *Педагогически Форум*, *11*(3), 45–54. DOI: 10.15547/PF.2023.018

Oxfam. (2015). *Education for Global Citizenship*. A guide for schools. Oxford, Oxfam.

Ozga, J. (2009). Governing education through data in England: From regulation to self-evaluation. *Journal of Education Policy*, *24*(2), 149–162. DOI: 10.1080/02680930902733121

Palfreyman, D. M., & Al-Bataineh, A. (2018). 'This is my life style, Arabic and English': Students' attitudes to (trans) languaging in a bilingual university context. *Language Awareness, 27*(1-2), 79–95. DOI: 10.1080/09658416.2018.1431244

Parker, R., Chainey, J., Goundar, P., Richardson, S., Dabrowski, A., Berry, A., & Scoular, C. (2023). *Being and becoming global citizens: Measuring progress toward SDG 4.7. Phase I: Monitoring teacher and school readiness to enact global citizenship in the Asia-Pacific region.* https://doi.org/DOI: 10.37517/978-1-74286-718-2

Parsons, M., Fisher, K., & Crease, R. P. (2021). Environmental Justice and Indigenous Environmental Justice. *Decolonising Blue Spaces in the Anthropocene: Freshwater management in Aotearoa New Zealand*, 39–73. https://doi.org/DOI: 10.1007/978-3-030-61071-5_2

Peltier, S. M. (2021). The Child Is Capable: Anishinaabe Pedagogy of Land and Community. *Frontiers in Education, 6*, 689445. DOI: 10.3389/feduc.2021.689445

Pidgeon, M. (2016). More than a checklist: Meaningful Indigenous inclusion in higher education. *Social Inclusion (Lisboa), 4*(1), 77–91. DOI: 10.17645/si.v4i1.436

Piller, I. (2016). *Linguistic diversity and social justice: An introduction to applied sociolinguistics*. Oxford University Press. DOI: 10.1093/acprof:oso/9780199937240.001.0001

Pollitt, C., & Bouckaert, G. (2017). *Public management reform: A comparative analysis—into the age of austerity*. Oxford University Press.

Poon, J. (2017). Engaging sustainability good practice within the curriculum design and property portfolio in the Australian higher education sector. *International Journal of Sustainability in Higher Education, 18*(1), 146–162. DOI: 10.1108/IJSHE-09-2015-0149

Portolés, L., & Martí, O. (2020). Teachers' beliefs about multilingual pedagogies and the role of initial training. *International Journal of Multilingualism, 17*(2), 248–264. DOI: 10.1080/14790718.2018.1515206

Porto, M. (2019). Affordances, complexities, and challenges of intercultural citizenship for foreign language teachers. *Foreign Language Annals, 52*(1), 141–164. DOI: 10.1111/flan.12375

Powell, W. W., & Snellman, K. (2004). The knowledge economy. *Annual Review of Sociology, 30*(1), 199–220. DOI: 10.1146/annurev.soc.29.010202.100037

Prabagar, S., Mothkur, R., Swetha, C. B., Kurian, A., Pavithra, K., & Aarthi, S. (2025). Generative AI for inclusive special education. In *Advances in Computational Intelligence and Robotics* (pp. 205–236). IGI Global Scientific Publishing. DOI: 10.4018/979-8-3373-3745-6.ch008

Pratiwi, P. H., Dwiningrum, S. I. A., & Sumunar, D. R. S. (2023). Integrated disaster risk management in the education process in schools. *Journal of Integrated Disaster Risk Management, 13*(1). Advance online publication. DOI: 10.5595/001c.91284

Purvis, B., Mao, Y., & Robinson, D. (2019). Three pillars of sustainability: In search of conceptual origins. *Sustainability Science, 14*(3), 681–695. DOI: 10.1007/s11625-018-0627-5

Putra, H. S., Sugiharto, S., Da Silva, A. M., & Syahrizal, T. (2025). Pedagogical Translanguaging in Language Education: Constructing Translanguaging Space to Reduce Cognitive Load in Support of Sustainable Development Goals (SDGs). *International Journal of Language Education, 9*(3), 656–675. DOI: 10.26858/ijole.v1i1.77403

Qablan, A., Alblooshi, K. M., & Alkaabi, F. A. (2023). Education for sustainable development (ESD) and school leadership. In Abdallah, A. K., & Alkaabi, A. (Eds.), *Restructuring Leadership for School Improvement and Reform* (pp. 378–398). IGI Global., DOI: 10.4018/978-1-6684-7818-9.ch019

Qablan, A., & Al-Qaderi, S. (2009). How to Change University Faculty Members' Attitudes and Behavior in the Context of Education for Sustainable Development. *Applied Environmental Education and Communication, 8*(3-4), 184–194. DOI: 10.1080/15330150903269407

Qablan, A., & DeBaz, T. (2015). Facilitating Elementary Science Teachers Implementation of Inquiry-Based Science Teaching. *Teacher Development, 19*(1), 3–21. DOI: 10.1080/13664530.2014.959552

Quilaqueo, D., & Torres, H. (2024). Aboriginal Country-centred learning. *Curriculum Perspectives, 44*(2), 205–216. Advance online publication. DOI: 10.1007/s41297-024-00233-2

Rahman, M. M. (2024). Practice and perceptions of classroom management: A study at the tertiary level. *Asian Journal of Contemporary Education, 8*(2), 62–76. DOI: 10.55493/5052.v8i2.5098

Rahman, T., Islam, K., & Rahman, A. (2025). Social science teachers' perceptions and practices of global citizenship education: A case study of primary schools in Bangladesh. *Journal of Interdisciplinary Studies in Education, 14*(4), 49–72. DOI: 10.32674/c7dz0p23

Rai, J., & Beresford-Dey, M. (2023). School leadership in the United Arab Emirates: A scoping review. *Educational Management Administration & Leadership*. Advance online publication. DOI: 10.1177/17411432231218129

Ram, A. T., Netto, S. G., & Mohanan, S. M. (2024). *NEP 2020 vision: Embedding Indian Knowledge System in contemporary Teacher Training and Curriculum Development*. Gurukul International Multidisciplinary Research Journal., DOI: 10.69758/GIMRJ/2412IV02V12P0014

Ravikumar, & Aarthi, S. (2025a). Natural Language Processing in conversational AI tutors. In *Advances in Computational Intelligence and Robotics* (pp. 275–306). IGI Global Scientific Publishing. DOI: 10.4018/979-8-3373-4217-7.ch011

Ravikumar, R. N., & Aarthi, S. (2025c). The role of educational developers in AI integration at scale. In *Advances in Computational Intelligence and Robotics* (pp. 289–322). IGI Global Scientific Publishing. DOI: 10.4018/979-8-3373-4129-3.ch011

Ravikumar, R. N., Aarthi, S., Ruzimova, F., & Mamadiyarov, Z. (2025). Training and development in AI-driven education scaling solutions. In *Advances in Computational Intelligence and Robotics* (pp. 267–298). IGI Global Scientific Publishing. DOI: 10.4018/979-8-3373-2185-1.ch010

Ravikumar, R. N., & Aarthi, S. (2025b). AI-powered academic libraries. In *Role of Library Resources for Higher Education in the Digital Era* (pp. 19–56). IGI Global Scientific Publishing., DOI: 10.4018/979-8-3373-3005-1.ch002

Redvers, N., Schultz, C., Vera Prince, M., Cunningham, M., Jones, R., & Blondin, B. (2020). Indigenous perspectives on education for sustainable healthcare. *Medical Teacher, 42*(10), 1085–1090. DOI: 10.1080/0142159X.2020.1791320 PMID: 32657230

Reid, A. J., Eckert, L. E., Lane, J. F., Young, N., Hinch, S. G., Darimont, C. T., Cooke, S. J., Ban, N. C., & Marshall, A. (2021). "Two-Eyed Seeing": An Indigenous framework to transform fisheries research and management. *Fish and Fisheries, 22*(2), 243–261. DOI: 10.1111/faf.12516

Reihana, K. R., Wehi, P. M., Harcourt, N., Booth, P., Murray, J. M., & Pomare-Pieta, M. (2021). Indigenisation of conservation education in New Zealand. *Pacific Conservation Biology, 27*(4), 493–504. DOI: 10.1071/PC20060

Reis, S. M., Baum, S. M., & Burke, E. (2014). An operational definition of twice-exceptional learners: Implications and applications. *Gifted Child Quarterly*, *58*(3), 217–230. DOI: 10.1177/0016986214534976

Richmond, D., Sibthorp, J., & Wilson, C. (2019). Understanding the role of summer camps in the learning landscape: An exploratory sequential study. *Journal of Youth Development*, *14*(3), 42. DOI: 10.5195/jyd.2019.780

Rizvi, F., & Lingard, B. (2010). *Globalizing education policy*. Routledge.

Rogier, D. (2012). *The effects of English-medium instruction on language proficiency of students enrolled in higher education in the UAE*. University of Exeter.

Ronksley-Pavia, M. (2015). A model of twice-exceptionality: Explaining and defining the apparent paradoxical combination of disability and giftedness in childhood. *Journal for the Education of the Gifted*, *38*(3), 318–340. DOI: 10.1177/0162353215592499

Sabilah, A. I. (2024). Revitalisasi fasilitas PAUD Pelangi guna meningkatkan kualitas pendidikan usia dini. *Jurnal Sains Teknologi Dalam Pemberdayaan Masyarakat*, *3*(1), 43–48. DOI: 10.31599/kg02s789

Sachs, J. D. (2015). *The age of sustainable development*. Columbia University Press. DOI: 10.7312/sach17314

Sadler, T. D. (2011). Situating Socio-scientific Issues in Classrooms as a Means of Achieving Goals of Science Education. In Sadler, T. D. (Ed.), *Socio-scientific Issues in the Classroom: Teaching, Learning and Research* (pp. 1–9). Springer Netherlands., DOI: 10.1007/978-94-007-1159-4_1

Sant, E., & González-Valencia, G. (2018). Global Citizenship Education in Latin America. In The Palgrave Handbook of Global Citizenship and Education (pp. 67–82). London: Palgrave Macmillan. Retrieved from https://www.researchgate.net/publication/325270842_Sant_E_Gonzalez-Valencia_G_2018_Global_Citizenship_Education_in_Latin_America_In_The_Palgrave_Handbook_of_Global_Citizenship_and_Education_pp_67-82_London_Palgrave_Macmillan

Şeremet, M. (2016). Geographical education in Turkey: Challenges and opportunities. *Geography (Sheffield, England)*, *101*(3), 146–155. DOI: 10.1080/00167487.2016.12093998

Shahjahan, R. A., & Seinn, A. P. (2025). Towards 'decolonizing' engineering education: An integrative review. *Cogent Education*, *12*(1), 1–15. DOI: 10.1080/2331186X.2025.2541075

Sharma, S. (2025). Indigenous Knowledge Systems and sustainable Socio-economic development in Northeast India: An analysis through the lens of Astalakshmi and Viksit Bharat. *International Journal For Multidisciplinary Research, 7*(5), 58933. Advance online publication. DOI: 10.36948/ijfmr.2025.v07i05.58933

Shotton, H. J., Lowe, S. C., & Waterman, S. J. (2023). *Beyond the asterisk: Understanding Native students in higher education*. Taylor & Francis.

Shrivastava, P., & Kennelly, J. J. (2013). Sustainability and place-based enterprise. *Organization & Environment, 26*(1), 83–101. DOI: 10.1177/1086026612475068

Shulman, L. S. (1986). Those who understand: Knowledge growth in teaching. *Educational Researcher, 15*(2), 4–14. DOI: 10.3102/0013189X015002004

Siemund, P., Al-Issa, A., & Leimgruber, J. R. E. (2021). Multilingualism and the role of English in the United Arab Emirates. *World Englishes, 40*(2), 191–204. DOI: 10.1111/weng.12507

Simpson, L. B. (2014). Land as pedagogy: Nishnaabeg intelligence and rebellious transformation. *Decolonization, 3*(3), 1–25.

Simpson, L. B. (2017). *As we have always done: Indigenous freedom through radical resistance*. U of Minnesota Press. DOI: 10.5749/j.ctt1pwt77c

Singh-Pillay, A., & Madlala, A. (2026). Legitimising Indigenous knowledge systems as a pathway to sustainable development. *Discover Education, 5*(1), 87. Advance online publication. DOI: 10.1007/s44217-025-01062-9

Skerrett, A., & Vlach, S. K. (2022). Advancing a sociocultural approach to decolonizing literacy education: Lessons from a youth in a "postcolonial" Caribbean geography. *Research in the Teaching of English, 57*(1), 67–88. DOI: 10.58680/rte202232002

Smith, G. A. (2002). Place-based education: Learning to be where we are. *Phi Delta Kappan, 83*(8), 584–594. https://www.jstor.org/stable/20440205. DOI: 10.1177/003172170208300806

Smith, G. A., & Sobel, D. (2010). *Place-and community- based education in schools*. Routledge.

Smith, L. T. (1999). *Decolonizing methodologies research and indigenous peoples*. University of Otago Press.

Sobel, D. (2004). 2005). In *Place-based education: Connecting classrooms & communities* (2nd ed.). The Orion Society.

Sobel, D. (2004). Place-based education: Connecting classroom and community. *Nature & Listening*, *4*(1), 1–7.

Spillane, J. P., Reiser, B. J., & Reimer, T. (2002). Policy Implementation and Cognition: Reframing and Refocusing Implementation Research. *Review of Educational Research*, *72*(3), 387–431. DOI: 10.3102/00346543072003387

Sposab, K., & Rieckmann, M. (2024). Development of Sustainability Competencies in Secondary School Education: A Scoping Literature Review. *Sustainability (Basel)*, *16*(23), 10228. DOI: 10.3390/su162310228

Staley, B., & Freeman, L. A. (2017). Digital storytelling as student-centred pedagogy: Empowering high school students to frame their futures. *Research and Practice in Technology Enhanced Learning*, *12*(1), 21. DOI: 10.1186/s41039-017-0061-9 PMID: 30595726

Star, J., Ringaert, L., & Larcombe, L. (2025). Qualitative methods case study: Using MAXQDA in Indigenous HIV journey mapping research. *International Journal of Qualitative Methods*, *24*(16094069251356668), 16094069251356667. Advance online publication. DOI: 10.1177/16094069251356667

Starkey, H. (2022). Challenges to global citizenship education: Nationalism and cosmopolitanism. In *Global Citizenship in Foreign Language Education* (pp. 62–78). Routledge. DOI: 10.4324/9781003183839-5

Steenbeek, H., van Geert, P., & van Dijk, M. (2022). A dynamic systems approach to the development of giftedness and talent. In Sternberg, R. J., & Ambrose, D. (Eds.), *Conceptions of giftedness and talent* (pp. 401–421). Palgrave Macmillan., DOI: 10.1007/978-3-030-56869-6_22

Steiner-Khamsi, G. (2016). Cross-national policy borrowing: Understanding reception and translation. *Asia Pacific Journal of Education*, *34*(2), 153–172. DOI: 10.1080/02188791.2013.875649

Stein, S. (2021). Critical Internationalization Studies at an Impasse: Making Space for Complexity, Uncertainty, and Complicity in a Time of Global Challenges. *Studies in Higher Education*, *46*(9), 1771–1784. DOI: 10.1080/03075079.2019.1704722

Sterling, S. (2001). *Sustainable education: Re-visioning learning and change*. Green Books for the Schumacher Society.

Sterling, S. (2011). Transformative learning and sustainability: Sketching the conceptual ground. *Learning and Teaching in Higher Education*, *5*(1), 17–33. DOI: 10.1177/1469787411429184

Sterling, S. (2024). *Learning and sustainability in dangerous times: The stephen sterling reader*. Agenda Publishing., DOI: 10.2307/jj.13473649

Stuckey, M., Hofstein, A., Mamlok-Naaman, R., & Eilks, I. (2013). The meaning of 'relevance' in science education and its implications for the science curriculum. *Studies in Science Education*, *49*(1), 1–34. DOI: 10.1080/03057267.2013.802463

Stumm, S. (2025). Epistemological implications of a system—Theoretical understanding for sustainability models. *Systems Research and Behavioral Science*. Advance online publication. DOI: 10.1002/sres.3149

Subotnik, R. F., Olszewski-Kubilius, P., & Worrell, F. C. (2011). Rethinking giftedness and gifted education: A proposed direction forward based on psychological science. *Psychological Science in the Public Interest : A Journal of the American Psychological Society*, *12*(1), 3–54. DOI: 10.1177/1529100611418056 PMID: 26168418

Subotnik, R. F., Olszewski-Kubilius, P., & Worrell, F. C. (2019). The talent development framework: Overview and components. In Subotnik, R. F., Olszewski-Kubilius, P., & Worrell, F. C. (Eds.), *The psychology of high performance: Developing human potential into domain-specific talent* (pp. 13–44). American Psychological Association., DOI: 10.1037/0000120-000

Suchman, M. C. (1995). Managing Legitimacy: Strategic and Institutional Approaches. *Academy of Management Review*, *20*(3), 571–610. DOI: 10.2307/258788

Suliman, O. (2000). *A descriptive study of the educational system in the United Arab Emirates*. (Doctoral dissertation, University of Southern California, 2000). Retrieved from ProQuest Dissertations and Theses database.

Sumida Huaman, E. (2023). Earth worlds and Indigenous dream-making: A reflection on teaching for beauty, repair, and balance. *Revista Española de Educación Comparada*, *43*(43), 158–173. DOI: 10.5944/reec.43.2023.36610

Sund, L., & Pashby, K. (2020). Delinking global issues in northern Europe classrooms. *The Journal of Environmental Education*, *51*(2), 156–170. DOI: 10.1080/00958964.2020.1726264

Sy, J. M. C. (2022). Till the land, defend the land: reflections on the critical place-based pedagogy of the Alternative Learning Center for Agricultural and Livelihood Development, Surigao del Sur. *AlterNative (Auckland, N.Z. : 2005)*, *18*(3), 402–411. DOI: 10.1177/11771801221115925

Tafese, M. B., & Kopp, E. (2025). Embedding social sustainability in education: A thematic review of practices and trends across educational pathways from a global perspective. *Sustainability (Basel)*, *17*(10), 4342. DOI: 10.3390/su17104342

Tahat, K., Mansoori, A., Qablan, A., & Tahat, D. (2024). Analyzing the newspapers' coverage of sustainable development issues in the United Arab Emirates. *Online Journal of Communication and Media Technologies*. Advance online publication. DOI: 10.30935/ojcmt/14710

Tang, F., & Calafato, R. (2021). Multilingual, bilingual, and monolingual Arabic teachers' development of learner self-regulation and language awareness in the Emirates. *Foreign Language Annals*, *54*(1), 233–254. DOI: 10.1111/flan.12515

Tata, E. F., Yongabi, K. A., Dzelamonyuy, E., Forbang, N. I., & Geh, E. A. (2023). Decolonizing African science: Efforts, challenges, and future directions. *International Journal of Science and Research Archive*, *10*(1), 823–833. DOI: 10.30574/ijsra.2023.10.1.0840

Teton Science Schools. (n.d.). *Preparing educators for the place-based education classroom.* https://www.tetonscience.org/preparing-educators-for-the-place-based-education-classroom/

The National. June 22, 2025. Arabic lessons daily in all UAE private schools for kindergarten pupils. https://www.thenationalnews.com/news/uae/2025/06/22/arabic-lessons-daily-in-all-uae-private-schools-for-kindergarten-pupils/

The University of Zambia Special Education and Sociology of Education, D. of E. P., & Kaziya, K. (2025). Harnessing Indigenous Knowledge Systems to transform mathematics instruction in rural Zambian schools: A case study from Kalomo District. *INTERNATIONAL JOURNAL OF SOCIAL SCIENCE AND EDUCATION RESEARCH STUDIES, 05*(03). DOI: 10.55677/ijssers/V05I03Y2025-13

Thomas, S. (2023). English in the United Arab Emirates. *World Englishes*, *42*(3), 344–370. DOI: 10.1111/weng.12560

Thompson, S. M., & Chapman, N. (2025). Learning Along the GreenWay: An Experiential, Transdisciplinary Outdoor Classroom for Planetary Health Education. *Sustainability (Basel)*, *17*(9), 4143. DOI: 10.3390/su17094143

Ticheloven, A., Blom, E., Leseman, P., & McMonagle, S. (2019). Translanguaging challenges in multilingual classrooms: Scholar, teacher and student perspectives. *International Journal of Multilingualism*, *18*(3), 491–514. DOI: 10.1080/14790718.2019.1686002

Times of India. (2023a, June 24). NEP 2020: Bringing experiential learning to the heart of education. *The Times of India.* https://timesofindia.indiatimes.com/education/news/nep-2020-bringing-experiential-learning-to-the-heart-of-education/articleshow/10124564.cms

Tomlinson, C. A. (2014). *The differentiated classroom: Responding to the needs of all learners* (2nd ed.). ASCD.

Torris-Hedlund, M. A., Palakiko, D.-M., & Komene, E. (2025). Asserting Indigenous academic sovereignty through assemblage, Indigenization, and cultural safety. *Creative Nursing, 31*(3), 243–253. DOI: 10.1177/10784535251357218 PMID: 40636965

Townend, G., & Brown, R. (2016). Challenging assumptions about the social and emotional adjustment of gifted students: A meta-analytic review. *Journal for the Education of the Gifted, 39*(4), 295–321. DOI: 10.1177/0162353216671059

Troudi, S., & Jendli, A. (2011). Emirati students' experiences of English as a medium of instruction. In Al-Issa, A., & Dahan, L. S. (Eds.), *Global English and Arabic: Issues of language, culture, and identity* (pp. 23–48). Peter Lang.

Truth and Reconciliation Commission of Canada. (2015). *Honouring the truth, reconciling for the future: Summary of the final report of the Truth and Reconciliation Commission of Canada*.

Tuck, E., & Yang, K. W. (2012). Decolonization is not a metaphor. *Decolonization, 1*(1), 1–40. DOI: 10.25058/20112742.n38.04

Tupas, R. (2019). Entanglements of colonialism, social class, and unequal Englishes. *Journal of Sociolinguistics, 23*(5), 529–542. https://doi-org.uaeu.idm.oclc.org/10.1111/josl.12384. DOI: 10.1111/josl.12384

UNESCO. (2014). *Shaping the future we want: UN decade of education for sustainable development (2005–2014) final report*. United Nations Educational, Scientific and Cultural Organization. https://unesdoc.unesco.org/ark:/48223/pf0000230171

UNESCO. (2015). *Global Citizenship Education: Topics and Learning Objectives*. UNESCO.

UNESCO. (2017). *Education for sustainable development goals : Learning objectives*. UNESCO. https://unesdoc.unesco.org/images/0024/002474/247444e.pdf

UNESCO. (2020). *Education for sustainable development: A roadmap*. United Nations Educational, Scientific and Cultural Organization (UNESCO). DOI: 10.54675/YFRE1448

UNESCO. (2022). *United Arab Emirates: Climate change communication and education, Education Profiles*. https://education-profiles.org/northern-africa-and-western-asia/united-arab-emirates/~climate-change-communication-and-education?utm_source=chatgpt.com

UNESCO. (2023). *Declaration on the common agenda for education and climate change at COP28*. https://www.unesco.org/sites/default/files/medias/fichiers/2023/12/Declaration-on-education-and-climate-change-en.pdf

UNESCO. (2023). *New report and guidelines for indigenous data sovereignty in artificial intelligence developments*. UNESCO. https://www.unesco.org/en/articles/new-report-and-guidelines-indigenous-data-sovereignty-artificial-intelligence-developments

United Arab Emirates Government. (2017). *UAE and 2030 Agenda for Sustainable Development: Excellence in Implementation*. Federal Competitiveness and Statistics Authority. https://fcsc.gov.ae/wp-content/uploads/2026/03/2018-UAE-and-the-2030-Agenda-for-Sustainable-Development-Excellence-in-Implementation.pdf

United Nations Educational, Scientific and Cultural Organization. (2015). Education 2030: Incheon Declaration and Framework for Action for the implementation of Sustainable Development Goal 4. UNESCO. https://unesdoc.unesco.org/images/0024/002456/245656e.pdf

United Nations General Assembly. (2007). United Nations Declaration on the Rights of Indigenous Peoples (A/RES/61/295). United Nations. https://undocs.org/A/RES/61/295

United Nations. (2007). United Nations declaration on the rights of Indigenous peoples. United Nations.https://www.un.org/development/desa/indigenouspeoples/wpcontent/uploads/sites/19/2018/11/UNDRIP_E_web.pdf

United Nations. (2015) Transforming Our World. The 2030 Agenda for Sustainable Development. https://sdgs.un.org/2030agenda

United Nations. (2015). *Transforming our world: The 2030 agenda for sustainable development*. https://sdgs.un.org/2030agenda

Vadivel, B., Alam, S., Nikpoo, I., & Ajanil, B. (2023). The impact of low socio-economic background on a child's educational achievements. *Education Research International*, *2023*, 1–11. DOI: 10.1155/2023/6565088

Van De Wetering, J., Leijten, P., Spitzer, J., & Thomaes, S. (2022). Does environmental education benefit environmental outcomes in children and adolescents? A meta-analysis. *Journal of Environmental Psychology*, *81*, 101782. DOI: 10.1016/j.jenvp.2022.101782

van Geert, P. (2023). Dynamic systems, process and development. In Koops, W., & van IJzendoorn, M. H. (Eds.), *The development of childhood* (pp. 87–112). Cambridge University Press., DOI: 10.1017/9781108779650.006

Vander Ark, T. (2018, May 9). Get kids in the community and change the world. *Forbes.* https://www.forbes.com/sites/tomvanderark/2018/05/09/get-kids-in-the-community-and-change-the-world/

Verger, A., Fontdevila, C., & Parcerisa, L. (2019). Reforming governance through policy instruments: How and to what extent standards, tests and accountability in education spread worldwide. *Discourse (Abingdon), 40*(2), 248–270. DOI: 10.1080/01596306.2019.1569882

Vogel, S., & García, O. (2017). *Translanguaging*. Oxford Research Encyclopedia of Education., DOI: 10.1093/acrefore/9780190264093.013.181

Wade-Lyles, T. A. (2016). Integration of place-based education into science classes from prekindergarten through grade 5 [Doctoral dissertation, Walden University].

Wals, A. E. J. (2010). *Message in a bottle: Learning our way out of unsustainability.* Wageningen University & Research (WUR).

Wals, A. E. J. (2011). Learning our way to sustainability. *Journal of Education for Sustainable Development*, *5*(2), 177–186. DOI: 10.1177/097340821100500208

WAM. (2024, November 3). *MoE, MOHESR cancel EmSAT exam for Grade 12 students.* Retrieved from https://www.wam.ae/en/article/b60yota-moe-mohesr-cancel-emsat-exam-for-grade-students

Weber, M. (1978). *Economy and society: An outline of interpretive sociology* (Roth, G., & Wittich, C., Eds.). Vol. 2). University of California Press.

Wheeler, H. C., Danielsen, F., Fidel, M., Hausner, V., Horstkotte, T., Johnson, N., Lee, O., Mukherjee, N., Amos, A., Ashthorn, H., Ballari, Ø., Behe, C., Breton-Honeyman, K., Retter, G.-B., Buschman, V., Jakobsen, P., Johnson, F., Lyberth, B., Parrott, J. A., & Vronski, N. (2020). The need for transformative changes in the use of Indigenous knowledge along with science for environmental decision-making in the Arctic. *People and Nature*, *2*(3), 544–556. DOI: 10.1002/pan3.10131

Whyte, K. (2020). Too late for indigenous climate justice: Ecological and relational tipping points. *Wiley Interdisciplinary Reviews: Climate Change*, *11*(1), e603. Advance online publication. DOI: 10.1002/wcc.603

Whyte, K. P. (2018). Indigenous science (fiction) for the Anthropocene: Ancestral dystopias and fantasies of climate change crises. *Environment and Planning. E, Nature and Space*, *1*(1-2), 224–242. DOI: 10.1177/2514848618777621

Wiek, A., Withycombe, L., & Redman, C. L. (2011). Key competencies in sustainability: A reference framework for academic program development. *Sustainability Science*, *6*(2), 203–218. DOI: 10.1007/s11625-011-0132-6

Wigfield, A., & Eccles, J. S. (2020). Chapter Five - 35 years of research on students' subjective task values and motivation: A look back and a look forward. In Elliot, A. J. (Ed.), Vol. 7, pp. 161–198). Advances in Motivation Science. Elsevier., https://doi.org/https://doi.org/10.1016/bs.adms.2019.05.002

Wijsman, K., Bogner, K., Avelino, F., Kalfagianni, A., & Weis, C. U. (2026). Towards a lexicon for just sustainability transitions: Thinking with capabilities, epistemology, intersectionality, spatiality, and temporality. *Environmental Innovation and Societal Transitions*, *59*, 101104. Advance online publication. DOI: 10.1016/j.eist.2026.101104

Williams, L. (2024). Indigenous intergenerational resilience and lifelong learning. *Sustainability (Basel)*, *16*(11), 4494. Advance online publication. DOI: 10.3390/su16114494

Wilson, C., Akiva, T., Sibthorp, J., & Browne, L. P. (2019). Fostering distinct and transferable learning via summer camp. *Children and Youth Services Review*, *98*, 269–277. DOI: 10.1016/j.childyouth.2019.01.017

Wilson, S. (2020). *Research is ceremony: Indigenous research methods*. Fernwood publishing.

Winebrenner, S. (2003). Teaching strategies for twice-exceptional students. *Intervention in School and Clinic*, *38*(3), 131–137. DOI: 10.1177/10534512030380030101

Wiyarsi, A., Çalik, M., Priyambodo, E., & Dina, D. (2024). Indonesian prospective teachers' scientific habits of mind: A cross-grade study in the context of local and global socio-scientific issues. *Science & Education*, *33*(5), 1257–1283. DOI: 10.1007/s11191-023-00429-4 PMID: 37359261

Woldegiorgis, E. T. (2025). Navigating the decolonial frontier: Leadership perspectives on epistemic justice in South African higher education. *International Journal of Educational Research*, *134*, 102816. DOI: 10.1016/j.ijer.2025.102816

World Commission on Environment and Development. (1987). *Our common future*. Oxford University Press.

Xu, C., Zhu, K., & Liu, S. (2023). Classroom management Strategies in secondary language education. *Journal of Education Humanities and Social Sciences*, *8*, 1649–1654. DOI: 10.54097/ehss.v8i.4539

Yele, W. B., & Specioza, A. (2024). Examining the access, quality, and relevance gaps in Liberia's educational policy environment. *International Journal for Multidisciplinary Research*, *6*(3), 20482. Advance online publication. DOI: 10.36948/ijfmr.2024.v06i03.20482

Yemini, M., Engel, L., & Ben Simon, A. (2025). Place-based education: A systematic review of literature. *Educational Review*, *77*(2), 640–660. DOI: 10.1080/00131911.2023.2177260

Yilmaz, T. (2021). Translanguaging as a pedagogy for equity of language minoritized students. *International Journal of Multilingualism*, *18*(3), 435–454. https://doi-org.uaeu.idm.oclc.org/10.1080/14790718.2019.1640705. DOI: 10.1080/14790718.2019.1640705

Yunkaporta, T. (2019). *Sand talk: How indigenous thinking can save the world*. Text Publishing.

Zeyer, A. (2022). Teaching two-eyed seeing in education for sustainable development: Inspirations from the science environment health pedagogy in pandemic times. *Sustainability (Basel)*, *14*(10), 1–12. DOI: 10.3390/su14106343

Zindi, B., & Sibanda, M. M. (2023). Challenges impeding the implementation of effective continuous teacher professional development programs in Eastern Cape Department of Education. *Business Ethics and Leadership*, *7*(4), 46–61. DOI: 10.61093/bel.7(4).46-61.2023

Zinga, D., & Styres, S. (2018). Decolonizing curriculum: Student resistances to anti-oppressive pedagogy. *Power and Education*, *175774381881056*. Advance online publication. DOI: 10.1177/1757743818810565

Zoghbor, W. S. (2023). Negotiating the Arabic and English space in UAE higher education: The ownership of English in the English-medium instruction context. In *English as a Medium of Instruction on the Arabian Peninsula* (pp. 89–111). Routledge. DOI: 10.4324/9781003183594-8

Zurba, M., Petriello, M. A., Madge, C., McCarney, P., Bishop, B., McBeth, S., Denniston, M., Bodwitch, H., & Bailey, M. (2021). Learning from knowledge co-production research and practice in the twenty-first century: Global lessons and what they mean for collaborative research in nunatsiavut. *Sustainability Science*, *17*(2), 449–467. DOI: 10.1007/s11625-021-00996-x

About the Contributors

Ahmad Qablan is a professor of science education at the College of Education at UAEU. He earned his Ph.D. degree in science and sustainability education from Florida State University, USA, in 2005 and worked for more than 15 years at the College of Educational Sciences at Hashemite University in Jordan, as well as 6 years as a senior programs manager at Queen Rania Teacher Academy, Jordan. I also worked as a visiting professor of science education at the University of Alberta, Canada, for 5 years. He has conducted several educational research projects with UNESCO, UNICEF, the EU, and USAID. He published over 100 books, chapters in books, and journal articles with local and international publishing houses.

Asma Abdallah is an Assistant Professor of Educational Leadership at Sharjah Education Academy, with extensive experience across higher education, school leadership, and educational consultancy in the UAE and Jordan. She previously served as a Visiting Assistant Professor at United Arab Emirates University and Assistant Professor at Abu Dhabi University, alongside her work as a school inspector and professional development consultant with ADEK's Irtiqa program. Her research focuses on teacher licensing, inclusive education, AI in education, and leadership development. She has authored and co-edited several scholarly books with IGI Global, including Building a Unified Teacher Licensing System: Policies, Education Reforms, and Cultural Integration (2025). She has published widely in Scopus- and Web of Science-indexed journals such as Heliyon, BMC Psychology, and Frontiers in Psychology. Dr. Abdallah holds a Ph.D. in Educational Leadership from the University of Jordan. She has been recognized with the Jordanian Independence Medal, First Class, awarded by His Majesty King Abdullah II for her contributions to education. She is an active reviewer for international journals and maintains a strong research profile with a Scopus h-index of 15.

Ahmed Alkaabi is an associate professor in the Learning and Educational Leadership Department—College of Education at United Arab Emirates University.

He is currently serving as the Assistant Dean for Student Affairs. Previously, Dr. Alkaabi held key roles, including Director of the Emirates Institute for Learning Outcomes Assessment at UAEU and Coordinator of the Master of Educational Innovation Program. His educational qualifications include a Ph.D. in Educational Administration and Policy with an emphasis on supervision from the University of Georgia–Athens, USA (2019).He was the recipient of two distinguished academic awards: the Ray Bruce Award in 2017 for his dedicated work and projects in the field of instructional supervision, and the Faculty Award in 2019 for his accomplishments in the doctoral program at the University of Georgia. Dr. Alkaabi is certified in Interdisciplinary Qualitative Studies and as a school inspector. His research interests fall within the spectrum of school leadership, specifically in the areas of instructional supervision, supervisoy leadership, leader and teacher evaluation, professional development, and school inspection.

***

Rasha Abdullah is a PhD candidate at the Leadership and Policy Studies in Education Program, College of Education, United Arab Emirates University.

Dina Al Khawaldeh is a Ph.D. student in Language and Literacy Education at United Arab Emirates University (UAEU). She holds a Master of Education in Curriculum and Methods of Instruction from the College of Education at UAEU and a Bachelor's degree in English Language, Literature, and Cultural Studies from The Hashemite University, Jordan. She has extensive experience as an English teacher in private schools across the United Arab Emirates. Her research interests include literacy education, multiliteracies, language acquisition, ESL/EFL, reading and writing development, and language arts.

Huseynova Abira Amanovna is the Regional Representative of the Commissioner for Human Rights (Ombudsman) of the Oliy Majlis of the Republic of Uzbekistan in the Bukhara Region. She holds the degree of Doctor of Philosophy and the academic title of Professor. Her professional and scholarly work focuses on human rights protection, legal advocacy, public administration, and the promotion of democratic values. She has extensive experience in monitoring rights compliance, addressing citizens' appeals, and contributing to policy-oriented research in law and governance. Huseynova Abira Amanovna is actively involved in academic and public initiatives aimed at strengthening human rights awareness and institutional accountability.

Azizjonova Asilabonu is a faculty member in the Department of Pedagogy and Psychology at Termez University of Economics and Service, Termez, Uzbekistan. She is engaged in teaching and academic activities related to pedagogy and educational psychology. Her professional interests include learner-centered teaching strategies, psychological aspects of learning, student development, and modern instructional methodologies in higher education. She actively contributes to curriculum development and academic mentoring, with a focus on enhancing teaching effectiveness and supporting students' cognitive and emotional growth. Her work reflects a commitment to improving educational quality through research-informed pedagogical practices.

Yuldoshev Rustam Elmurodovich is a faculty member in the Department of Primary Education at Karshi State University, Karshi, Uzbekistan. He is engaged in teaching and academic activities focused on primary education pedagogy, curriculum development, and learner-centered instructional practices. His academic interests include innovative teaching methods, classroom management, learner assessment, and inclusive education in primary schools. He is committed to strengthening foundational literacy and numeracy skills through effective pedagogical approaches. Yuldoshev Rustam Elmurodovich actively contributes to academic and institutional initiatives aimed at enhancing the quality of primary teacher education.

Raju Ahammed Fakir completed Bachelor of Education (B. Ed) and Master of Education (M. Ed) from the Institute of Education and Research (IER), University of Dhaka. He is passionate about research on social issues including human rights education, social justice, child rights, and citizenship education.

Nour Hammo is a PhD student in Language and Literacy Education at the United Arab Emirates University. Her love for languages and cultures shapes her interest in literacy, multilingualism, and translanguaging practices in education.

Sovanna Huot, currently enrolled as a PhD candidate in Political Science at the University of Delhi, New Delhi, India, is also a vice principal at a Government Public High School of Cambodia, where he holds the responsibilities of the school's managerial and academic institutional leadership. An accomplished scholar in a vast array of fields, he has undergone a two-year teacher training course from 2010-2012 and has earned bachelor's degrees in English and Educational Administration. Moreover, he holds two master's degrees, which he earned from a highly selective scholarship program, in TEFL and a Master of Education, in School Leadership, Upper Secondary School Principalship, from the National Institute of Education (NIE), Phnom Penh, 2021. His areas of focus are on educational reform and the practice-policy nexus of digit pedagogy, alongside glonacal (global–national–local) education systems. His scholarly work encompasses both political science and education; he has formal training in both fields and has published in both, including Scopus-indexed articles (Author ID: 59316244400).

Ali Ibrahim holds a Ph.D. in Administrative and Policy Studies in Education from the University of Pittsburgh, USA. He is a Professor at the United Arab Emirates University. He is also the co-founder of the Gulf Comparative Education Society (GCES), a not-for-profit organization that fosters academic, professional, and educational dialogue from a comparative perspective with a focus on the Arabian Gulf region. Dr. Ibrahim has published extensively in highly ranked and indexed journals. His research interests include school organizational culture, education policy, school leadership, education reform, and teacher professionalism in the Arab Gulf states.

Khairul Islam, PhD., is a Professor and has been working as a faculty member at the Institute of Education and Research (IER), University of Dhaka, Bangladesh since 2010. He holds a PhD degree in the area of right-based pedagogy from the Faculty of Education, Monash University, Melbourne, Australia. He was awarded British Government Commonwealth Scholarship in 2012 and completed his MA in Global and Citizenship Education from the University of York, United Kingdom. He is experienced in qualitative research in the field of Social Science Education. His publications include peer-reviewed book chapters and articles in the area of Social Science Education. His research interests include right-based pedagogies, child rights and child protection, citizenship and human rights education, teacher education, and action research.

Ali Kocalar is a faculty member at Marmara University's Atatürk Education Faculty, specializing in Human Geography and Geography Education. His academic research primarily focuses on political geography, migration studies, and environmental issues, with a strong emphasis on pedagogical field studies.

Lovleen Marwaha, Ph.D., is a Professor at the Department of Zoology, School of Bioengineering and Biosciences, Lovely Professional University, Phagwara, India. She has completed her M.S. and Ph.D. in Cytogenetics from the Department of Zoology, Punjab University, Chandigarh. She has worked on the evaluation of pesticide genotoxicity using insect genomes, specifically on different mosquito species and Drosophila melanogaster. She has carried out an intensive study on the correlation between pesticide exposure limit and inductive genotoxic consequences on Dipteran larval salivary polytene chromosomes. Furthermore, she has studied chromosomal and gene mutations in mosquito and fruit fly genomes due to pesticide exposure. Currently, she is working on Apiculture and Sericulture-related topics.

She has many research papers published in National and International Journals of repute. Presently, she is working on insecticide resistance in mosquitoes, pesticide genotoxicity evaluation, honey bee development, honey processing, silk fiber, and pearl quality improvement. She has published three authored books on Apiculture and on Mobile Genetic Elements. Forthcoming books include Worker Honey Bees, Apiculture Products, Basic Concepts of Genetics, and Introduction to Sericulture.

Ashraf Moustafa is a faculty member of Special and Gifted Education Department at the United Arab Emirates University, He has clinical and educational experience working with children with Sensory Impairment, L.D, Autism, ADHD and gifted students. His research interests are in Sensory Impairment and Assessment for children with Sensory Impairment, developmental disorders and learning difficulties.

Ravikumar R N is currently a Postdoctoral researcher at Universiti Malaysia Perlis, Malaysia. Also, working as an Associate Professor in the Department of Computer Engineering at Marwadi University, Rajkot, Gujarat. He has 13+ years of national and international teaching experience. He completed his PhD at Amity University, Rajasthan, M.E at VMRF Deemed University, B.E at Anna University, Chennai. He is the co-founder of the Research Activities Club at Marwadi University and trains students on research activities, which leads to the publication of 8 Scopus/ SCI journals, 60+ Book chapters, 30+ Conference papers, 16+ patents, and 1 book. His specialization is ML, DL, Federated Learning, Recommendation Systems, and Optimization Techniques.

Arifur Rahman is the faculty at the Institute of Education and Research, University of Dhaka, Dhaka, Bangladesh. He conducts social science education courses at both undergraduate and graduate levels. His research interests include social science education, early childhood education, higher education, teaching-learning, and policy-making. He published several research articles in peer-reviewed journals.

Tanjima Rahman completed her Bachelor of Education (B. Ed) and Master of Education (M. Ed) from the Institute of Education and Research (IER), University of Dhaka. She is very passionate about research on social issues. Particularly, her research area covers global citizenship education, social justice, and child rights.

Aarthi S is a dedicated Assistant Professor in Computer Science with over six years of teaching experience. She is currently pursuing her Ph.D. at Marwadi University, Gujarat, and holds an M.Phil. and MCA from Bharathiar University, Tamil Nadu. Throughout her academic career, Aarthi has specialized in programming languages such as C++, Java, and C, and contributed to research in recommender systems and AI-related data privacy. She has published journal articles, contributed to book chapters, and actively participated in technical workshops. Known for her student mentorship and engagement in academic events, Aarthi remains committed to fostering innovation and excellence in computer science education.

Rustamov Shuxratovich is a Doctor of Psychological Sciences (DSc) and Associate Docent at Bukhara State University, Uzbekistan. He is an experienced academic and researcher in the field of psychology, with expertise spanning educational psychology, personality development, and psychological assessment. His scholarly work focuses on applying psychological theories to educational and social contexts, with particular attention to mental well-being, learning behavior, and cognitive development. He has contributed to research, teaching, and postgraduate supervision, and actively participates in academic initiatives aimed at advancing psychological science and evidence-based educational practices.

Phearun Tep, a Cambodian scholar, holds a degree in English from the Department of English at the University of Delhi. With immense experience teaching English literature, working as a translator, and authoring scholarly works in Cambodia, he is a committed researcher with interests in South Asian literature, Translation Studies, Indian Literary Theory & Criticism, fiction, cultural and post-conflict studies, and identity representation. His research examines how historical trauma, cultural memory, and narrative imagination inform human experience, contributing to literary scholarship and interdisciplinary studies.

Zahid Thoker is a serious research Scholar at Lovely Professional University who is working on Insects (Entomology), specifically on Honey Bees. He graduated with his Master's degree in Zoology from Chandigarh University, Gharuan, Mohali, where he developed a very solid academic grounding in life sciences. Zahid has already contributed greatly to academia; one of his research papers was published, one review paper, and over 25 book chapters in reputable national and international volumes. His academic publication is an indication of the desire to promote science, especially in the biological sciences. As a visionary who desires to make a difference in the research world and in society, Zahid hopes to broaden his academic and professional experiences with integrity and excellence. He is also strongly driven by the goal of benefiting society with his research work and making his parents proud by leaving his mark on the field of science and Technology.

Yusuf Üstün is an academic at Marmara University's Atatürk Education Faculty, specializing in geography education, regional planning, and human geography.

Antuni Wiyarsi works as Professor of Chemistry Education at the Faculty of Mathematics and Natural Science, Yogyakarta State University, Indonesia. Her research interest and experiences in context-based learning, Socio-scientific issues, pedagogical content knowledge, and science teacher education, that granted much research fundings both national and international. She also acts as international/ national journal reviewer and editor, lecturer, and trainer for chemistry and science teacher training program under the ministry of education programs.

Noha Zoelfakar is a PhD student in the Leadership and Policy Studies in Education program at the United Arab Emirates University (UAEU).

Index

C

D

E

G

I

L

M

N

O

P

S

T

www.ingramcontent.com/pod-product-compliance
Lightning Source LLC
LaVergne TN
LVHW060005031026
849161LV00025B/1490